D0164186

# Construction Safety Manual

## Construction Books from McGraw-Hill

BIANCHINA • *Forms & Documents for the Builder*

BOLT • *Roofing the Right Way, 3/e*

BYNUM, RUBINO • *Handbook of Alternative Materials in Residential Construction*

CLARK • *Retrofitting for Energy Conservation*

DOMEL • *Basic Engineering Calculations for Contractors*

GERHART • *Everyday Math for the Building Trades*

HACKER • *Residential Steel Design and Construction*

HUTCHINGS • *National Building Codes Handbook*

HUTCHINGS • *OSHA Quick Guide*

HUTCHINGS • *Construction Claims for Residential Contractors and Builders*

JAHN • *Residential Construction Problem Solver*

KOREJWO • *Kitchen Installation, Design and Remodeling*

MILLER, BAKER • *Carpentry and Construction 3/e*

PHILBIN • *Painting, Staining and Refinishing*

POWERS • *Kitchens: Professional's Illustrated Design & Remodeling Guide*

POWERS • *Bathrooms: Professional's Illustrated Design & Remodeling Guide*

SCHARFF AND THE STAFF OF ROOFER MAGAZINE • *Roofing Handbook*

SCHARFF AND THE STAFF OF WALLS & CEILINGS MAGAZINE • *Drywall Construction Handbook*

SHUSTER • *Structural Steel Fabrication Practices*

TRELLIS • *Documents, Contracts and Worksheets for Home Builders*

VERNON • *Professional Surveyor's Manual*

WOODSON • *Be a Successful Building Contractor, 2/e*

## Dodge Cost Books from McGraw-Hill

MARSHALL & SWIFT • *Dodge Unit Cost Book*

MARSHALL & SWIFT • *Dodge Repair & Remodel Cost Book*

MARSHALL & SWIFT • *Dodge Heavy Construction Unit Cost Book*

MARSHALL & SWIFT • *Dodge Electrical Cost Book*

# Construction Safety Manual

Dave Heberle

McGraw-Hill
New York San Francisco Washington, D.C. Auckland Bogotá
Caracas Lisbon London Madrid Mexico City Milan
Montreal New Delhi San Juan Singapore
Sydney Tokyo Toronto

*McGraw-Hill*

*A Division of The* ***McGraw-Hill*** *Companies*



1 2 3 4 5 6 7 8 9 0 DOC/DOC 9 0 3 2 1 0 9 8

ISBN 0-07-034454-X (P)
ISBN 0-07-034339-X (H)

*The sponsoring editor for this book was Zoe G. Foundotos, the editing supervisor was Penny Linskey, and the production supervisor was Pamela Pelton. It was set in New Century Schoolbook by Paul Scozzari of McGraw-Hill's Professional Book Group Hightstown composition unit.*

**Illustrations by Jessica Olszewska Heberle.**

*Printed and bound by R. R. Donnelley & Sons Company.*

This book is printed on recycled acid-free paper containing a minimum of 50% recycled, de-inked paper.

# Contents

# Introduction

It's a cold, sunny mid-November morning. There's a foot of snow on the ground but it hasn't bothered the builders and contractors who are busily working on seven new homes in an upscale residential development. From a quarter-mile away I'm picking my way down a part gravel, part frozen mud road that curves into the back of the subdivision, and I can clearly hear construction sounds echoing through the dry air: hammers driving nails, the knocking of lumber against lumber, the whine of power saws and drills, an air compressor revving up, diesel engines throbbing, vehicle doors slamming shut, and occasional shouting and loud talking. As I draw near this cluster of single-family houses which are relatively close together, I see that they're in varying stages of construction, but most are under roof already.

The builders all have signs advertising themselves and their latest efforts, and together they represent a cross section of the area's better home builders.

Only 10 feet away, at the back of a pickup truck, a worker in low-cut black basketball sneakers stands on the frozen mud that will eventually be the home's front yard. Leaning against the truck's tailgate, he is cutting the ends of several two-by-six planks with a power saw. He wears no glasses while cutting, and a cloud of sawdust swirls about his face. Across the street I see another worker partly inside the unroofed second level of a garage. He is standing on something (I can't see what) that positions his head and shoulders just above the top of the wall, where he is driving nails at face height, with short wrist-action strokes that strike hammer-to-nail only a few inches away from his unprotected eyes.

At yet another partially built dwelling I notice a worker standing on the very top of an 8-foot aluminum stepladder. He is installing a piece of aluminum soffit. The stepladder is only ten inches away from an electrical drop line.

On an adjacent site I see workers bricking the outside of the front entrance. Their scaffold is constructed so that the front doorway can still be used while the overhead brick work goes on. Plumbers and electricians are entering the dwelling from beneath the scaffold while two masons above continue to lay brick and perform related activities. There's no midrail or toe boards on the scaffold. None of the workers who are walking beneath or near the scaffold are wearing hard hats. At the far end of the same house, several wall openings at floor level appear to be awaiting glass sliding doors that may eventually lead to a deck. These spaces are wide open, without any kind of protective guardrails or barriers.

Today a multitude of government regulations apply to construction worksites, as part of a major federal program to eliminate unsafe work habits and practices like those described above. Today worker safety and health are important to the federal government, and they're important to company management. Yet this wasn't always so. Back in the 1800s and early 1900s, American manufacturing plants and construction companies were notorious for injuring their employees. Workers' hands and arms were frequently caught in pulleys, belts, rollers, and presses. Falling objects caused fatalities and numerous injuries. Chemicals splashed about. Molten metal burned through work clothes with relentless regularity. Employees fell from all heights, were buried in cave-ins, and were asphyxiated in confined spaces. Workers were electrocuted, crushed, chemically abused, overheated, overworked, and overstressed. Coal miners were striken with black-lung disease, meat cutters rarely made it out of their companies with all of their fingers. Asbestos killed more miners, as well as insulation industry employees. Workers lost their hearing and their sight gradually or all at once.

And it continued that way for years and years, until a coal-mine explosion that killed 78 miners and the widespread diagnosis of black-lung disease finally led to the passage of the Coal Mine Health and Safety Act of 1969. Once started, the legislation to protect American workers quickly spread to more comprehensive forms. Congress passed the Occupational Safety and Health Act (see Chap. 5), the most comprehensive legislation on workplace safety ever enacted. The Act created the Occupational Safety and Health Administration (OSHA) in 1970.

Since then, companies and entire industries have been paying a lot more attention to the health and safety of their work forces.

Welcome to *Construction Safety Manual.* Within these pages you'll read about how to run a safe (or safer) operation afield. The goals of this book? On a personal level, if the book saves you or one of your employees a single ambulance ride, or a single anxious visit to a physician or emergency room, it's done its job. Within the larger scheme of things, it's meant to do that many times over, by supplying workable guidelines for a no-nonsense system of worksite safety.

## Why Safety Management?

There are many reasons for establishing an effective safety management program, but the most important include:

1. *The prevention of injuries and illnesses.* Pure and simple, this is by far and away the overriding reason for effective safety management. The physical pain, mental stress, and eventual financial hardships that injured, ill, or disabled employees and their families experience are reasons enough.

2. Additional reasons for managing safety are *to avoid direct and indirect costs that work-related injuries, illnesses, and property damages incur.*

*Direct costs* are expenses *directly* associated with a work-related injury or illness. They include the following costs:

- Medical expenses
- Medications
- Worker's compensation benefits
- Rehabilitation costs

*Indirect costs* are *all other expenses* resulting from work-related incidents or illnesses. They include the following costs:

- Production losses or delays
- Property or product damage
- Training
- Supervisory time
- Administrative time

How many ways can a safety incident disrupt normal operations and lead to additional costs? Plenty. What about time taken in getting or giving medical help? Or the work interruption at the time of the incident? It's only natural that everything but care-for-the-injured stops. There's time spent on the necessary investigation, searching for immediate and basic causes, plus developing corrective and preventive measures so similar incidents won't reoccur. Further down the road, there's time and effort spent processing paperwork and payments. Meanwhile, someone's got to work in the injured employee's spot. That means time spent training or retraining replacement workers. No matter how quickly less-experienced replacement workers learn their duties, some loss of productivity is likely. According to reliable studies, if the direct costs of an injury total, say, $1,500, then the estimated dollar amount required for additional indirect expenses for the same incident will fall somewhere between $6,000 and $13,500, or four to nine times higher.

Overall, an unsafe operation yields a high cost in physical and mental suffering, inconvenience, and dollars. It's a cost that's ultimately wrung out of the company's profits, taken directly from the bottom line. It's a cost that makes it more difficult for your company to compete with other builders, in the open market.

As if we haven't already said enough, there is one other major reason for managing safety:

3. *To prevent injuries or illnesses which may result in regulatory investigations, citations, and criminal penalties—including fines and incarceration.* That doesn't include fallout from the unflattering publicity that inevitably accompanies those activities.

Some employers, despite their awareness of the moral responsibility and sound business sense of maintaining a safe operation, focus entirely on compliance with the rules. They're determined to let regulations dictate their every move. These members of management do only what they believe the regulations require. Unfortunately, a company can still have high injury rates even if every safety regulation is followed to the letter. Regulations alone cannot address many of the human factors of safety, especially attitudes and behaviors. Moreover, even the most comprehensive and current regulations won't address all the situations you'll encounter in the worksite.

There are publications—huge, thick binders—in which OSHA and other regulatory requirements are translated and interpreted line by line into simple English. Sometimes, as the ink is drying on their pages, the rules are changing. So supplements are required to keep subscribers current. Similar updates are often available through OSHA offices and local or state home builder's associations. And recently, OSHA has gone to the Worldwide Web with user-friendly "expert adviser" programs that effectively function just as a safety consultant might—by identifying pertinent rules and regulations that apply to situations in question. At this writing, the OSHA advisers can be found on the OSHA Web site at www.osha.gov. Look for fewer pamphlets and books from them in the future, and more free and easy access information on the Web.

## Preparing Your Company's Safety Management Plan

Before you can set up a program, you've got to have a plan. There are a lot of ready-made programs out there for the asking: huge, formal written programs in large binders with information on every conceivable aspect of safety, referencing OSHA and other regulatory agencies, and cross-referencing all kinds of procedures and rules. Some of them are good; and most of them simply take and embellish what OSHA has prescribed; but all of them have to be customized to meet your needs, because no single program fits every situation.

A review of more programs than the average person would want to see reveals a number of components which are common to the most successful versions. Of course, smaller companies can have written programs that may be less formal, but every system must be reviewed frequently with the idea of being flexible enough to adapt to changing conditions and needs. These components are:

1. *A written safety plan.* OSHA's 1926.20 (b) (1), under Accident prevention responsibilities, states that "It shall be the responsibility of the employer to initiate and maintain such programs as may be necessary to comply with this part." OSHA's interpretation of this is that *the company's safety management plan—including all of its mandatory components—must be prepared in a written form,* and that (b) (2) "Such programs shall provide for frequent and regular inspections of the job sites, materials, and equipment to be made by competent persons designated by the employers." That means that *the written plan must include provisions for competent persons to frequently get out there and inspect the worksites.* The written plan does not belong behind a company owner's desk, accumulating dust. It's got to be an actual working plan.

Other components include:

2. *A comprehensive safety policy.* The company safety policy should be simple enough to read in less than a minute. It should be a statement of policy

that in a nutshell says nothing is worth being unsafe about. It's got to address training, procedures, and management support.

3. *Management support and direction.* This often tells the tale of whether a safety program will float or sink. Management, starting with the owner, president, or other recognized leader, must lead by example and handle safety with as much attention and enthusiasm as he or she would treat any other component of the business such as quality or budgeting or marketing.

4. *Safety responsibilities and accountability.* Safety isn't going to just happen. Individuals have to know what they're responsible for, and what roles they play. Then they've got to be held accountable for their behavior. The key individual at the jobsite is the foreman or supervisor or crew leader—whoever is in immediate charge.

5. *Safe behavior development.* Traditionally, OSHA and other regulatory agencies look for unsafe equipment and conditions, and cite those deficiencies. Unfortunately, as we discussed in the previous section, much of that attention fails to address the cause of up to 95 percent of safety incidents—unsafe behavior.

6. *Job site audits.* These are best done at least weekly (formal), and daily (informal) on certain aspects of the job, perhaps just notes in a notebook, not necessarily a formal checklist. Particular tasks can be looked at—such as framing, flooring, dry walling, excavating, or mobile equipment operation.

7. *Incident investigations.* Incident investigations should be completed on every safety case from near misses up. A good safety management system will learn from incident investigations, not just mechanically do them. Don't merely gather information. Find out *why* an incident occurred, and *what can be done to prevent it from happening again.* Such information must be passed to people who could experience similar exposures. Information learned from incidents must be shared with others.

8. *Employee training.* There's new-hire orientation. It all starts there. Overall attitude of the company toward safety must be conveyed early and often. Reviewing the safety policy, discussing everyone's roles, and reflecting on how important safety is. It shouldn't just be a safety video or a thick binder you ask the employee to read. Real people, from top management to the foremen or supervisors, should be involved. Avoid just passing lists of rules and regulations out, or mechanically reading them. Instead, employ specific job training with equipment and methods. Use supervisory safety training. Schedule safety meetings and training and tool box sessions. Don't take it for granted that people know. Refresh yearly or as often as needed. Keep your supervisors trained. Are enough competent individuals available to properly inspect all your company's worksites, with individual emphasis on fire protection, sling and wire rope inspections, cranes and hoists, assured electrical equipment grounding programs, scaffolding, fall protection, aerial lift operations, and excavations?

9. *First-Aid and Health Planning.* Have a plan for first-aid procedures, and make arrangements for emergency services. Some builders elect to invite members of emergency responder groups to safety meetings for discussions on how to contact the responders and what response times and services can be

expected. Communications should be tested and tried in advance of an actual incident. A trial demonstration is also a good idea.

10. *Site Security Plan.* Limit access of unauthorized individuals. Secure tools and materials on the site. Make the site fire-resistant.

11. *Mobile Equipment Safety Plan.* Are only those employees who have been trained to operate specific pieces of equipment allowed to operate them? Are there vehicle use procedures? Seatbelt policy? Do individuals know how to report a vehicular accident?

12. *Materials Safety Plan.* Is there preplanning of equipment moves and materials deliveries? Housekeeping policy?

13. *Program Measurement and Review.* How do you know whether your safety program is working? Are you seeing injury or near-miss reductions? Are your employees displaying safe behaviors? Are your safety costs being reduced? If you're not getting results, you've got to look back and see what's missing or what's not working. You've got to tweak the program from time to time. It's no good to have the program just for its own sake. You want results. Pull out your records, the audits, the lunch box or tool box minutes, training sheets, and see whether things are actually happening.

## How to Get the Most from This book

Some safety management books are intended to be everything-but-the-kitchen-sink reference works, to be shelved in walnut bookcases and referred to once or twice a year. Others, intended to be revised almost yearly, are full of regulations that change as fast as shifting sand. And some—this book included—are meant to be worn out with constant use, passed from one individual to another, sometimes splayed open, face-down on a copy machine, other times collecting dog-eared pages and dirty fingerprints, or wrinkles from sleet that's blowing in through a partly open pickup truck window. This book is meant to be used. There's nothing within its pages that shouldn't be shared with anyone in your company.

To get the most from this book, finish this Introduction and take a quick peek at the Table of Contents to see what's covered. Then read one chapter after another (they're not difficult) at a leisurely pace. If you feel that a particular chapter or section has no application to your operation—if you never use a miniloader or skid-steer loader, for instance—just skim through that part of the mobile equipment chapter so you realize it's available if needed in the future. Resist the urge to refer only to subjects as related problems surface at the worksite. That approach will keep you working from behind, instead of planning in advance—which is so necessary for achieving prevention. By the time you complete a first-read, you'll have a good understanding of how the key elements of a safety management program fit together.

During a second-read, mark the sections that apply to your situation for later reference. Use a red felt-tipped pen or easy-to-see highlighting marker. Feel free to make notes in the margins, to underline key passages, or fold down the page corners as reminders. Again, this book is meant as a working tool—one that gets handled a lot. It can't save you injuries, time, money, and worry by sitting idle on a desk or bookshelf. The more you use it, and the more familiar the company's employees become with it, the better off your safety program can be.

## This Book's Organization

There's no doubt that a comprehensive safety management program covers a lot of topics. This book tries to make sense of the big picture by breaking down the program into five parts:

- Basic Principles of Safety Management
- Setting Up an Effective Safety Management System
- Personal Protective Equipment
- Recognizing and Dealing with Jobsite Hazards
- Using Safety to Help Promote Your Business

Part 1, *Basic Principles of Safety Management,* presents a basic safety model—how safe and unsafe behaviors figure in the realm of near misses, first-aid incidents, recordable incidents, lost-workday incidents, and worse. You'll see that mathematical laws of probability can be put to work for (or against) your safety program. Chapter 5 discusses OSHA, and reviews its relationship to safety management.

Part 2, *Setting up an Effective Safety Management System,* includes information about your company safety policy, management support, employee participation, employee training, safety meetings, safety audits, incident investigations, and general safety rules that all complanies should enforce. There are discussions about correcting at-risk behavior, as well as about reinforcing safe behavior—something that isn't done often enough by most companies. The science of ergonomics is reviewed in this part, and you'll learn some warning signs to watch for at the worksite—to help prevent the onset of carpel tunnel, tendinitis, and other cumulative-injury disorders. The preparation of specific safe procedures, each called a "safe job analysis," can be used for risky and routine tasks that your company's employees perform. A review of setting up plans for routine and emergency medical services at the worksite is also touched upon.

Part 3 is about *Personal Protective Equipment.* It reviews clothes, foot protection, eye and face protection, head protection, hearing protection, respiratory protection, hand protection, and fall protection.

Part 4 is about *Recognizing and Dealing with Jobsite Conditions and Hazards.* This section includes discussions about hazardous energy control, confined-space entry, hazardous chemicals and communications, hot works, fire prevention, material handling, hand tools, portable power tools and other equipment, signs, signals and barricades, housekeeping, and environmental concerns. It reviews mobile equipment from pickup trucks to front-end loaders and bulldozers. It touches upon site clearing, excavations and trenches, material storage, ladders and staircases, door and window openings and glass, and site security.

Part 5 presents *Using Safety to Help Promote Your Business,* and reviews designing safety features into your homes or products, seeking advice from local and national safety organizations, and presenting your successes to trade journals and/or the local media.

*One Cautionary Note:* Although there is a lot of information in this book, it's not all included to expressly keep you in compliance with OSHA or any other regulatory agency. Additional publications are devoted to line-by-line regulation readings or interpretations so that home builders and other contractors can operate within legal compliance. *But consider that as the ink was drying on this page, some safety laws and regulations were being dropped, modified, proposed, or made law.* That means that when you need the answers to specific regulatory questions, you should go directly to the latest available authoritative source—whether it's your area OSHA office, or through the online services OSHA and other regulatory agencies offer. But always remember, operating in compliance really means doing the legal minimums—following minimal guidelines. *Simply being in compliance won't guarantee a successful safety management program.* Generally speaking, when you have employee safety at heart, achieving compliance will in almost every case come naturally, like taking reasonable precautions you would take anyway. A sound safety program will include attention to the laws and regulations in your area, and can generally be strengthened by participating in home builders' or other trade associations and organized groups in your area. In other words, there's more to safety than what can be presented in this or any other book or publication. Safety management is a living, breathing thing; the parts are forever in flux, with new and improved methods to be learned, and new tools, equipment, and materials, and new and modified laws and regulations, too. But if you have the safety of your employees at heart, and demonstrate so with your actions, that's 90 percent of the battle, and the odds will be in their and your favor.

Back at home now, I think about the contractors working to construct those seven new homes. They're giving their best efforts to produce quality dwellings. They're working in an extremely competitive marketplace. They're employing tradespersons and keeping subcontractors busy. They're doing their best to run small businesses and make a profit. Yet for the most part, they appear to be leaving worksite safety to chance.

If that's what your company has been doing, it's time to stop, take a deep breath, and change directions. Start planning your safety program, and start making changes.

Part

# 1

# Basic Principles of Safety Management

*Part 1 deals with the basics. First comes a safety model, to explain how accidental safety incidents and injuries come about and how we can affect their numbers. The model has resulted from what large companies have learned about safety over the years from thousands and thousands of near misses, first-aid incidents, recordable incidents, and fatalities. Next come discussions about the individual incident types, such as which types of safety cases qualify as near misses and why employees should report them. Beyond near misses are first-aid, recordable, and lost-workday cases—each representing varying levels of severity differentiated from each other simply through luck or chance. The last portion of this part reviews OSHA's recordkeeping guidelines and discusses OSHA's interest in safety management.*

Chapter

# 1

# A Safety Model

## Quick Scan

1. To a certain extent, getting injured "accidentally" is a matter of probability.
2. On average, a certain number of unsafe acts will yield near-miss (close calls) and property-damage incidents.
3. On average, a certain number of near-miss and property-damage incidents will result in one incident requiring first-aid treatment.
4. On average, a certain number of first-aid incidents will eventually result in a more serious recordable injury.
5. Similarly, on average, a certain number of recordable injuries will tip the scales toward one lost-time injury.
6. Unfortunately, having enough lost-time incidents may eventually lead to a fatality.
7. Once a safety incident (or accident) is set in motion, the severity of the outcome is largely due to sheer luck or chance.

In order to understand the entire safety management process, a simple model will help. Over the years numerous geometrical shapes have served as models to describe and illustrate the "safety universe." Here, our model resembles a simple triangular pine tree with a very wide trunk and root ball.

Although this book has been printed in black and white, for clarity's sake, please obtain three highlighting markers: one green, one yellow, and one red. Those colors will help emphasize the three major parts of the tree as we're building our model throughout the chapter.

We'll start with the bottom section of our tree—the trunk and root ball (see Fig. 1.1). The trunk and root ball, the parts of the tree upon which the rest is supported, represent the underlying tenet or belief that S*afety is everyone's responsibility.* Safety is not *only* the responsibility of your crew or crews. It's not *only* the responsibility of a foreman or supervisor. And it's not *only* your responsibility. Safety at (or away from) the worksite is everyone's responsibility. If it isn't accepted as such throughout the company, then no matter how persistent the effort, maximum safety results will remain out of reach.

Directly above the trunk, in the lower branches or body of the tree (see Fig. 1.2), are three statements. One is that *Safety is manageable.* Indeed, large companies have proven this so. Just as numerous aspects of a business can be controlled, such as labor relations, material costs, subcontractor services, and quality of construction, so can safety be effectively managed. You probably wouldn't have selected this book if you hadn't already agreed that it's a good idea to manage safety, and throughout the introduction we've hashed out a number of reasons why actively managing safety is so critical. Indeed, safety *must* be actively managed; in this chapter you'll see why it can't be left to chance.

Next comes the belief that *All accidental injuries and illnesses are preventable.* Here is a stumbling point for some, who will inevitably point to a singular or hypothetical case of being in the wrong place at the wrong time, or to how something *could* happen that is totally beyond the control of the participants—resulting in an injury for which there appears to be no reasonable or meaningful corrective or preventive action. Well, forget about trying to anticipate those apparent exceptions. Instead, concentrate on the other 99.9 percent of safety incidents or conditions that *are* preventable. That's where you're going to get results by managing safety, and those results are going to pay you and your employees back in spades.

The third and last statement in the lower branches is one of philosophy: that nothing we do on the job is worth getting injured for, and, consequently, that *No work is worth taking safety risks for.* Okay, here someone invariably says, "But what if my son or brother or father is working with me and he passes out in the bottom of a gas-filled trench, and...." Yes, there are things we'd sacrifice or take risks for, but not in the line of our routine work. Who wants to risk cutting off a thumb to finish framing a wall before dark? Who wants to risk falling from an icy roof to install that last piece of flashing? This lower portion of the tree, due to its positive contribution toward safety management, should be outlined in green.

The next level of the model includes classifications that are frequently tossed about by management trainers in industrial or manufacturing environments (see Fig. 1.3). But these safety components apply to home building and related trades as well. The first one is *total employee involvement and support.*

**Figure 1.1** Safety model, part 1.

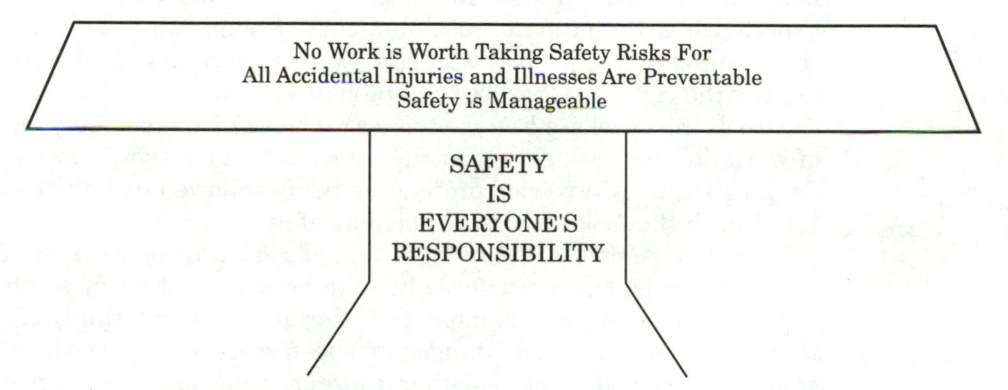


**Figure 1.2** Safety model, part 2.

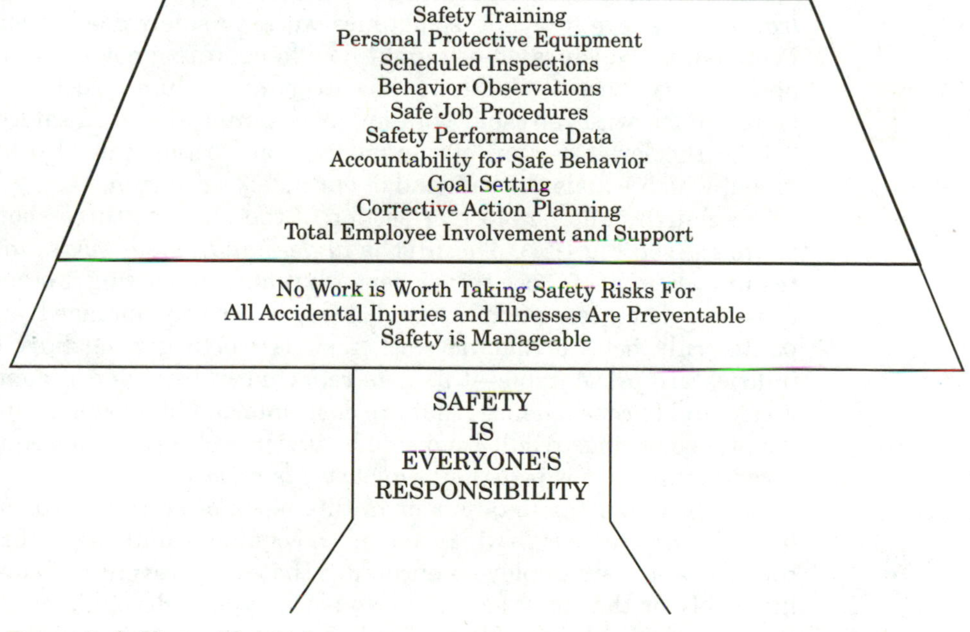


**Figure 1.3** Safety model, part 3.

And by total employee, that means you, too—management and employees alike. It's the only way to make safety everyone's responsibility. The next items are *corrective action planning* and *goal setting*. Both of these take considerable thinking and advance planning, with input from employees, from management, and sometimes from outside agencies and other sources. Another is *accountability for safe behavior*. That's simple. If no one is held accountable for choosing or following safe ways to work, then those ways of doing things may just not happen. Employees must know what's expected of them when it comes to safe behavior. Next comes *safety performance data*. That means understanding the safety statistics pertinent to your operation, such as how many near misses, first-aid incidents, and recordable incidents your employees are experiencing in relation to total numbers of hours worked. Topics of much of the training materials deal with *safe job procedures,* or simply how to do a job or task the right way, without taking needless risks.

Also at this level are *behavior observations,* which are simply audits or slices of work life that look closely at the safety aspects of tightly focused behaviors or work tasks. Are proper procedures being followed or have things deteriorated until the tasks are being done unsafely?

Yet another important concept is that of *scheduled inspections*. Tools, equipment, methods, and procedures have to be looked at every so often to make sure that they're in good shape, that they're the safest things available, and that they haven't started falling apart, been cracked or warped, or had a guard shaken or rusted loose. Another concept is *personal protective equipment* (PPE). It's knowing what to wear when, how to inspect it, how to wear it, and how to get into the habit so it's impossible to forget.

The last important safety behavior in this section of the model is *safety training*. Believe it or not, sometimes working safely doesn't come naturally. People must be educated or taught how to operate a powerful drill motor...or how to carry concrete blocks...or how to operate a dump truck. Then they must be reminded with refresher training. New employees need safety orientation.

Like the lower section of our model, this section can also be outlined in green, which equals the safe end of our safety spectrum.

The fourth building block or section of the tree has three short statements or items (see Fig. 1.4). The first is *understanding safe ways to work,* which results when all of those things beneath it and supporting it occur. When safety is everyone's responsibility, and safety is actively managed, and all participants truly believe that no work task is worth getting hurt for, and that injuries are preventable—this gets reflected in behavior of doing things correctly, and in consequences that are desirable and profitable to all involved. No doubt, good things result from safe behavior and correct consequences. So the second statement is just that: *Safe work is rewarded.*

The third and final statement in this block or section is *roadblocks to safe behavior are removed,* which means recognizing and highlighting obstacles that you and your employees encounter that may prevent individuals from acting safely or that may lead them to act unsafely. Recognizing the hazards is the first step to being able to avoid them and develop corrective action plans for their reduction or elimination.

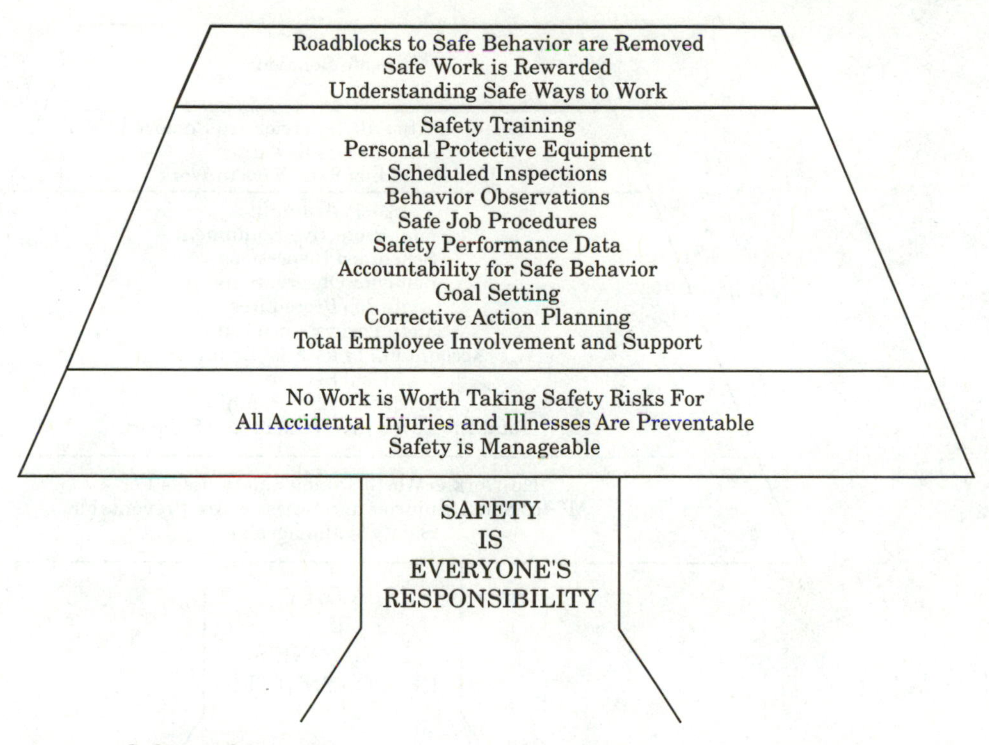


**Figure 1.4** Safety model, part 4.

Those four sections make up the green portion of our triangular tree—the broad base in which most of our actions should be happening. It would be nice to simply forget about the rest of the tree, because that's not where we want to be.

But we can't. Unfortunately, for some companies, and some employees, that's where a lot—too much—of the action is.

Please outline the next part—the center portion of the tree (see Fig. 1.5)—yellow for *warning,* or yellow for *unsafe behavior.* Or yellow for *slow down and think about what you're doing.* To expand on the unsafe behaviors section (see Fig. 1.6), unsafe behaviors include taking dangerous shortcuts, hurrying, overexertion, taking unnecessary risks, poor housekeeping, violating or ignoring safety rules or procedures, reporting to work tired, not paying attention, using poor judgment, and lack of proper training. Unsafe behavior, conditions, or incidents may, as a whole, be considered symptoms of deficient management systems. In other words, you and your employees will be only as safe as your system allows—only as safe as your worksites are being managed.

From the yellow level of unsafe behavior, our safety model ascends into *danger* into the red (see Fig. 1.7), which is what the remaining upper layers of the tree should be outlined in. Sharing the bottom section of this treetop are two statements, the first being *near misses.* A near miss is a perfect name for an incident which could have resulted in injury but didn't. Could have been, but wasn't. The adjacent section is labeled *property damages.* No injury here, either, but it has tangible, quantifiable physical damage. Something is broken,

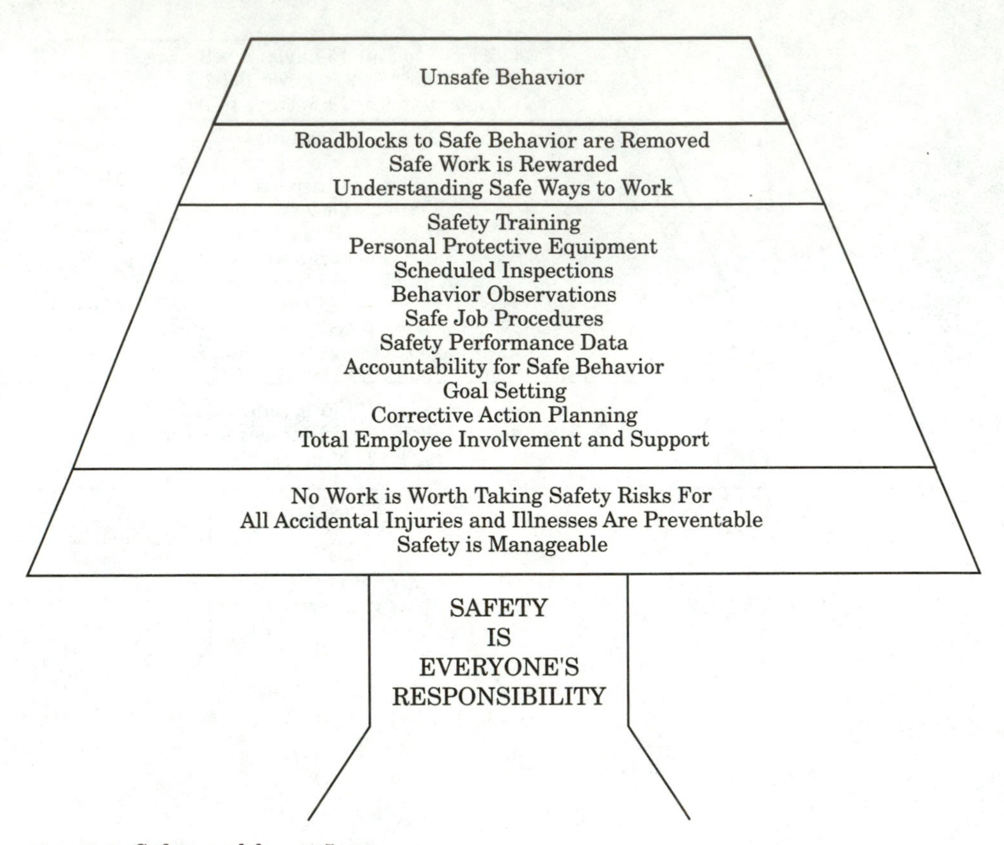


**Figure 1.5** Safety model, part 5.

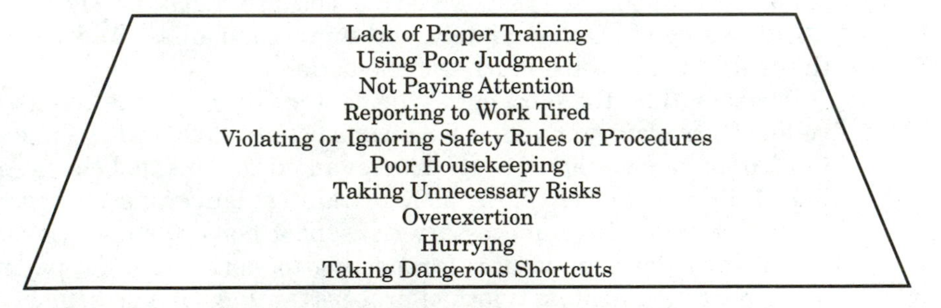


**Figure 1.6** Safety model, expanded unsafe behavior section.

a portion of work is lost, damages are done. Money is lost, time is lost. An injury could have happened, but didn't. Again, these are outlined in red.

The next levels of our safety model (see Fig. 1.8) first include a seemingly harmless level of *first-aid incidents,* or incidents in which some injury or illness presents itself that needs medical attention, however minor. Yet the simple existence of a first-aid case is not good. In fact, numerous first-aid incidents are an ominous sign, a sign that enough are building up to support the next level of the tree. And here it's worth noting that our model can also, from the yellow zone upward, equate to a somewhat proportional representation of the numbers and types of incidents, upward from unsafe behavior. In other words, let your worksite experience enough unsafe behaviors, and the odds are you'll be having first-aid cases. Then, have enough of them and you're bound to travel into the next level of red, labeled *recordable incidents.*(See Fig. 1.9)

Recordable incidents are a recognized measurement of injuries or illnesses that require more than first aid to treat. Some result in an injury following which the individual will have to work a restricted duty, and will be able to perform only certain parts of his or her job. Recordables and the incidents which make up the next layer, the *lost-workday incidents* (See Fig. 1.10), where the individual is injured badly enough to lose work, are considered more serious (and less likely to occur) than recordable or first-aid incidents. Naturally, the number of lost-workday incidents is related to and somewhat dependent on the number of recordables; which, in turn, is somewhat dependent upon

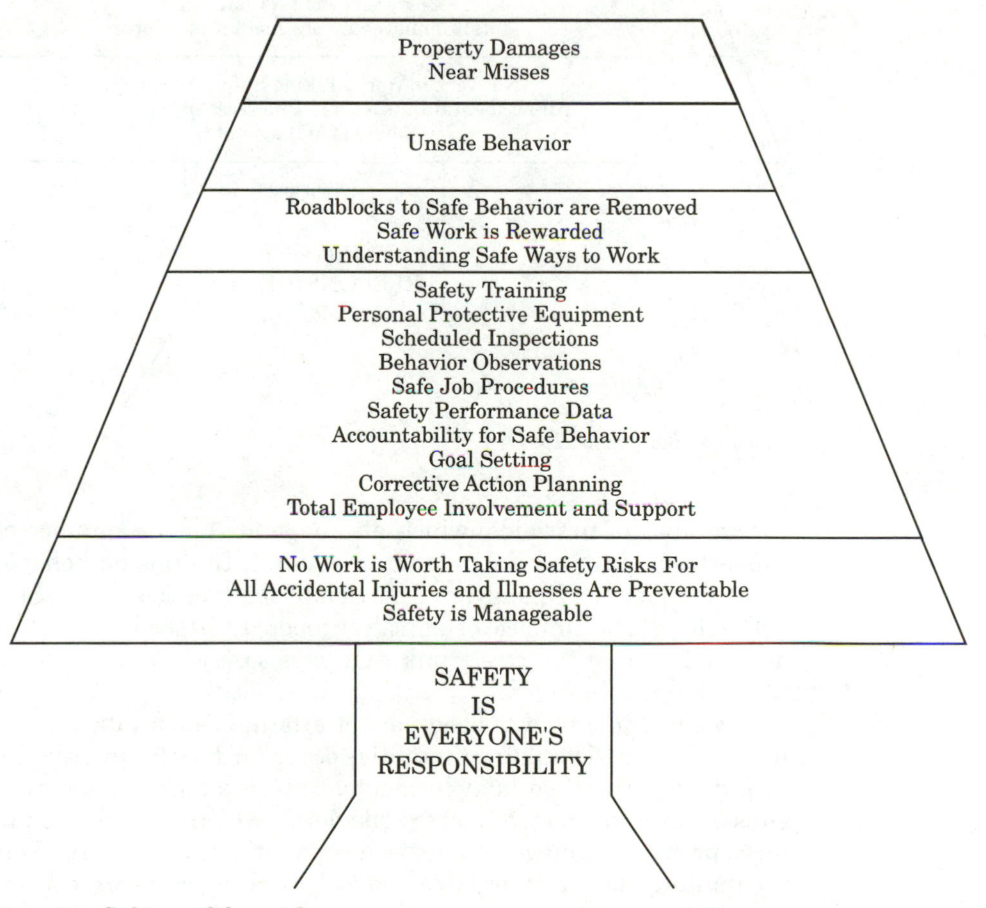


**Figure 1.7** Safety model, part 6.

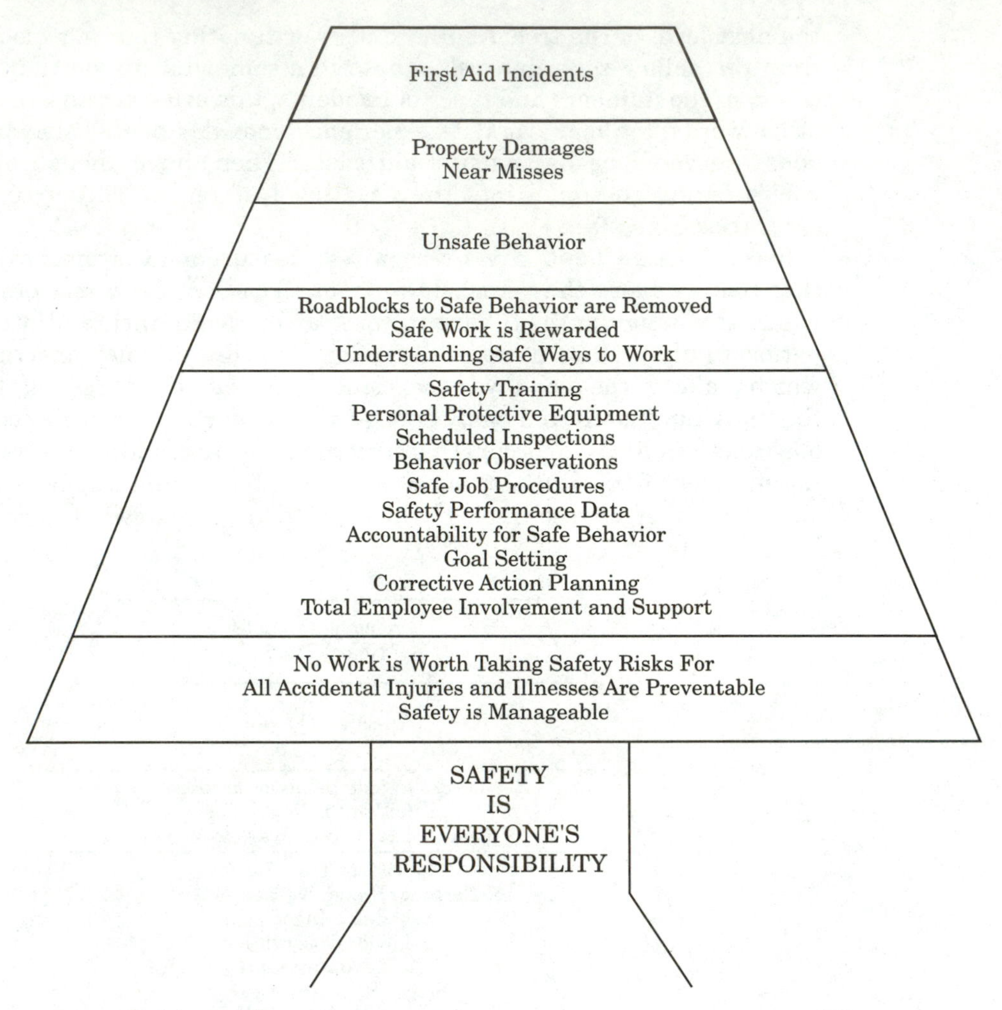


**Figure 1.8** Safety model, part 7.

the number of first-aids; which also depends on the number of near misses, property-damage incidents, and, in general, the unsafe behavior that's going on at the jobsite. This layer or portion of the tree is also in the red.

Finally, at the pinnacle of our safety model tree (see Fig. 1.11), and as far away as possible from the tree trunk that says *safety is everyone's responsibility,* is the single word *fatality.*

Anyone who has ever been part of a fatality or a fatality investigation will never forget it. Thankfully, fatalities occur far less frequently than do first-aid, recordable, or lost-workday incidents. But, on a true-but-sad note, if your work-sites experience enough unsafe behaviors—which, in turn, lead to enough near-miss, property-damage and first-aid incidents, your sites are likely to eventually accumulate enough recordable and lost-workday cases to result in a fatality.

That's our safety model. As you see, a lot of it is related to probability. Do enough of the right things—enough of the safe behaviors—and you won't have the unsafe behaviors that pave the way to property damages, near misses, first-aids, recordables, and worse. Allow too many of the wrong things—the unsafe behaviors—and bad things are likely to happen.

This doesn't mean, though, that a single incident incurred by a company all year couldn't end up as a fatality. The safety tree is only a model—a representation of what happens on average. Statistics are not everything, and that's why safety *must* be everyone's responsibility. You can't rest comfortably on past safety performance. Nor can you totally predict incident outcomes or levels of severity. Theoretically, you, with your single ticket, can win the state

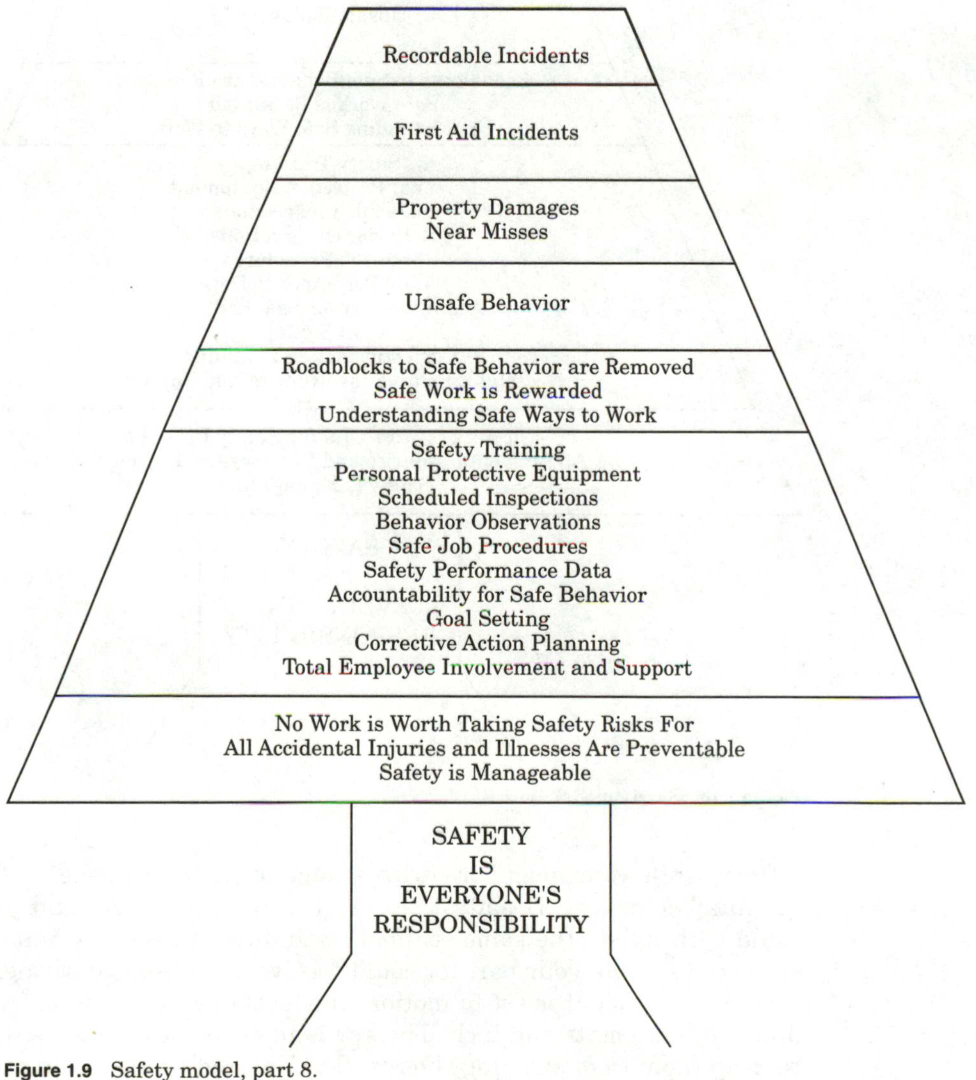


**Figure 1.9** Safety model, part 8.

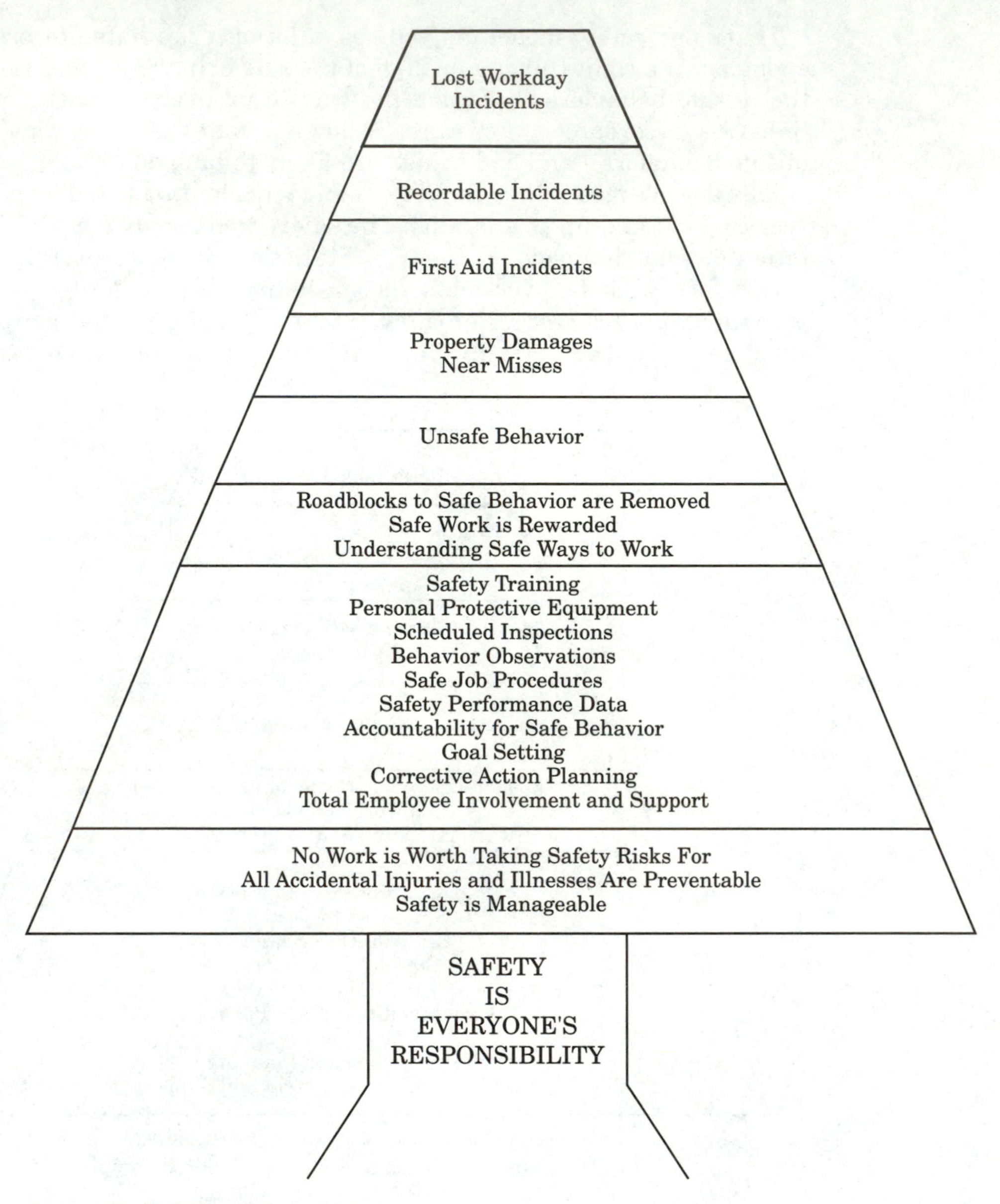


**Figure 1.10** Safety model, part 9.

lottery, while your neighbor—with a huge bucketful of tickets—has not a single number drawn. Despite the laws of probability, you and your coworker could both fall off the same scaffold, with different results. You could just get shaken up, while your partner could receive a broken collarbone.

Once an accident is set in motion, the level of severity of its outcome can be and is often a matter of luck. The key is to keep your employees in the green, to keep them from entering those yellow and red levels of the safety model.

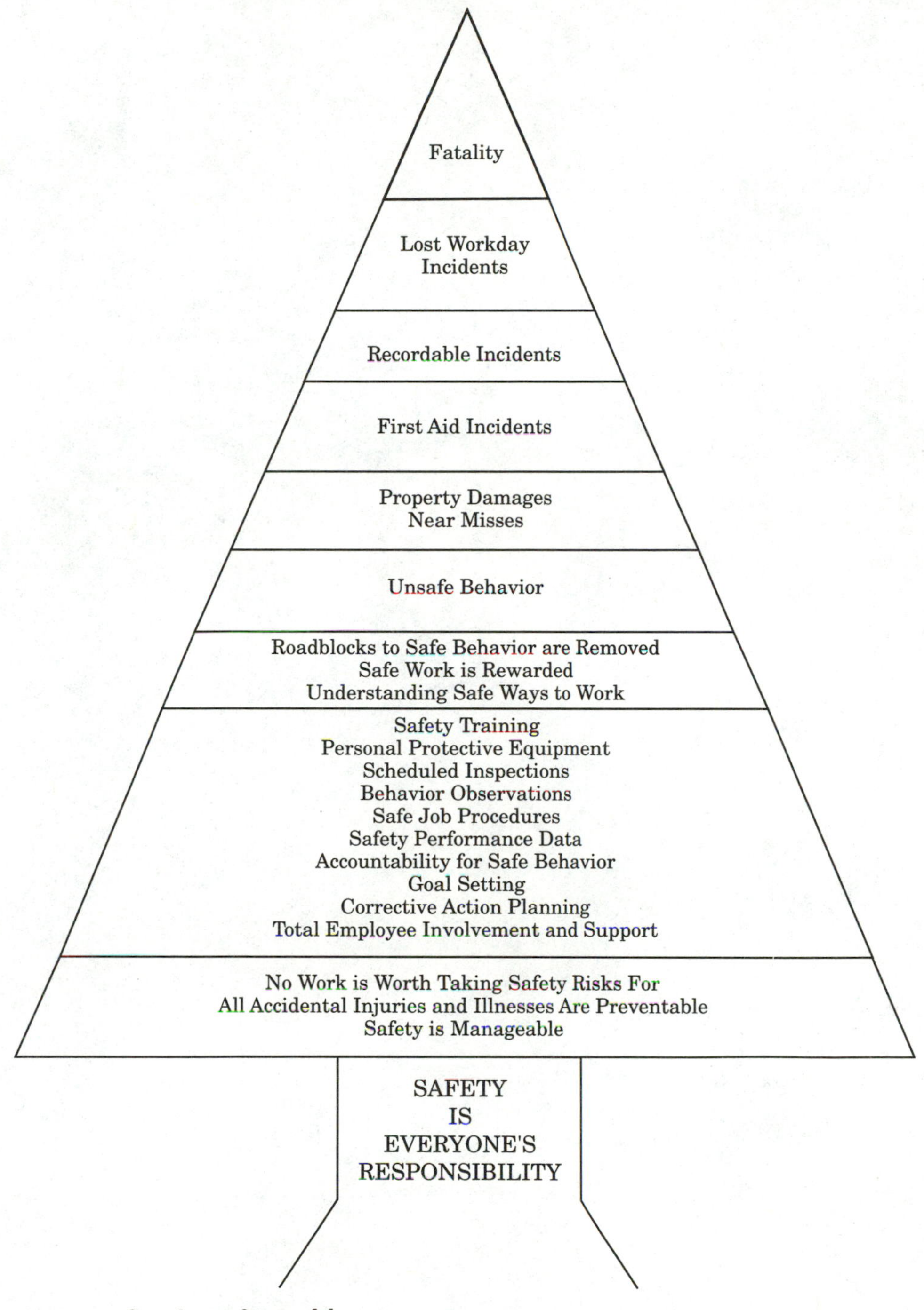


**Figure 1.11** Complete safety model.

Chapter

# 2

# Safe and Unsafe Behavior

## Quick Scan

1. Attitudes are internal ways of thinking or believing that are arrived at through an accumulation of experiences and personal views; once established, they're difficult to change.
2. Behaviors are external, observable acts or activities; unlike attitudes, behaviors can be changed in a relatively short time.
3. The way to modify someone's attitude toward a certain end is to change that person's behavior.
4. The way to turn a person's unsafe behavior into safe behavior (and eventually improve his or her accompanying attitude) is through punctual, certain applications of appropriate consequences.
5. Negative consequences must follow unsafe behaviors to discourage such behaviors from being repeated in the future.
6. Positive consequences should follow individual safe behaviors to reinforce the likelihood of those specific behaviors being repeated in the future.

Why is it that, no matter how many times they're told, some employees continue to perform tasks unsafely? While it's impossible to come up with a correct answer every time, certain guidelines and principles do exist, and once known they can be employed by management to help tip the scales of behavior toward the safe side.

This chapter reviews how behaviors are related to safety at the worksite. Unlike attitudes, which are internal ways of thinking or believing, behaviors are external, observable acts or activities. While attitudes may be *expressed* through observable actions, body language, facial expressions, and speech, the attitudes themselves remain internal and are difficult to deal with directly.

Because between 85 and 95 percent of all work-related injuries and illnesses can be associated with behavior, a basic understanding of why employees choose to behave unsafely (or safely) is critical. You need to know why employees take unnecessary risks, and only then can you deliberately work on either *convincing* those employees to change their ways, or on *making* those employees choose safe behaviors. Naturally, it's always best to have workers select safe behavior options on their own accord.

Before going any further, it bears mentioning that although most accidental injuries are attributable in one way or another to at-risk behaviors, those behaviors ultimately occur *when supervision does not take an active enough role in safety management.* That cannot be stressed strongly enough. Consultants and safety experts have a grand time telling company owners, managers, and anyone else that the safety or injury level a company experiences is pretty much *what the company's management is managing for.* In other words, safetywise, a company usually gets what it deserves. That makes sense because a company's management is responsible for planning, organizing, training, establishing rules, supervising, and carrying out or seeing to most other components of the safety process. If people are getting hurt, that typically points to deficiencies somewhere within the safety management system.

Often, a company's management and employees will deceive themselves into believing that their problems mainly result from unsafe conditions. In most cases, however, a close review of the situation reveals that contrary to this belief, unsafe *behaviors* are truly responsible for whatever unsafe conditions exist. For example, an employee working with drywall sheets reaches into a canvas tool bag for a tape measure, and is deeply sliced on the left thumb by a utility knife that was left out of its sheath. A different employee had last used the razor-sharp utility knife and failed to secure its sheath before returning it to the canvas tool bag. To the employee who got injured, the unsheathed knife was indeed a hazardous condition. Certainly, he or she didn't create the hazard. But the root cause of the injury was actually the unsafe behavior performed by the uninjured employee who previously used the knife, *when he or she failed to sheath the knife* before returning it to the tool bag. And, taking the behavior perspective a bit further, should the individual who received the cut have carefully looked into the tool bag before reaching in? Will he or she in the future?

Another example of behavior versus condition is a scaffold carelessly built so that some of its planks are positioned with little or no overlapping, with some of them not tied or fastened down. An injury resulting from such a scaffold's *condition* can instead be attributed to several unsafe *behaviors:* to the scaffold's haphazard construction, and also to the failure to inspect the scaffold before its initial use.

To analyze the difference between safe and unsafe behaviors, individuals specializing in behavioral approaches to safety management have studied what elicits both kinds of activities. And guess what? Often there's not much difference between the motivations for either.

Behaviors, again, are external and observable, and can be stopped or changed midstream, with an immediacy that's plain to see. They're a reflection of one's attitudes and beliefs. An individual act is an instantaneous event, while behavior encompasses a whole series of activities over a period of time. If we wish to influence a change in a person's behavior, it can most easily be done by changing the behavior itself, not by going after the internal attitude of the person performing the behavior.

## Behavioral Components

Although numerous individuals, from psychologists to safety specialists, use various systems to describe the behavioral process, they all key on three different but related components: *activators* or *antecedents, behaviors* (with their activities and results), and *consequences.*

### Activators or antecedents

These components always trigger or prompt behaviors. An activator is any person, place, or thing that comes before a behavior and encourages that behavior to take place, or that encourages a person to perform a task or series of tasks in a certain manner. Every behavior has an activator or antecedent. Activators or antecedents communicate information. They're effective in triggering behavior because they're invariably paired with consequences or results. *Activators* influence a behavior *before* the behavior occurs. *Consequences* influence the likelihood that the behavior will be repeated *after* the behavior in question occurs. A consequence for one person could be an activator for another individual. Without consequences, activators have short-lived effects, because no specific result can be counted upon. To be effective in triggering a certain behavior, an activator must identify a specific behavior and what the outcome of that behavior will be in advance. Activators usually provide some kind of instruction or communication. They tell us what to do or not to do, how to do it, and when to do it. Activators for safe behavior are commonly provided as operating procedures, warning signs or displays, or alarms—they've got to be specified clearly to prompt a desired safe behavior.

Effective activators have the following characteristics:

- They identify specific observable behaviors, so there's no doubt as to what needs to be done.
- They specify the outcome of the behavior.

Even the best activators, however, are not long-term solutions to maintaining safe behaviors. Hazard warnings on tools, for example, are activators—but how many individuals take them seriously? So are operating procedures you may train your employees to follow—has anyone ever violated them? And so are signs and labels that are clearly written. Other activators include goals and objectives, safe behavior of coworkers, a supervisor's instructions, and even certain worksite conditions. Take housekeeping, for example. Studies have shown that safety and quality definitely improve when high standards of housekeeping are maintained.

### Behaviors

Behaviors immediately follow activators. Behaviors have two identifiable components—*activities* and *results.* Human nature typically guides us on a course of action which will bring us the greatest amount of pleasure or satisfaction for the least amount of pain or effort. *Activities* are behaviors undertaken to yield specific results. *Results* are products of activities that are observed after the behavior is completed. It's important to distinguish both of these components of behavior since the results component of unsafe behavior is often mistaken for an unsafe condition. These conditions must be recognized for what they really are, and be corrected as unsafe *behaviors.*

### Consequences

Consequences, as parents of practically any teenager will confirm, are the key to managing behavior. Consequences—or what happens once a result occurs from an activity—largely determine whether a behavior will be performed or repeated in the future. If a 17 year old comes home at 1 A.M., an hour after curfew, one consequence could be taking the car keys away for a few days. Unfortunately, the ineffective use and application of consequences can result in poor quality, low productivity, and less desirable worksite safety performance. Consequences determine if a behavior will be repeated, or decrease the possibility of it being repeated. Always consider a consequence from the perspective of the person who will benefit or suffer from it.

Three factors govern a consequence's strength as a motivator: *significance, timing,* and *consistency.* The consequences which have the most power to influence behavior are those that are delivered in a positive, immediate, and certain fashion. Too often we apply what we perceive as a positive consequence to a behavior that we want to encourage, while, in reality, that same consequence may be perceived as negative by the recipient. Consider the following:

*Recognition for an outstanding quantity of excellent work production.* This might backfire because the recipient's fellow workers feel that they, too, will be expected to produce more-than-average amounts of work.

*Discipline for having an accident.* This can be counterproductive because it may discourage the reporting of safety incidents and injuries, while encouraging the hiding of information that could otherwise help make the company's safety management program stronger.

*Supplying poor-quality or ill-fitting safety equipment.* Inferior-quality, poorly fitting safety equipment will probably not be worn willingly.

*Supervisors ignoring unsafe behavior.* If supervisors don't follow a company's own safety rules, neither will the workers.

There are four basic reasons for unsafe behavior:

1. Employees do not know the desired behavior.
2. Employees do not know how to perform the desired behavior.

3. Obstacles prevent employees from performing the desired behavior.
4. Employees choose not to perform the desired behaviors.

The first three are fairly cut and dry, and, within reason, are relatively simple to address. The fourth is related to consequences. *The main reason that employees choose not to do something is that there are stronger, positive consequences for not doing it.* An employee may opt to attempt to clear a jam in a power fastening tool on the run instead of locking it out because of being expected to finish so the crew can start on another scheduled construction site the following day. The motivation for hurrying to clear the jam is strong because the employee considers the criticism sure to come for not completing the work a far greater evil than any possibility of being criticized or disciplined for fixing the jam in an unsafe manner.

Think about the common behavior of wearing safety glasses. Now list all of the positive and negative outcomes or consequences of wearing safety glasses at the worksite:

| | |
|---|---|
| Pos. | The employee's eyes won't be injured. |
| Pos./Neg. | Recognition from peers and supervisors for safe behavior. |
| Neg. | Extra time required to use and maintain the eyewear. |
| Neg. | Scratches or lens defects may distort vision. |
| Neg. | Discomfort. |
| Neg. | Safety glasses may fog in warm, humid environments. |
| Neg. | Inconvenience. |
| Neg. | Peer pressure from other employees who do not wear the glasses. |

A single behavior can thus result in numerous conflicting consequences, all of which may not be favorably regarded by the employee. The consequences perceived as being the most important by the employee will largely control that employee's behavior. In other words, what the company considers to be a careless action—the unjamming of a fully charged fastening tool—may, in the employee's way of looking at things, be the most logical behavior to select. If management doesn't readjust the consequences of how that task is done in order to produce safe behavior, so that the employee controls the possibility of an unexpected activation of the power tool before unjamming it (which is the essence of hazardous energy control), then the unsafe behavior option will continue.

## Positive and Negative Consequences

A positive consequence either provides something an individual likes or removes something the individual doesn't like. Positive consequences are reinforcers; they maintain behaviors. Negative consequences provide something the employee doesn't like or take away something the employee does like. Negative consequences punish a behavior. Negative consequences are used to stop unsafe behaviors.

How can consequences be used to manage behavior? How strongly a consequence influences behavior depends on three factors:

1. How soon the consequence is received following the behavior
2. How certain the employee is that the consequence will be received
3. Whether the consequence is positive or negative

The key to understanding how strong a consequence will be depends upon how the *employee,* not the company, perceives the consequence. In other words, if you were threatened with no MTV for a month, you might not care less. Say that to the typical teenager, however, and it could carry considerably more threat. The second consequence in the preceding list of examples on using safety glasses—recognition from peers and supervisors for safe behavior—can be positive in the eyes of the company, but negative in the eyes of peers who may be jealous of the employee or who may tease the employee for being a "company man." It's happened.

*Consequences that are positive, certain, and provided soon after the behavior are the most powerful in maintaining a behavior.*

### Timeliness—sooner or later

Consequences that are delivered immediately following a behavior strongly influence whether that behavior will be repeated. Consider an example of a carpenter driving a common nail into an electric line behind the wall and receiving a nasty shock. The consequence of the behavior—pain—is immediate, and the carpenter jumps back, pulling that hand away. It's likely that the carpenter will remember this shocking experience and its definite, immediate nature. The negative consequence occurred because the carpenter didn't take the time to make sure there were no electrical lines behind that area, and the threat of a similar consequence, or shock, is likely enough to make him or her more careful the next time.

Consequences for behavior must be provided as soon as possible after the behavior. The longer the delay in receiving consequences, the more difficult it is for the individual to associate the consequence with a specific behavior.

That is why group awards, by themselves, are not effective in promoting safe behavior. These awards are usually provided after some milestone—for example, so many safe hours worked—has been reached, which usually comes much later than most of the actual safe time in question. It's difficult for an employee to identify specific safe behaviors that have contributed to such an achievement.

### Certainty—certain or uncertain

If the individual is certain that either positive or negative consequences will be received for a particular behavior, that strongly influences the behavior. Referring back to the carpenter shock example, the carpenter realizes that a shock will be inevitable if a "hot" line is hit with a nail.

Everyone would follow safe procedures if they were injured every time they behaved unsafely. But, since actually being injured is an uncertain outcome of unsafe behavior, the fear of being injured is not a strong negative consequence. Usually there are other more certain, positive consequences for unsafe behavior, such as additional break time, completing tasks more quickly, or reaching levels of personal comfort that offset the chance consequences of being injured.

### Positive or negative

Whether the consequence is perceived to be positive or negative also determines if the behavior will be repeated. The word *perceived* must be emphasized, since it is the employee's perception of the consequence, not the company's, that really counts. Remember, a positive consequence provides something the employee wants, or removes something the employee doesn't want.

This can be a very frustrating element of behavior management. A good example of failing to recognize the employee's perception of consequences is disciplining an employee for lockout violations with a 3-day suspension during deer-hunting season. While the supervisor may feel that adequate punishment has been administered, the employee may perceive the temporary suspension as a benefit, that it's great to be off during hunting season.

## Consequence Analysis

*Consequence analysis* is a fancy term for a method used to identify the consequences influencing a behavior. It can be a useful tool for analyzing problem behaviors in order to provide strong positive consequences for the safe behavior. The consequence analysis procedure is as follows:

**Step 1: Specify the behavior to be analyzed.** It is important to identify a specific behavior. The analysis won't be effective if the behavior isn't stated in specific, observable terms.

**Step 2: Identify all possible consequences for the behavior.** Remember, consequences will be both positive and negative. View the possible consequences from the employee's perspective.

**Step 3. Determine the factors associated with each consequence.** Each consequence will be positive or negative, sooner or later, and certain or uncertain. Remember to always take the employee's viewpoint when calling the factors.

**Step 4. Rank the consequences to determine the ones that carry the most weight in controlling the behavior.** Use the following consequence list to help determine which consequences will be the most powerful. Remember that *positive*, *certain*, and *sooner* consequences are strongest.

1. Positive, certain, sooner
2. Negative, certain, sooner
3. Positive, certain, later
4. Negative, certain, later
5. Positive, uncertain, later
6. Negative, uncertain, later

This ranking explains why people continue certain risk-taking behaviors. It partially explains why I, like millions of others, am so many pounds above my ideal weight range. I realize that having hot buttered popcorn, loaded with salt, time and again is not good for me. On the other hand, it's not going to affect me tonight, or tomorrow night, or next week (I don't think). Gaining weight is an insidious process that often takes years. So you might say that, to me, the consequences of eating buttered popcorn are *negative, uncertain,* and *later*—right there at the bottom of the consequence list. And that's why I continue to eat it. And that's why people continue to smoke cigarettes, chew tobacco, and drive without buckling their seatbelts. The negative outcomes of those behaviors are not certain, nor are they likely to be sooner.

On the other hand, the consequence for downing a spoonful of rat poison would rank number 2 on the list: *negative, certain,* and *sooner.*

For a heroin addict, the ranking shows why it's so hard to quit. For as self-destructive as the behavior *eventually* becomes, along the way the feelings received from the drugs are ranked—by the addict—as *positive, certain,* and *sooner.*

**Step 5: Provide positive consequences or remove negative consequences to maintain safe behaviors.** You may also be able to provide a strong negative consequence to stop a certain behavior, but the goal of supervisors should be to provide positive consequences to motivate safe behavior.

For more specific information on behavior management, see Chaps. 13 and 14 on correcting at-risk behavior and reinforcing positive behavior, respectively.

Chapter

# 3

# Near-Miss and First-Aid Incidents

## Quick Scan

1. Train all employees on near-miss incident reporting.
2. Instruct all employees to report and investigate near misses in a timely fashion.
3. Reinforce the reporting of near misses in a positive manner, and share information learned from the resulting investigations with all employees.
4. Complete near-miss incident corrective actions and report their status to all employees.
5. Do the same for all first-aid incidents.
6. Minimizing near-miss and first-aid incidents is really the key to reducing the overall frequency of worksite injuries.
7. There is generally a predictable mathematical ratio, based on historical data specific to your company, between the numbers of near-miss and first-aid incidents occurring on your worksites and the numbers of more serious incidents taking place in the same areas.

It's a few minutes before 12:30 on a hot, humid afternoon. A front-end loader operator and a construction crew of three masons, four carpenters, and an electrician have just eaten lunch and are sitting outside the doorless garage, relaxing. One carpenter steps away from the others and lies down in the grass in shade provided by the huge rear wheels of the loader. At 12:45, everyone but the reclining (and now sleeping) carpenter heads back to work. The loader operator climbs into the machine, starts it up, and allows the loader to lurch forward down a slight grade while beginning to build up air pressure. The carpenter lying in the grass screams and leaps up at the sound of the diesel engine and is glanced on the forehead by the machine's steel counterweight as it pulls away. It is just a scrape, and everyone else has a good laugh, but the carpenter wonders privately what would have happened if the loader had been parked on a downgrade instead, and it had drifted backwards.

A near miss is an incident that does not result in injury, but that either (1) has the potential for serious bodily harm, or (2) results in property or product damage.

Near misses, or close calls, are common at many worksites. They don't result in injury—but they may cause property damage. If, say, an employee had been in a slightly different position or place, or the equipment or product placement had been to the left or right, serious injury and/or damages *could* have resulted. A lot depends on sheer luck and circumstance.

The loader operator mentioned in the preceding situation should have walked around the machine before climbing into the cab. Doing so would have prevented the near miss. But after the incident occurred, as simple as it was, it should have been investigated and reported—so that all crew members on the worksite would realize what went wrong and understand how a similar situation could be prevented in the future. Just the simple act of discussing the event could eventually prevent a similar occurrence from happening—and, perhaps, the next time the loader *could* be parked on a downgrade.

For the little effort it takes to review a near miss with affected employees, it doesn't make sense not to. Plus, once a number of individuals discuss a near miss, helpful suggestions are likely to be made. What about the loader's parking brake? Why wasn't that set? And what about the judgment of the carpenter who laid down behind the tire in the first place? Once written up, the same incident should be shared with other company workers and even used as a future training topic for new employees.

Near misses, while not desirable, still provide opportunities to help strengthen and refine the effectiveness of safety programs. Why investigate a near miss? Why even talk about something *that didn't happen*? Because near misses are like flags on the playing field of safety. A near miss gives management and employees a chance to investigate a potentially dangerous circumstance and correct problem situations before the same thing or something worse happens again. That's important because the next time the outcome could be quite different, with far more serious consequences. After all, while the fact that an incident occurs can be attributed to one or more errors or causes, the severity of the outcome is largely a matter of chance.

## Reporting Near Misses

This chapter emphasizes the importance of near-miss reporting for a number of reasons, most having to do with recognizing unsafe behavior and worksite hazards. But it's also true that individuals who study safe and unsafe worksite behaviors are in agreement that, unfortunately, most near misses are not reported and are rarely included in the safety data that companies use to help evaluate their own safety programs. Why aren't more near misses reported? There are many reasons, including the following:

*Lack of meaningful action resulting from similar near-miss incidents that had been reported previously.* Maybe past incidents were reported but never got any attention or results. Soon employees develop a "no-one-ever-does-anything" attitude.

*Fear of discipline.* This is especially true with newer employees who may not know how a supervisor or crew leader will react to a reported near miss. That's why a discussion of near misses and incident investigations should be mandatory for new employee safety orientation.

*Macho reputations to maintain.* Individuals may be too embarrassed to report something that they feel resulted from a foolish mistake they made. Unless employees have a true understanding of why near misses are reported, employees may try to cover up close calls instead.

*A lack of understanding of how important studying and learning from near misses can be to preventing future incidents.* This necessitates at least basic training on safety management fundamentals—the safety tree model and its components.

*The annoyance of filling out paperwork or sitting through incident reviews.* This can be a serious problem that no one is aware of. Some construction crew members—like certain employees in many of the trades—may be a whiz at their jobs but have deficiencies in writing ability due to circumstances beyond their control. Don't assume that everyone can fill out an incident form or sit through an incident review without help. Make sure, in a tactful way, that peer or other assistance is provided, especially to individuals for whom English is a second language.

*The potential interruption or spoiling of a safety record.* The verdict is still out on the benefits of using safety incentives as rewards for individuals, crews, or teams to achieve low or improved workplace injury rates. For our purposes, consider that tangible goals and incentives have a meaningful place, as long as they're not relied upon exclusively, or even too heavily. Also, some safety consultants suggest that too much peer pressure toward achieving tangible rewards encourages the hiding of incidents instead of reporting them.

Effective handling of near misses requires that they're reported immediately, so corrective action can be decided upon and taken right away. It's recommended that a shortened incident form or *near-miss report* (see Fig. 3.1) be used, because there isn't as much data required as is needed when an injury or illness results.

## Near-Miss Examples

A three-person crew is scheduled to complete siding the front and one side of a two-story new-construction house on a Friday. Bad weather forced them to quit

**NEAR-MISS REPORT**

Date: ______________ Time: ______________

Location: ______________

Reported By (Optional): ______________

What Happened: ______________

______________

______________

______________

______________

______________

______________

______________

______________

Action Taken to Prevent Repeat of Incident: ______________

______________

______________

______________

______________

Received by: ______________

Date: ______________

Figure 3.1 Near-miss report form.

early the day before, and the following Monday they are supposed to start at a new site, across town. They agree that they need to pick up the pace, so a decision is made to use an assortment of lightweight, easy-to-handle boards and planks on their scaffold; to avoid taking the extra time to fasten them down; and that guardrails won't be necessary because they aren't going to be up there long, and everyone knows how to be careful enough, so nothing could happen.

They also realize that their supervisor is at another site and isn't likely to come around until the end of the day.

So they start. The first four hours, things go smoothly. The time they save from just throwing the scaffold together is used to gain on the siding installation. Everything is fine until one employee decides, after finishing a section, to move to the opposite end of the scaffold to help the other two. The employee isn't aware of a certain section that can't be stepped on. The other two employees know about this section, and have discussed with each other the necessity to avoid it, but they saw no reason to inform their partner—who was not supposed to be there anyway. Before they can shout a warning, their partner steps on an end piece of plywood that teeters downward, and slides down behind it, through the platform. Luckily, the worker reaches out and grabs a support bar that holds—preventing a 12-foot fall to the dwelling's concrete driveway. The other two employees pull the worker to safety. Afterward, the crew decides to finish the scaffold the right way. They complete the siding project a few hours past the regular quitting time, but never mention the near miss to their supervisor, or to any other employees in the company.

As an opportunity to learn from, and as a means to help prevent future incidents, their experience is lost.

On the other hand, one of your company's employees, a truck driver, has been delivering small loads of 2B gravel to the worksite, dumping them on several sides of the foundation. The last load goes near the far side of the house, where the driver backs up, lifts the bed to dump the gravel, then pulls ahead 10 feet or so and attempts to lower the empty bed. The bed, however, won't go down. The driver inspects the controls. They seem okay. Then the driver hears several coworkers screaming to back up. The driver finally looks where they are pointing and sees that the bed is being held in the air by an energized line from the temporary electric service. Luckily, the driver is able to back up and get free from the dangerously stretched line. All the individuals involved, however, know that it could easily have become deadly in a hurry.

The employee, despite plenty of good-natured ribbing from coworkers, makes out a near-miss report and turns it in to the supervisor (see Fig. 3.2). Afterward, and throughout the week, brief tool-box discussions with each of the company's worksite crews emphasize the importance of using extreme caution when working near electrical components with mobile equipment. While the crews are able to laugh at their fellow employee's "misfortune," at least they are left with a renewed appreciation for the importance of knowing where overhead lines are when operating mobile equipment.

In short, a near miss indicates that something is seriously wrong. When reported, it gives us an opportunity to investigate the situation and highlight or correct the behaviors or conditions before the same thing happens again.

Experience has proven that if accident causes are not removed, similar accidents will occur again and again.

By keeping silent about a near miss, participants may avoid having to deal with it. But try to explain that to a coworker who ends up in a wheelchair when someone knew that a hazard existed but was too lazy or proud to talk about it.

**NEAR MISS REPORT**

**Date:** 3/12/98 **Time:** 10$^{30}$ AM

**Location:** 3845 CANTERBURY DRIVE

**Reported By (Optional):** HAROLD

**What Happened:** After spreading three loads of gravel, one each on the front, east, and west sides of the foundation, I backed the fourth load near the oak tree in the back, raised the dump bed, dumped my load and pulled ahead to lower the bed, but it wouldn't go down. It was held in the air by the temporary electric service line. I pulled ahead and let the bed down. The line was stretched, but not broken. Rich reported it to the electric co.

**Action Taken to Prevent Repeat of Incident:** We all talked about it and I told Alex what happened, so he will be careful with his truck, too.

**Received by:** Richard Thomas

**Date:** 3/12/98

**Figure 3.2** Completed near-miss form.

Controlling near-miss incidents is one good way to reduce the overall frequency of accidents. Companies that are leaders in the area of loss control typically place a great deal of importance on reporting near-miss incidents. The vital part, though, is to apply corrective action right away. The only way that can happen effectively is if a near miss is reported immediately after it occurs. That way we can learn as much as possible, as soon as possible.

## First-Aid Incidents

A first-aid incident is a one-time treatment (plus any follow-up visit for the purpose of observation) of minor scratches, cuts, burns, splinters or similar injuries which may call for minor attention but do not ordinarily require medical care. Companies not familiar with OSHA recording and reporting rules frequently confuse the classification of first-aid cases, often misnaming them recordable incidents instead. This can occur when someone totally unfamiliar with the difference is responsible for keeping the safety books. The company's safety numbers can thus be inflated to appear worse than they really are—which will inadvertently bring about higher insurance premiums and insurance modifier rates.

I've often seen companies that list an injury as recordable solely because the treatment was provided by an emergency-room physician. In truth, it doesn't matter *who* provided the treatment; rather, it's *what* was required.

The following treatments are generally considered to be first aid:

- Application of antiseptics during a first visit to medical personnel
- Treatment of first-degree burns
- Application of bandages
- Use of elastic bandages during a first visit to medical personnel
- Removal of a foreign body from the eye if not embedded and only irrigation is involved
- Removal of a foreign body from a wound if not complicated
- Use of nonprescription medications or use of a single dose of prescription medication on a first visit for minor injury or discomfort
- Use of soaking therapy during an initial visit to medical personnel or in order to remove a bandage
- Use of hot or cold compresses or heat therapy during a first visit to medical personnel
- Use of a whirlpool bath during a first visit to medical personnel

First-aid cases, like near misses, are there to be learned from. They're positioned within the third (and red-colored) part of the safety tree model, as verifiable injuries to employees. Safety experts recognize that, averaged over hundreds or thousands of hours worked by company employees, a certain ratio between employee first-aid cases and the more serious recordable cases exists. Let's say that, based on over 65,000 hours worked by your company's crews last year, your company's ratio of first-aid injuries to recordable injuries was 7 to 1: that means that for every safety incident that resulted in one of your employees receiving first-aid treatment, at least one incident entailed medical treatment *beyond* first aid, or met other recordable injury or illness criteria. This is one of the important principles discussed in Chap. 2.

Keeping all of this in mind, can you see that it makes sense to concentrate on preventing first-aid incidents? Beyond prevention, also consider that someone on each worksite crew should be versed in basic first aid, so attention can be given to small injuries—to keep them from becoming recordable injuries at a later date. Also, there needs to be a plan for reporting and accessing emergency medical aid and additional help. You'll read more on this in Chap. 17.

Chapter

# 4

# Recordable Injuries, Illnesses, and Recordkeeping

## Quick Scan

1. Contractor companies must maintain records of certain work-related (occupational) injuries and illnesses.
2. All work-related illnesses are recordable per OSHA.
3. Work-related injuries are recordable per OSHA that:

- Require medical treatment beyond first aid
- Result in loss of consciousness
- Involve restriction of work or motion
- Result in a transfer to another job

4. Refer to Fig. 4.7 on page 42 for a simple guide to recordability of cases under the Occupational Safety and Health Act.
5. At this writing, contractors required to keep injury and illness records should use OSHA 200 logs and OSHA 101 supplemental records forms.
6. Obtain a copy of the OSHA booklet (at this writing) "Recordkeeping Guidelines for Occupational Injuries and Illnesses."

To individuals just going after the meat and potatoes of construction worksite safety, this chapter may at first be difficult to digest. It reads somewhat OSHA-like, being a discussion about the fine lines that exist between worksite injuries and illnesses, and between serious injuries which are recordable on OSHA logs and not-so-serious injuries. It's an important chapter, though. It helps provide a way to measure the severity and frequency of workplace injuries and illnesses, while reviewing some of the recordkeeping requirements prescribed by OSHA. Once the logic for recordkeeping and recordable incident decision making is understood, the concepts in this chapter become far simpler than they first appear.

A common term used by safety professionals and insurance agents is *recordable rate,* meaning a way to measure or express the frequency at which a group or class of employees is incurring work-related injuries or illnesses. The *recordable* part comes from the insistence by OSHA that most companies *record* work-related injuries and illnesses on an official log as a kind of ongoing tally. At the current writing, the official OSHA log referred to is the OSHA 200 log (see Figs. 4.1, 4.2, 4.3, and 4.4). This recording of injuries and illnesses is required for a variety of reasons. First of all, before OSHA came about there was no centralized, comprehensive way that companies collected work-related injury and illness data, and there was no accurate means for anyone to monitor national, statewide, and individual industry occupational health and safety trends and problems. Oh, some individual states and private organizations collected information, but the overall totals and national data were grossly inaccurate due to the many reporting gaps and estimates in the system. To yield uniform reporting of injuries and illnesses, OSHA ruled that most employers of 11 or more employees must maintain records of certain occupational injuries and illnesses as those exposures occur. Note that the "11 or more employees" is a *cumulative number for the year*—meaning that although fewer than 10 individuals may have worked for the company at any one time, if the total number of employees throughout the year totals more than 10 different employees, that's enough to kick in the mandatory recordkeeping requirements. That holds true for companies throughout the country—including those under the jurisdiction of state-run programs.

Most of the recordkeeping requirements center around the use of the OSHA 200 log and the OSHA 101 form (see Figs. 4.5 and 4.6). The OSHA 200 log is a large spreadsheet-type form on which information is entered about work-related injuries and illnesses (refer to Figs. 4.1 through 4.4). Although OSHA says it's okay to use a different form if the requested information is present, you may as well just use the 200 anyway. In short, recordable injuries and illnesses must be entered on the log within 6 days of their occurrence. Logs must be maintained at the company's locale "for five years following the end of the calendar year to which they relate." Each year, from February 1 until March 1, the company must post a copy of the log's annual summary totals and information about workplace injuries and illnesses for the year so that the employees can see them.

The OSHA 101 form is kept as a supplement to the OSHA 200 log, as a more detailed account of recordable injuries and illnesses. It contains information sections about the employer, about the affected employee, about the accident or exposure to occupational illness, about the illness or injury, and about any other detail related to medical treatment or to the incident. Both the OSHA 200 log and the OSHA 101 form must be made available to OSHA "without delay and at reasonable times for examination" by OSHA or Department of Labor representatives.

Bureau of Labor Statistics
Log and Summary of Occupational
Injuries and Illnesses

| NOTE: This form is required by Public Law 91-596 and must be kept in the establishment for *5 years*. Failure to maintain and post can result in the issuance of citations and assessment of penalties. *(See posting requirements on the other side of form.)* | | | RECORDABLE CASES: You are required to record information about every occupational death; every nonfatal occupational illness; and those nonfatal occupational injuries which involve one or more of the following: loss of consciousness, restriction of work or motion, transfer to another job, or medical treatment (other than first aid). *(See definitions on the other side of form.)* | | |
|---|---|---|---|---|---|
| Case or File Number | Date of Injury or Onset of Illness | Employee's Name | Occupation | Department | Description of Injury or Illness |
| Enter a nonduplicating number which will facilitate comparisons with supplementary records. | Enter Mo./day. | Enter first name or initial, middle initial, last name. | Enter regular job title, not activity employee was performing when injured or at onset of illness. In the absence of a formal title, enter a brief description of the employee's duties. | Enter department in which the employee is regularly employed or a description of normal workplace to which employee is assigned, even though temporarily working in another department at the time of injury or illness. | Enter a brief description of the injury or illness and indicate the part or parts of body affected.<br><br>Typical entries for this column might be: Amputation of 1st joint right forefinger; Strain of lower back; Contact dermatitis on both hands; Electrocution--body. |
| (A) | (B) | (C) | (D) | (E) | (F) |
| | | | | | PREVIOUS PAGE TOTALS → |
| | | | | | |
| | | | | | |
| | | | | | |
| | | | | | |
| | | | | | |
| | | | | | |
| | | | | | |
| | | | | | |
| | | | | | |
| | | | | | |
| | | | | | |
| | | | | | |
| | | | | | |
| | | | | | |
| | | | | | |
| | | | | | |
| | | | | | |
| | | | | | |
| | | | | | TOTALS (Instructions on other side of form.) → |

OSHA No. 200

FOLD

**Figure 4.1** OSHA 200 log, page 1.

**U.S. Department of Labor**

For Calendar Year 19 ______ Page ___ of ___

| Company Name | Form Approved O.M.B. No. 1220-0029 |
|---|---|
| Establishment Name | See OMB Disclosure Statement on reverse. |
| Establishment Address | |

| Extent of and Outcome of INJURY | | | | | | Type, Extent of, and Outcome of ILLNESS | | | | | | | | | | | | |
|---|---|---|---|---|---|---|---|---|---|---|---|---|---|---|---|---|---|---|
| Fatalities | Nonfatal Injuries | | | | | Type of Illness | | | | | | | Fatalities | Nonfatal Illnesses | | | | |
| Injury Related | Injuries With Lost Workdays | | | | Injuries Without Lost Workdays | CHECK Only One Column for Each Illness *(See other side of form for terminations or permanent transfers.)* | | | | | | | Illness Related | Illnesses With Lost Workdays | | | | Illnesses Without Lost Workdays |
| Enter DATE of death. Mo./day/yr. | Enter a CHECK if injury involves days away from work, or days of restricted work activity, or both. | Enter a CHECK if injury involves days away from work. | Enter number of DAYS *away from work.* | Enter number of DAYS of *restricted work activity.* | Enter a CHECK if no entry was made in columns 1 or 2 but the injury is recordable as defined above. | Occupational skin diseases or disorders | Dust diseases of the lungs | Respiratory conditions due to toxic agents | Poisoning (systemic effects of toxic materials) | Disorders due to physical agents | Disorders associated with repeated trauma | All other occupational illnesses | Enter DATE of death. Mo./day/yr. | Enter a CHECK if illness involves days away from work, or days of restricted work activity, or both. | Enter a CHECK if illness involves days away from work. | Enter number of DAYS *away from work.* | Enter number of DAYS of *restricted work activity.* | Enter a CHECK if no entry was made in columns 8 or 9. |
| (1) | (2) | (3) | (4) | (5) | (6) | (7) (a) | (b) | (c) | (d) | (e) | (f) | (g) | (8) | (9) | (10) | (11) | (12) | (13) |

INJURIES

ILLNESSES

Certification of Annual Summary Totals By ______________ Title ______________ Date ______________

FOLD

OSHA No. 200 **POST ONLY THIS PORTION OF THE LAST PAGE NO LATER THAN FEBRUARY 1.**

**Figure 4.2** OSHA 200 log, page 2.

Public reporting burden for this collection of information is estimated to vary from 4 to 30 (time in minutes) per response with an average of 15 (time in minutes) per response, including the time for reviewing instructions, searching existing data sources, gathering and maintaining the data needed, and completing and reviewing the collection of information. If you have any comments regarding this estimate or any other aspect of this information collection, including suggestions for reducing this burden, please send them to the OSHA Office of Statistics and/or the Department of Labor, Office of IRM Policy, Room N-1301, 200 Constitution Avenue, N.W. Washington, D.C. 20210

**Instructions for OSHA No. 200**

**I. Log and Summary of Occupational Injuries and Illnesses**

Each employer who is subject to the recordkeeping requirements of the Occupational Safety and Health Act of 1970 must maintain for each establishment a log of all recordable occupational injuries and illnesses. This form (OSHA No. 200) may be used for that purpose. A substitute for the OSHA No. 200 is acceptable if it is as detailed, easily readable, and understandable as the OSHA No. 200.

Enter each recordable case on the log within six (6) workdays after learning of its occurrence. Although other records must be maintained at the establishment to which they refer, it is possible to prepare and maintain the log at another location, using data processing equipment if desired. If the log is prepared elsewhere, a copy updated to within 45 calendar days must be present at all times in the establishment.

Logs must be maintained and retained for five (5) years following the end of the calendar year to which they relate. Logs must be available (normally at the establishment) for inspection and copying by representatives of the Department of Labor, or the Department of Health and Human Services, or States accorded jurisdiction under the Act. Access to the log is also provided to employees, former employees and their representatives.

**II. Changes in Extent of or Outcome of Injury or Illness**

If, during the 5-year period the log must be retained, there is a change in an extent and outcome of an injury or illness which affects entries in columns 1, 2, 6, 8, 9, or 13, the first entry should be lined out and a new entry made. For example, if an injured employee at first required only medical treatment but later lost workdays away from work, the check in column 6 should be lined out, and checks entered in columns 2 and 3 and the number of lost workdays entered in column 4.

In another example, if an employee with an occupational illness lost workdays, returned to work, and then died of the illness, any entries in columns 9 through 12 should be lined out and the date of death entered in column 8.

The entire entry for an injury or illness should be lined out if later found to be nonrecordable. For example: an injury which is later determined not to be work related, or which was initially thought to involve medical treatment but later was determined to have involved only first aid.

**III. Posting Requirements**

A copy of the totals and information following the fold line of the last page for the year must be posted at each establishment in the place or places where notices to employees are customarily posted. This copy must be posted no later than ***February 1 and must remain in place until March 1.***

Even though there were no injuries or illnesses during the year, zeros must be entered on the totals line, and the form posted.

The person responsible for the ***annual summary totals*** shall certify that the totals are true and complete by signing at the bottom of the form.

**IV. Instructions for Completing Log and Summary of Occupational Injuries and Illnesses**

Column A – **CASE OR FILE NUMBER.** Self-explanatory.

**Column B** – **DATE OF INJURY OR ONSET OF ILLNESS.**
For occupational injuries, enter the date of the work accident which resulted in injury. For occupational illnesses, enter the date of initial diagnosis of illness, or, if absence from work occurred before diagnosis, enter the first day of the absence attributable to the illness which was later diagnosed or recognized.

**Columns C through F** – Self-explanatory.

**Columns 1 and 8** – **INJURY OR ILLNESS-RELATED DEATHS.**
Self-explanatory.

**Columns 2 and 9** – **INJURIES OR ILLNESSES WITH LOST WORKDAYS.**
Self-explanatory.

Any injury which involves days away from work, or days of restricted work activity, or both must be recorded since it always involves one or more of the criteria for recordability.

**Columns 3 and 10** – **INJURIES OR ILLNESSES INVOLVING DAYS AWAY FROM WORK.** Self-explanatory.

**Columns 4 and 11** – **LOST WORKDAYS—DAYS AWAY FROM WORK.**
Enter the number of workdays (consecutive or not) on which the employee would have worked but could not because of occupational injury or illness. The number of lost workdays should not include the day of injury or onset of illness or any days on which the employee would not have worked even though able to work.
**NOTE:** For employees not having a regularly scheduled shift, such as certain truck drivers, construction workers, farm labor, casual labor, part-time employees, etc., it may be necessary to estimate the number of lost workdays. Estimates of lost workdays shall be based on prior work history of the employee AND days worked by employees, not ill or injured, working in the department and/or occupation of the ill or injured employee.

**Columns 5 and 12** – **LOST WORKDAYS—DAYS OF RESTRICTED WORK ACTIVITY.**
Enter the number of workdays (consecutive or not) on which because of injury or illness:

(1) the employee was assigned to another job on a temporary basis, or
(2) the employee worked at a permanent job less than full time, or
(3) the employee worked at a permanently assigned job but could not perform all duties normally connected with it.

The number of lost workdays should not include the day of injury or onset of illness or any days on which the employee would not have worked even though able to work.

*U.S. Government Printing Office: 1995 — 387-186/22983

**Figure 4.3** OSHA 200 log, page 3.

Columns
6 and 13 – INJURIES OR ILLNESSES WITHOUT LOST WORKDAYS. Self-explanatory.

Columns 7a
through 7g – TYPE OF ILLNESS.
Enter a check in only *one* column for each illness.

TERMINATION OR PERMANENT TRANSFER—Place an asterisk to the right of the entry in columns 7a through 7g (type of illness) which represented a termination of employment or permanent transfer.

V. **Totals**

Add number of entries in columns 1 and 8.
Add number of checks in columns 2, 3, 6, 7, 9, 10, and 13.
Add number of days in columns 4, 5, 11, and 12.
Yearly totals for each column (1-13) are required for posting. Running or page totals may be generated at the discretion of the employer.

If an employee's loss of workdays is continuing at the time the totals are summarized, estimate the number of future workdays the employee will lose and add that estimate to the workdays already lost and include this figure in the annual totals. No further entries are to be made with respect to such cases in the next year's log.

VI. **Definitions**

OCCUPATIONAL INJURY is any injury such as a cut, fracture, sprain, amputation, etc., which results from a work accident or from an exposure involving a single incident in the work environment.
NOTE: Conditions resulting from animal bites, such as insect or snake bites or from one-time exposure to chemicals, are considered to be injuries.

OCCUPATIONAL ILLNESS of an employee is any abnormal condition or disorder, other than one resulting from an occupational injury, caused by exposure to environmental factors associated with employment. It includes acute and chronic illnesses or diseases which may be caused by inhalation, absorption, ingestion, or direct contact.

The following listing gives the categories of occupational illnesses and disorders that will be utilized for the purpose of classifying recordable illnesses. For purposes of information, examples of each category are given. These are typical examples, however, and are not to be considered the complete listing of the types of illnesses and disorders that are to be counted under each category.

7a. **Occupational Skin Diseases or Disorders**
Examples: Contact dermatitis, eczema, or rash caused by primary irritants and sensitizers or poisonous plants; oil acne; chrome ulcers; chemical burns or inflammations; etc.

7b. **Dust Diseases of the Lungs (Pneumoconioses)**
Examples: Silicosis, asbestosis and other asbestos-related diseases, coal worker's pneumoconiosis, byssinosis, siderosis, and other pneumoconioses.

7c. **Respiratory Conditions Due to Toxic Agents**
Examples: Pneumonitis, pharyngitis, rhinitis or acute congestion due to chemicals, dusts, gases, or fumes; farmer's lung; etc.

7d. **Poisoning (Systemic Effect of Toxic Materials)**
Examples: Poisoning by lead, mercury, cadmium, arsenic, or other metals; poisoning by carbon monoxide, hydrogen sulfide, or other gases; poisoning by benzol, carbon tetrachloride, or other organic solvents; poisoning by insecticide sprays such as parathion, lead arsenate; poisoning by other chemicals such as formaldehyde, plastics, and resins; etc.

7e. **Disorders Due to Physical Agents (Other than Toxic Materials)**
Examples: Heatstroke, sunstroke, heat exhaustion, and other effects of environmental heat; freezing, frostbite, and effects of exposure to low temperatures; caisson disease; effects of ionizing radiation (isotopes, X-rays, radium); effects of nonionizing radiation (welding flash, ultraviolet rays, microwaves, sunburn); etc.

7f. **Disorders Associated With Repeated Trauma**
Examples: Noise-induced hearing loss; synovitis, tenosynovitis, and bursitis; Raynaud's phenomena; and other conditions due to repeated motion, vibration, or pressure.

7g. **All Other Occupational Illnesses**
Examples: Anthrax, brucellosis, infectious hepatitis, malignant and benign tumors, food poisoning, histoplasmosis, coccidioidomycosis, etc.

MEDICAL TREATMENT includes treatment (other than first aid) administered by a physician or by registered professional personnel under the standing orders of a physician. Medical treatment does NOT include first-aid treatment (one-time treatment and subsequent observation of minor scratches, cuts, burns, splinters, and so forth, which do not ordinarily require medical care) even though provided by a physician or registered professional personnel.

ESTABLISHMENT: A single physical location where business is conducted or where services or industrial operations are performed (for example: a factory, mill, store, hotel, restaurant, movie theater, farm, ranch, bank, sales office, warehouse, or central administrative office). Where distinctly separate activities are performed at a single physical location, such as construction activities operated from the same physical location as a lumber yard, each activity shall be treated as a separate establishment.

For firms engaged in activities which may be physically dispersed, such as agriculture; construction; transportation; communications; and electric, gas, and sanitary services, records may be maintained at a place to which employees report each day.

Records for personnel who do not primarily report or work at a single establishment, such as traveling salesmen, technicians, engineers, etc., shall be maintained at the location from which they are paid or the base from which personnel operate to carry out their activities.

WORK ENVIRONMENT is comprised of the physical location, equipment, materials processed or used, and the kinds of operations performed in the course of an employee's work, whether on or off the employer's premises.

**Figure 4.4** OSHA 200 log, page 4.

Bureau of Labor Statistics
Supplementary Record of
Occupational Injuries and Illnesses

**U.S. Department of Labor**

| **This form is required by Public Law 91-596 and must be kept in the establishment for *5 years.* Failure to maintain can result in the issuance of citations and assessment of penalties.** | Case or File No. | Form Approved O.M.B. No. 1220-0029 |
|---|---|---|

**Employer**

1. Name
2. Mail address *(No. and street, city or town, State, and zip code)*
3. Location, if different from mail address

**Injured or Ill Employee**

4. Name *(First, middle, and last)* — Social Security No.
5. Home address *(No. and street, city or town, State, and zip code)*
6. Age — 7. Sex: *(Check one)* Male ☐ Female ☐
8. Occupation *(Enter regular job title, not the specific activity he was performing at time of injury.)*
9. Department *(Enter name of department or division in which the injured person is regularly employed, even though he may have been temporarily working in another department at the time of injury.)*

**The Accident or Exposure to Occupational Illness**

If accident or exposure occurred on employer's premises, give address of plant or establishment in which it occurred. Do not indicate department or division within the plant or establishment. If accident occurred outside employer's premises at an identifiable address, give that address. If it occurred on a public highway or at any other place which cannot be identified by number and street, please provide place references locating the place of injury as accurately as possible.

10. Place of accident or exposure *(No. and street, city or town, State, and zip code)*
11. Was place of accident or exposure on employer's premises? Yes ☐ No ☐
12. What was the employee doing when injured? *(Be specific. If he was using tools or equipment or handling material, name them and tell what he was doing with them.)*
13. How did the accident occur? *(Describe fully the events which resulted in the injury or occupational illness. Tell what happened and how it happened. Name any objects or substances involved and tell how they were involved. Give full details on all factors which led or contributed to the accident. Use separate sheet for additional space.)*

**Occupational Injury or Occupational Illness**

14. Describe the injury or illness in detail and indicate the part of body affected. *(E.g., amputation of right index finger at second joint; fracture of ribs; lead poisoning; dermatitis of left hand, etc.)*
15. Name the object or substance which directly injured the employee. *(For example, the machine or thing he struck against or which struck him; the vapor or poison he inhaled or swallowed; the chemical or radiation which irriatated his skin; or in cases of strains, hernias, etc., the thing he was lifting, pulling, etc.)*
16. Date of injury or initial diagnosis of occupational illness — 17. Did employee die? *(Check one)* Yes ☐ No ☐

**Other**

18. Name and address of physician
19. If hospitalized, name and address of hospital

| Date of report | Prepared by | Official position |
|---|---|---|

OSHA No. 101 (Feb. 1981)

**Figure 4.5** OSHA 101 log, page 1.

**SUPPLEMENTARY RECORD OF OCCUPATIONAL INJURIES AND ILLNESSES**

To supplement the Log and Summary of Occupational Injuries and Illnesses (OSHA No. 200), each establishment must maintain a record of each recordable occupational injury or illness. Worker's compensation, insurance, or other reports are acceptable as records if they contain all facts listed below or are supplemented to do so. If no suitable report is made for other purposes, this form (OSHA No. 101) may be used or the necessary facts can be listed on a separate plain sheet of paper. These records must also be available in the establishment without delay and at reasonable times for examination by representatives of the Department of Labor and the Department of Health and Human Services, and States accorded jurisdiction under the Act. The records must be maintained for a period of not less than five years following the end of the calendar year to which they relate.

Such records must contain at least the following facts:

1) ***About the employer***—name, mail address, and location if different from mail address.

2) ***About the injured or ill employee***—name, social security number, home address, age, sex, occupation, and department.

3) ***About the accident or exposure to occupational illness***—place of accident or exposure, whether it was on employer's premises, what the employee was doing when injured, and how the accident occurred.

4) ***About the occupational injury or illness***—description of the injury or illness, including part of body affected; name of the object or substance which directly injured the employee; and date of injury or diagnosis of illness.

5) ***Other***—name and address of physician; if hospitalized, name and address of hospital; date of report; and name and position of person preparing the report.

SEE *DEFINITIONS* ON THE BACK OF OSHA FORM 200.

**Figure 4.6** OSHA 101 log, page 2.

Simply put, recordable incidents can mean a number of different injury or illness categories, including work-related fatalities; work-related illnesses; work-related injuries which require medical treatment other than first aid; and work-related injuries which involve days away from work, restriction of work or motion, transfer to another job, or loss of consciousness. The formula for determining a recordable incident rate is likewise fairly simple, but it's best done with a calculator in hand because a large number is used in the ratio's numerator to equalize the rate so that operations with relatively small numbers of hours worked can be compared with larger companies in a meaningful way.

The formula for determining a recordable rate is as follows:

$$\text{Recordable rate} = \frac{\text{Number of recordables} \times 200{,}000}{\text{Number of hours worked}}$$

Let's say you have 11 employees who worked a total of 26,000 hours last year, and you had 3 recordable injuries. Your recordable rate computations would be as follows:

$$\text{Recordable rate} = \frac{3 \text{ recordables} \times 200{,}000}{26{,}000 \text{ hours}}$$

$$\text{Recordable rate} = \frac{600{,}000}{26{,}000} = 23.07$$

The 200,000 number represents the total hours 100 employees would work during a year (40 hours/week × 50 weeks × 100 employees = 200,000). The 23.07 means that the company is working with a recordable injury and illness rate that expects 23 out of each 100 workers to experience a recordable incident sometime throughout the year.

Recordable rates, along with *experience insurance modifiers* (EMRs), are what many companies use to gauge the effectiveness of a contractor's or company's safety program. You already know what recordable rates are. EMRs are numbers based on actual safety records that indicate how safe a particular company is in relation to the safety record for the company's overall industry. A modifier of 1 says the company's insurance rate costs are exactly the industry average. A modifier of 1.20 says the insurance cost will run an additional 20 percent, while a modifier of .80 could bring an insurance safety *savings* of 20 percent. When all is said and done, such businesses as financial institutions, real estate companies and major investors prefer to associate and work with the safer home construction companies—companies with lower recordable rates. That goes for the selection of subcontractors, too. Nowadays it's too risky to partner up with an unsafe contractor. In addition to the unsafe behavior that such a company's employees bring to the site, there's a chance that some of the bad habits of visiting workers could be tried by the main contractor's otherwise safe employees. What's more, with the high costs that accompany workplace illnesses and injuries, unsafe contractors and subcontractors are finding it increasingly difficult to make a profit.

## Guidelines to Injury and Illness Recordability

According to OSHA, recording an injury or illness under the OSHA system does not necessarily imply that management was at fault, that the worker was at fault, that a violation of an OSHA standard has occurred, or that the injury or illness is compensable under workers' compensation or other systems. The scope of recordability of the OSHA system is broader and more inclusive than that of most other recordkeeping systems. OSHA includes injuries and illnesses that may not be *compensable* in the workers' compensation context, or even *recordable* under individual company safety and health recordkeeping systems. The reason for this was "to make the system as simple as possible, and the alternative of developing a detailed list of exceptions for not recording specific injuries and illnesses was felt to impose far greater administrative and reporting burdens on most employers than requiring that a relatively small number of borderline cases be recorded." The simple and streamlined OSHA recording guidelines encourage a valid, consistent, and uniform recordkeeping system that's capable of producing reliable statistical information.

Use the OSHA guidelines for recordability, and don't try to develop your own criteria. The bible for recordkeeping guidelines is OSHA's "Recordkeeping Guidelines for Occupational Injuries and Illnesses," published by the U.S. Department of Labor, Bureau of Labor Statistics, effective April 1986, based on the Occupational Health and Safety Act of 1970 and 29 CFR 1904. That booklet contains guidelines for keeping the occupational injury and illness records necessary to fulfill your recordkeeping obligations.

As far as *who* has the authority to determine recordkeeping decisions, it is clear that OSHA gives it to the employers, or in this case, the contractor's management. Further, OSHA comments on its requirement of good faith policy:

> Although employers ultimately decide if and how a particular case should be recorded, their decision must not be an arbitrary one, but should be made in accordance with the requirements of the act, regulations, and instructions on the forms, and the guidelines in this report ["Recordkeeping Guidelines for Occupational Injuries and Illnesses"]. Information from medical, hospital, or supervisors' record should be reviewed along with other pertinent information, and the employee should be interviewed to determine his or her medical condition and ability to perform normal job duties.

What checks and balances exist in the system? From a worker's point of view, what makes an employer toe the line and follow the recordkeeping rules? OSHA built the system for accurate reporting when it required the involvement of all participants in the recordkeeping and reporting system. Consider the importance of collecting meaningful workplace injury and illness data from which safety improvements can be planned. Consider the mandatory yearly postings of OSHA logs for review by employees. Consider the periodic log and report inspections by OSHA. All of these are designed to verify the accuracy and validity of employer recordkeeping determinations. Additional incentives for companies to keep accurate injury and illness records are the stiff financial and criminal penalties (with possible imprisonment) for recordkeeping violation convictions.

## Recordability Decision Making

The recordability decision-making process regarding injury or illness exposures consists of five steps:

1. Determining whether a case (a death, injury, or illness) has occurred
2. Determining that the case is work related
3. Deciding whether the case is an injury or an illness
4. Recording the case as an illness, if appropriate
5. Deciding if an injury is a recordable case

A review of the five steps follows:

**Step 1: Determining whether a case (a death, injury, or illness) has occurred.** The recognition of an injury or illness is usually a simple matter. But some situations can be confusing. It may sound contrary to what you'd first think, but fault plays no role in determining recordability. OSHA makes no distinction between incidents that are compensable under workers' compensation laws, incidents caused by employer neglect, incidents that are preventable, or random incidents in which no one appears to be at fault. In other words, it doesn't matter if the employee, or the company, or no one was at fault, a recordable incident still gets recorded in any of those cases. For example, an employee falls backwards off a bench while eating lunch—if a resulting injury meets the medical criteria, it must be logged on the company's 200 log. Also, an employee needn't be involved in a specific job task to be injured at work, nor are cases recordable only if they occur during hours for which wages are paid. An employee arriving early or staying late at the worksite is still considered to be under the company's employ.

**Step 2: Determining that the case is work related—that it resulted from an event or exposure in the work environment.** Generally, injuries and illnesses that result from an event or exposure on the employer's premises are considered to be work related. In our case, the *employer's premises* extends to home construction sites. OSHA's definition includes "the primary work facility and other areas which are considered part of the employer's general work area." When an employee is off the employer's premises, work relationship must be established. This includes not only the primary facility, but also such areas as company storage facilities. In addition to physical locations, equipment or materials used in the course of an employee's work environment are included. See Fig. 4.7 for guidelines for establishing work relationship.

Injury exposures experienced by employees transporting, picking up, and unloading equipment or materials away from the jobsite or work premises would still be work related because the employees are engaged in work-related activities. A company-owned, -rented, or -borrowed truck, backhoe, or front-end loader is considered part of the work environment even though it is not part of the employer's premises or worksite. The work environment in these instances includes locations where employees are present due to the nature of their jobs or as a condition of their employment.

If an employee reports for work, but then performs a personal errand or task away from the work environment and receives an injury, that injury is not recordable. OSHA explains that work relationship must be established for employee activities off premises—it is not presumed. To be performing work-related activity off premises, the employee must have been doing some job, task, or service for the employer, or must have been present at the off-premises location in connection with his or her employment. If the employee is off the employer's premises, and leaves the normal area of operations entirely for his or her own purpose, then these activities would not be considered work related.

Is an injury occurring during the lunch break of an employee working off the employer's premises considered to be work related? It would be if it occurred in the off-premises work environment or if it was a work-related lunch. If individuals leave the work environment for lunch and are injured, those injuries are not recordable unless the lunch is in some way required by their jobs.

What about horseplay? Do employers have to record injuries that clearly result from horseplay or workplace violence, such as a robbery or dog bite? Yes. Again, injuries that occur during activities in the work environment are subject to recordability, without distinction between fault.

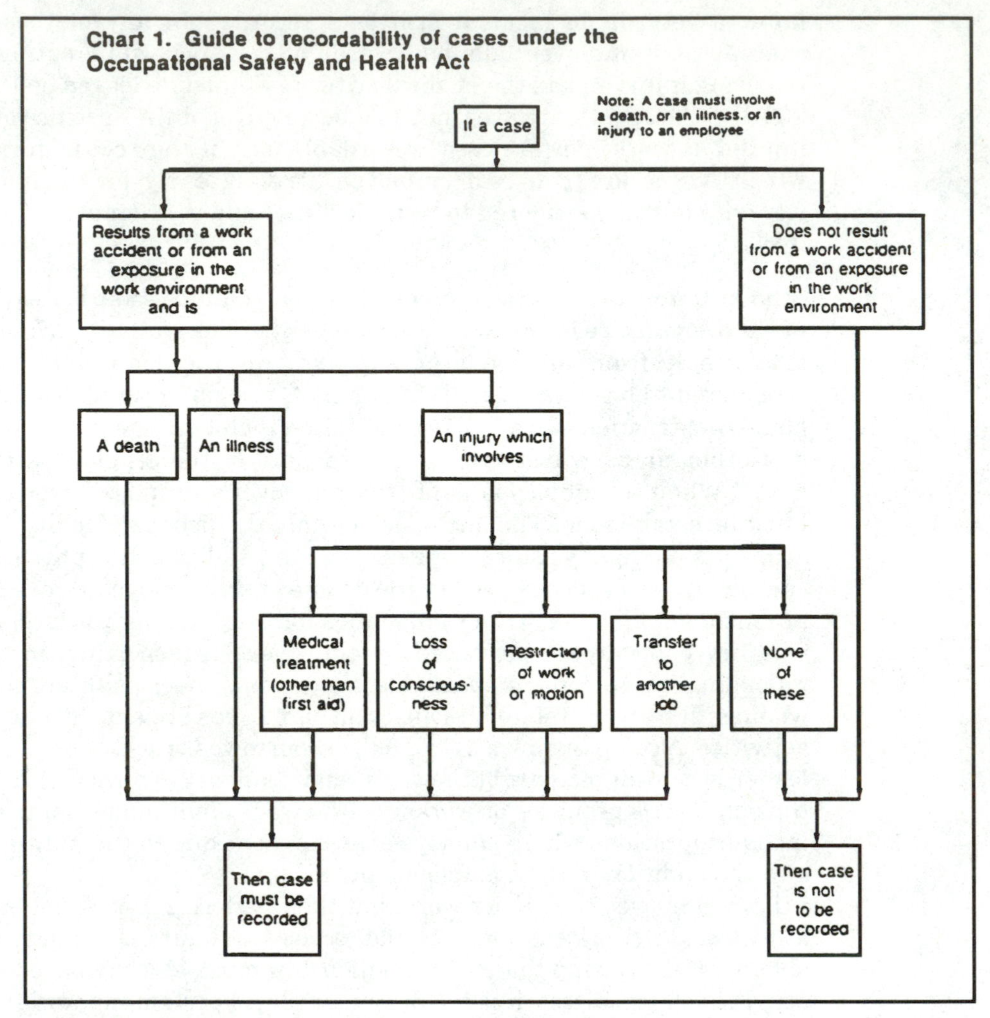


**Figure 4.7** OSHA guide to recordability.

Are there specific requirements for evaluating the occurrence of back or hernia cases? No. Back and hernia cases should be evaluated in the same manner as any other case. Questions concerning the recordability of these cases usually revolve around the impact of a previous back or hernia condition on the recordability of the case, or on whether or not the back injury or hernia was work related.

Preexisting conditions generally do not impact the recordability of cases under the OSHA system. For a back or hernia case to be considered work related, it must have resulted from a work-related event or exposure in the work environment. Employers may sometimes be able to distinguish between back injuries that *result* from an event in the work environment, and back injuries that are caused elsewhere and merely *surface* in the work environment. The former are recordable; the latter are not. Of course, this test should be applied to all injuries and illnesses, not just back and hernia cases.

An employee's back goes out while he or she is performing a routine activity at work. Assuming the employee was not involved in any stressful activity, such as lifting a heavy object, is the case recordable? Particularly stressful activity is not required. If such an event as a slip, trip, fall, or sharp twist occurred in the work environment *that caused or contributed to the injury,* the case would be recordable, assuming it meets the other requirements for recordability.

Must there be an identifiable event or exposure in the work environment for there to be a recordable case? What if someone experiences a backache, but cannot identify the particular movement which caused the injury? Usually, there will be an identifiable event or exposure to which the employer or employee can attribute the injury or illness. However, this is not necessary for recordkeeping purposes. If it seems likely that an event or exposure in the work environment either caused or contributed to the case, the case is recordable, even though the exact time or location of the particular event or exposure cannot be identified.

If the backache is known to have resulted from some non-work-related activity outside the work environment and has merely surfaced at work, then the employer need not record the case. In these situations, employers may want to document the reasons why they feel the case is not work related.

What about cases where the employee alleges that an injury or illness has occurred? Must employers record these cases without any medical verification? Medical verification is not required for recordability. However, employers have the ultimate responsibility for making good-faith recordkeeping determinations. If an employer doubts the validity of an employee's alleged injury or illness and there is no evidence supporting the allegation, the employer need not record the case.

Must occupational injuries and illnesses that are disputed be recorded? Within 6 workdays after receiving information that an injury or illness has occurred, the employer must determine whether the case is recordable. Questionable cases should be entered on the OSHA 200 log, and can be lined out at a later date if they are found to be not recordable.

**Step 3: Deciding whether the case is an injury or an illness.** Under OSHA, all work-related illnesses must be recorded, while only those injuries which require medical treatment other than first aid or involve loss of consciousness, restriction of work or motion, or transfer to another job are recordable. Whether a case involves an injury or illness is determined by the nature of the original event or exposure which caused the case, not by the resulting condition of the affected employee. *Injuries* are caused by instantaneous events in the work environment. Cases resulting from anything other than instantaneous events are considered *illnesses.* This concept of illnesses includes acute illnesses which result from exposures of relatively short duration.

An *occupational injury* is defined on the back of the log and summary form, OSHA 200, as "any injury such as a cut, fracture, sprain, amputation, etc., which results from a work accident or from an exposure involving a single incident in the work environment." Conditions resulting from animal bites, such as insect, dog, or snake bites, or from one-time exposure to chemicals are considered to be injuries.

An *occupational illness,* as defined on the OSHA 200 log, is "any abnormal condition or disorder, other than one resulting from an occupational injury, caused by exposure to environmental factors associated with employment. It includes acute and chronic illnesses or diseases which may be caused by inhalation, absorption, ingestion, or direct contact."

Some conditions could be classified as either an injury or an illness (but not both), depending on the nature of the event that produced the condition.

An infection from a laceration would be classified as an injury because the classification is based on the original event—the laceration—not on the subsequent developments.

Back cases can be tricky. Frequently, situations arise where an employee complains about back pain, but is unable to associate it with a single instantaneous event. OSHA says that most back cases should be classified as injuries—the ones for which identifiable events occur, and also for cases where a specific event cannot be identified, since back cases are usually triggered by some specific movement, such as a slip, trip, fall, or sharp twist.

*Carpal tunnel syndrome* is a condition involving compression of the median nerve in the wrist, which results in tingling, discomfort, and numbness in the thumb, index, and long fingers. Because work-related carpal tunnel syndrome cases almost always result from repetitious movement, they should be classified as occupational illnesses. The entry for these cases should be in column 7(f) of the log for disorders associated with repeated trauma.

OSHA provides a further example:

> A chemical worker contracted a mild case of dermatitis on both hands while working in a solution for several hours. The employee was sent to the doctor, who recommended application of a topical lotion (a commercial, non-prescription remedy). The employee bought a bottle of the lotion and treated the rash for a few days until it disappeared. There were no subsequent visits to the doctor. The rash did not prevent the employee from performing all the duties of the job.

The answer OSHA gives is:

> The case is a recordable occupational illness. The answer to this question is based on the distinction between an injury and an illness. If considered an injury, the case would not be recordable since no medical treatment was provided. However, since the case almost certainly did not involve a single instantaneous exposure, it should be classified as an occupational illness. Consequently, the kind of treatment given by the doctor (none in this case) is immaterial, since all occupational illnesses are recordable.

**Step 4: Recording the case as an illness, if appropriate.** Occupational illnesses must be diagnosed to be recordable. However, they do not necessarily have to be diagnosed by a physician or by other medical personnel. Diagnosis may be by a physician, a registered nurse, or a person who by training or experience is capable to make such a determination. Employers, employees, and others may be able to detect some illnesses, such as skin diseases or disorders, without the benefit of specialized medical training. However, a case that is more difficult to diagnose, such as silicosis, would require evaluation by properly trained medical personnel.

In addition to recording the occurrence of occupational illnesses, employers are required to record each illness case in one of the seven categories on the front of the log. The back of the log form contains a listing of types of illnesses or disorders and gives examples for each illness category. Recording and classifying occupational illnesses may be difficult for employers, especially the chronic and long-term latent illnesses. Many illnesses are not easily detected, and it is often difficult to determine whether an illness is work related. Also, employees may not report illnesses because the symptoms may not be readily apparent, or because they do not think the illness is serious or work related.

Are employee complaints of such common subjective symptoms as general malaise, headache, and/or nausea recordable as cases of illnesses if there are no indications that the symptoms are work related? No. Such subjective symptoms are not recordable if there is no apparent association with the employee's work environment. However, in evaluating these cases, employers should be aware that many subjective complaints, including feelings of malaise, headache, and nausea are symptomatic of a wide range of diseases, a number of which are occupational in origin. In this regard, employers should pay attention to the distribution of such subjective complaints with respect to time and place, particularly when such complaints are observed to occur among one or more groups of employees.

**Step 5: Deciding if an injury is a recordable case.** The following guidelines will help you determine if work-related injuries are recordable:

*Medical treatment other than first aid is required.*

- Injuries that must be treated *only* by a physician or licensed medical personnel. However, it's *the kind of treatment* which is, or should be, provided that's the determining factor, not the place or person providing the treatment. If a case comes under the definition of *medical treatment* rather than *first aid,* it's recordable. On the other hand, first-aid treatment is not recordable, even when given by a doctor. The key factor to consider is the type of treatment which is, or should be, provided, not the person administering it.

  If treatment was not given, but should have been, that's enough to meet the definition of recordability. Let's say a worker rips through the web of skin between the thumb and forefinger with a saw, but doesn't report it and takes care of the injury without treatment. Eventually, the wound heals, with considerable scarring and a reduced mobility of the thumb. This is a case that should be recorded because medical treatment was clearly required, but was not actually provided.

- Injuries that impair bodily functions, such as normal use of senses or limbs.
- Injuries that result in damage to physical structure of a nonsuperficial nature, such as a fracture. Some individuals may receive injuries in which a

hairline fracture occurs that is given no treatment and does not interfere with the employee's work activities. Such hairline fractures are still recordable, because they are considered serious and will generally require medical treatment or restriction of work or motion.

- Injuries that involve complications requiring follow-up medical treatment. Casts, splints, and orthopedic devices to immobilize are almost always medical treatments. The use of elastic bandages or nonconstraining devices, such as wristlets, are considered first-aid treatment regardless of how long or how often they are used.
- Follow-up visits to a doctor or medical facility for minor cuts or burns, if simply for observation or to change an adhesive or small bandage, are not recordable. The injury becomes recordable, though, if any medical treatment is provided.

Examples which are almost always recordable include the following:

- Treatment of *infection.* If, however, the employee has a minor scratch and the doctor administers a preventive tetanus shot anyway, the shot does not count as medical treatment. Rabies shots, however, are a different animal—they're vaccinations that are considered absolutely necessary, and they involve a series of injections far beyond what is considered first aid. On the other hand, medical advances may soon provide a less complicated vaccination for rabies, and it could then be considered first aid.
- Application of *antiseptics* during second or subsequent visits to medical personnel.
- Treatment of *second* or *third degree burns.*
- Application of *sutures* or *stitches.*
- Application of *butterfly adhesive dressings* or *steri strips* in lieu of sutures.
- Removal of *foreign bodies embedded in eye.*
- Removal of *foreign bodies from wound,* if procedure is *complicated* because of depth of embedment, size, or location.
- *Chipped* or *broken teeth,* because they ordinarily require medical treatment.
- *Use of prescription medicine* (except a single dose administered on a first visit for minor injury or discomfort). Just the issuance of the prescription is enough to make it recordable. Also, a single dose of each of two prescription medicines on a first visit, such as eyedrops for local anesthetic, and eyedrops to treat or prevent infection, can qualify as more than a single dose, and thus qualify as a recordable case.
- Use of *hot* or *cold soaking therapy* during a second or subsequent visit to medical personnel.
- Application of *hot* or *cold compresses* during a second or subsequent visit to medical personnel. The use of ice packs or hot compresses on a second visit

is considered medical treatment even if it is for a minor injury, such as a muscle strain.

- *Cutting away dead skin* (surgical debridement) is considered medical treatment even when a nurse treats or dresses a minor wound and records "minor wound debrided and dressed."
- Application of *heat therapy* during a second or subsequent visit to medical personnel.
- *Positive X-ray diagnosis* (fractures, broken bones, etc.).
- A single treatment by a chiropractor is considered first aid. A *series of treatments* is considered medical treatment.
- *Admission to a hospital* or equivalent medical facility *for treatment*; hospitalization for observation does not denote recordability.

*An employee loses consciousness as the result of a work-related injury or condition.*

*Restriction of work or motion results.* If the physician places an employee on restricted work duty—that is, unless the worker is at his or her permanently assigned job, is working a full shift, and is able to do each and every task required by that job—the injury qualifies as recordable for restricted duty.

*The employee must be transferred to another job as the result of a work-related injury or condition.* Injuries requiring employees to be transferred to another job are considered serious enough to be recordable regardless of the type of treatment provided.

## Fatalities

This is a topic we'd all rather do without. It has many implications, and triggers activities that inevitably make everyone at an affected company wish more time and effort had been spent doing the things that could have prevented the fatality from happening in the first place. In short, there are many ills a fatality brings, beyond, of course, the pain and loss to the deceased and his or her family.

A fatality will have a psychological effect on the company's management, on the employees, and, to a certain extent, on all of their families. A fatality will demand that considerable effort be spent completing reports to regulatory agencies and other related paperwork, and it could well result in citations, fines, or even criminal prosecution.

A fatality will certainly bring the wrong kind of publicity to a business. It can have a long-term psychological impact on all affected employees, as well.

A fatality, as seen in the safety model, is at the very peak of the treetop. It often results when recordable and lost-time injuries occur with a frequency that eventually tips the odds that one of them will break into the next higher and final level of severity—that of death.

But, unfortunately, a fatality doesn't always *have* to come from that kind of situation. You can have an excellent safety management program and still have a brief lapse where a situation develops and leads to a critical injury. It could be from an individual's lack of training on a particular task, or an employee could deliberately decide to act unsafely, for a variety of reasons. That's why your safety program must be designed to reward safe, not unsafe, behavior—to prevent someone from deciding to perform an unsafe act in the first place.

Fatalities have a way of staying with you for a long time. Suddenly, people throughout the affected company realize the importance of safety management, and the reason for all of that preparation that may have seemed boring or unnecessary at the time. Indeed, when the news hits and you're on your way to the accident scene; when you're hoping for the best, that it's just a close call—a warning to get things in order—but then you see the ambulance or emergency crews there; when you have to notify the victim's family and attend the funeral and answer dozens of questions with the incident investigation; when you're filling out the cold, stark paperwork, resting in the horrible stigma surrounding the situation—that's when you truly realize the reason for developing an effective program of safety management.

Chapter

# 5

# OSHA and Safety Management

## Quick Scan

1. Obtain the latest copy of the *Code of Federal Regulations,* Title 29, Part 1926 (see Fig. 5.1), which contains the text of OSHA standards for the construction industry.
2. Obtain a copy of *Selected Construction Regulations for the Home Building Industry* (see Fig. 5.2), which excerpts many standards applicable to residential construction and related trades from Title 29.
3. Familiarize yourself with OSHA's online website, at http://www.osha.gov, where OSHA regulations are now available.
4. Highlight the sections of OSHA standards that cover your work tasks and programs, and make sure you are at least meeting the minimum requirements.
5. Avail yourself and your company of OSHA's services. Work with your local office.
6. Keep current on OSHA regulations affecting your industry through the *Federal Register,* trade organizations and publications, safety consulting or publishing services, or OSHA's ever-growing bank of electronic information available on the Internet and through various fax and CD-ROM resources.

## OSHA

In its infancy, almost 30 years ago, *OSHA* stood for the *Occupational Safety and Health Act of 1970.* It was, and still is, a public law passed by Congress. Under the Act, the Occupational Safety and Health Administration (which, coincidentally, also has the same initials—OSHA) was created within the code of federal regulations

Labor

29

PART 1926

**Revised as of July 1, 1997**

**Figure 5.1** 29 CFR 1926 Code of Federal Regulations.

Selected Construction Regulations for the Home Building Industry

U.S. Department of Labor
Occupational Safety and Health Administration

1997

**Figure 5.2** OSHA's selected construction regulations for the home building industry.

Department of Labor. The initials stuck with the agency; thereafter, the law became known as simply "the Act," as it is still known today.

## OSHA's Purpose

OSHA was, as the agency itself states, created to do the following:

- Encourage employers and employees to reduce workplace hazards and to implement new or improved existing safety and health programs.
- Provide for research in occupational safety and health to develop innovative ways of dealing with occupational health and safety problems.
- Establish "separate but dependent responsibilities and rights" for employers and employees for the achievement of better safety and health conditions.
- Maintain a reporting and recordkeeping system to monitor job-related injuries and illnesses.
- Establish training programs to increase the number and competence of occupational safety and health personnel.
- Develop mandatory job safety and health standards and enforce them effectively.
- Provide for the development, analysis, evaluation, and approval of state occupational safety and health programs.

## What OSHA Does

OSHA undertakes activities required to meet its goals, as described in the following.

### Standards development

OSHA creates or promulgates legally enforceable specific standards alongside its general duty clause—which says that "each employer shall furnish...a place of employment which is free from recognized hazards that are causing or are likely to cause death or serious physical harm to employees."

### Publication of standards

Standards can be found in the *Federal Register* and in the *Code of Federal Regulations* (CFR), or in similar publications from states having OSHA-approved programs.

### Workplace inspections

OSHA has the authority to inspect any company under its jurisdiction. Details can be found in OSHA's booklet on inspections, OSHA Publication 2098 (see Fig. 5.3). The inspection process follows set rules, and includes the following items:

# OSHA Inspections

U.S. Department of Labor
Occupational Safety and Health Administration

OSHA 2098
1996 (Revised)

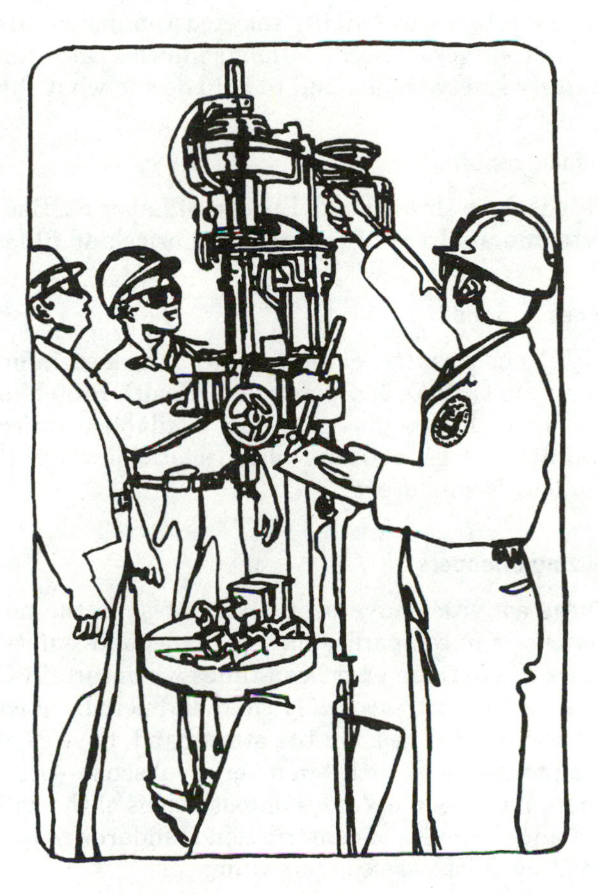

**Figure 5.3** OSHA's Booklet #2098, on inspections.

1. The OSHA inspector's credentials are presented.
2. An opening conference takes place at which the compliance officer explains why the inspection is being conducted.
3. The inspection tour is conducted, during which the inspector will speak with on-the-job employees and view whatever operations the inspector selects.
4. A closing conference between the compliance officer and the employer takes place when the inspection ends.

#### Recordkeeping

Although not all companies are required to forward their workplace injury and illness records to OSHA, selected companies are required to participate in annual surveys of occupational injuries and illnesses that are used to help identify safety trends and to help decide what the agency should focus on.

#### Generating statistical reports

This is done through the Bureau of Labor Statistics, with collections of report data and analyses of workplace injuries and illnesses.

#### Keeping employees informed

OSHA ensures that employers keep workers informed by requiring employers to put up OSHA's Job Safety and Health Protection workplace poster (see Fig. 5.4), by making copies of the Act available upon request, and by posting injury and illness logs, petitions, and violations where they will be seen by all of the company's employees.

#### Citing and penalizing offenders

These activities have been placing fear in the hearts of managers (especially managers in companies having inadequate safety management programs) for years. Have there ever been times when certain OSHA offices and compliance officers have *unreasonably* cited and penalized construction and manufacturing operations? Yes. On the other hand, have certain company managers ever tried to get away with—in a very real sense—occupational murder? Yes again. There have been excesses on both sides of the safety coin.

Some examples of construction standards frequently cited for violations and fined by OSHA are the following:

| | |
|---|---|
| 1926.10(a) | Protective Helmets |
| 1926.20(b)(1) | Accident Prevention Programs, Initiate & Maintain |
| 1926.20(b)(2) | Accident Inspection Programs, Inspections |
| 1926.21(b)(2) | Employer Safety Training and Education |
| 1926.25(a) | Scrap Lumber & Debris |

# JOB SAFETY & HEALTH PROTECTION

The Occupational Safety and Health Act of 1970 provides job safety and health protection for workers by promoting safe and healthful working conditions throughout the Nation. Provisions of the Act include the following:

## Employers

All employers must furnish to employees employment and a place of employment free from recognized hazards that are causing or are likely to cause death or serious harm to employees. Employers must comply with occupational safety and health standards issued under the Act.

## Employees

Employees must comply with all occupational safety and health standards, rules, regulations and orders issued under the Act that apply to their own actions and conduct on the job.

The Occupational Safety and Health Administration (OSHA) of the U.S. Department of Labor has the primary responsibility for administering the Act. OSHA issues occupational safety and health standards, and its Compliance Safety and Health Officers conduct jobsite inspections to help ensure compliance with the Act.

## Inspection

The Act requires that a representative of the employer and a representative authorized by the employees be given an opportunity to accompany the OSHA inspector for the purpose of aiding the inspection.

Where there is no authorized employee representative, the OSHA Compliance Officer must consult with a reasonable number of employees concerning safety and health conditions in the workplace.

## Complaint

Employees or their representatives have the right to file a complaint with the nearest OSHA office requesting an inspection if they believe unsafe or unhealthful conditions exist in their workplace. OSHA will withhold, on request, names of employees complaining.

The Act provides that employees may not be discharged or discriminated against in any way for filing safety and health complaints or for otherwise exercising their rights under the Act.

Employees who believe they have been discriminated against may file a complaint with their nearest OSHA office within 30 days of the alleged discriminatory action.

## Citation

If upon inspection OSHA believes an employer has violated the Act, a citation alleging such violations will be issued to the employer. Each citation will specify a time period within which the alleged violation must be corrected.

The OSHA citation must be prominently displayed at or near the place of alleged violation for three days, or until it is corrected, whichever is later, to warn employees of dangers that may exist there.

## Proposed Penalty

The Act provides for mandatory civil penalties against employers of up to $7,000 for each serious violation and for optional penalties of up to $7,000 for each nonserious violation. Penalties of up to $7,000 per day may be proposed for failure to correct violations within the proposed time period and for each day the violation continues beyond the prescribed abatement date. Also, any employer who willfully or repeatedly violates the Act may be assessed penalties of up to $70,000 for each such violation. A minimum penalty of $5,000 may be imposed for each willful violation. A violation of posting requirements can bring a penalty of up to $7,000.

There are also provisions for criminal penalties. Any willful violation resulting in the death of any employee, upon conviction, is punishable by a fine of up to $250,000 (or $500,000 if the employer is a corporation), or by imprisonment for up to six months, or both. A second conviction of an employer doubles the possible term of imprisonment. Falsifying records, reports, or applications is punishable by a fine of $10,000 or up to six months in jail or both.

## Voluntary Activity

While providing penalties for violations, the Act also encourages efforts by labor and management, before an OSHA inspection, to reduce workplace hazards voluntarily and to develop and improve safety and health programs in all workplaces and industries. OSHA's Voluntary Protection Programs recognize outstanding efforts of this nature.

OSHA has published Safety and Health Program Management Guidelines to assist employers in establishing or perfecting programs to prevent or control employee exposure to workplace hazards. There are many public and private organizations that can provide information and assistance in this effort, if requested. Also, your local OSHA office can provide considerable help and advice on solving safety and health problems or can refer you to other sources for help such as training.

## Consultation

Free assistance in identifying and correcting hazards and in improving safety and health management is available to employers, without citation or penalty, through OSHA-supported programs in each State. These programs are usually administered by the State Labor or Health department or a State university.

## Posting Instructions

Employers in States operating OSHA approved State Plans should obtain and post the State's equivalent poster.

*Under provisions of Title 29, Code of Federal Regulations, Part 1903.2(a)(1) employers must post this notice (or facsimile) in a conspicuous place where notices to employees are customarily posted.*

### More Information

Additional information and copies of the Act, OSHA safety and health standards, and other applicable regulations may be obtained from your employer or from the nearest OSHA Regional Office in the following locations:

| | |
|---|---|
| Atlanta, GA | (404) 347-3573 |
| Boston, MA | (617) 565-9860 |
| Chicago, IL | (312) 353-2220 |
| Dallas, TX | (214) 767-4731 |
| Denver, CO | (303) 844-1600 |
| Kansas City, MO | (816) 426-5861 |
| New York, NY | (212) 337-2378 |
| Philadelphia, PA | (215) 596-1201 |
| San Francisco, CA | (415) 975-4310 |
| Seattle, WA | (206) 553-5930 |

Washington, DC
1996 (Reprinted)
OSHA 2203

Robert B. Reich, Secretary of Labor

**U.S. Department of Labor**
Occupational Safety and Health Administration

**Figure 5.4** OSHA's Job Safety & Health Protection Workplace Poster.

| | |
|---|---|
| 1926.28(a) | Employer Required Wearing of Personal Protective Equipment |
| 1926.50(c) | Person Trained in First Aid |
| 1926.50(d)(1) | First-Aid Supplies |
| 1926.59(b)(1) | Chemical Hazard Communication Program |
| 1926.59(e)(1) | Written Hazard Communication Program |
| 1926.59(f)(5) | Labels, Hazardous Chemicals |
| 1926.59(g)(1) | Workplace, Material Safety Data Sheets |
| 1926.59(h) | Employee Information & Training, Hazardous Chemicals |
| 1926.95(a) | Providing Personal Protective Equipment |
| 1926.150(a)(1) | Fire Protection Program |
| 1926.150(c)(1) | Fire Extinguishers & Small Hose Lines |
| 1926.152(a)(1) | Approved Containers, Flammable & Combustible Liquids |
| 1926.300(b)(2) | Machine Guarding |
| 1926.304(f) | Standards, Woodworking Tools |
| 1926.350(a)(9) | Securing Compressed Gas Cylinders |
| 1926.403(b)(2) | Electrical Equipment, Installation & Use |
| 1926.404(b)(1)(i) | Branch Circuits, Ground Fault Protection |
| 1926.404(f)(6) | Grounding Path, Circuits, Equipment & Enclosures |
| 1926.405(a)(2)(ii)(J) | Extension Cords |
| 1926.405(g)(2) | Identification, Splices & Terminations, Flexible Cords & Cables |
| 1926.416(e)(1) | Electrical Cords & Cables |
| 1926.451(a)(3) | Erection of Scaffolding, Competent Person |
| 1926.451(a)(4) | Guardrails & Toeboards, Scaffolding |
| 1926.451(a)(13) | Planking, Scaffolding |
| 1926.451(d)(10) | Tubular Welded Frame Scaffolds, Guardrails |
| 1926.500(b)(1) | Guarding of Floor Openings & Floor Holes |
| 1926.501(b)(11) | Duty to Have Fall Protection |
| 1926.503(a)(1) | Training, Fall Protection |
| 1926.602(a)(9) | Audible Alarms, Earthmoving Equipment |
| 1926.602(a)(9) | Seat Belts, Motor Vehicles |
| 1926.651(c)(2) | Inspections, Excavations |
| 1926.652(a)(1) | Protective Systems, Excavations |
| 1926.1051(a) | Providing Stairways or Ladders |
| 1926.1052(c)(1) | Stairrails and Handrails |
| 1926.1053(b)(16) | Withdrawal from Service, Defective Ladders |
| 1926.1060(a) | Training Program, Ladders & Stairways |

### Providing consultation services

Consultation services are available upon request, primarily for smaller employers with more hazardous operations. The consultation services are delivered by state government agencies, at no cost to employers. When delivered at the worksite, consultation assistance includes an opening conference with the employer to explain the ground rules for consultation, a walk through the workplace to identify any specific hazards and to examine those aspects of the employer's safety and health program which relate to the scope of the visit, and a closing conference followed by a written report to the employer of the consultant's findings and recommendations.

Possible violations of OSHA standards will not be reported to OSHA enforcement staff unless the employer fails or refuses to eliminate or control worker exposure to any identified serious hazard or imminent danger situation.

### Supplying training and education

OSHA's area or local offices are all full-service centers that offer a variety of informational services, including compliance experts available for speaking engagements, sundry publications on safety and health issues, audiovisual aids on workplace hazards, and technical advice. The OSHA Training Institute in Des Plaines, Illinois provides basic and advanced training seminars and classes for public and private employers and employees. Training schedules (see Fig. 5.5) may be found on OSHA's website or through OSHA area offices.

## The Act's Coverage

OSHA's definition of an employer covered under the Act is any "person engaged in a business affecting commerce who has employees, but does not include the United States or any State or political subdivision of a State." That means the Act applies to employers and employees in manufacturing, construction, longshoring, agriculture, and similar pursuits.

But it does not cover the following:

- Self-employed persons (even if they're employed in construction work or manufacturing). The Act does not cover the activities of those who have no employees.
- Farms on which only immediate members of the farm employer's family are employed.
- Working conditions regulated by other federal agencies under other federal statutes (except any parts not covered as such).

## OSHA-Approved State Programs

OSHA covers the Unites States and its territories through numerous regional, district, and area offices (see Fig. 5.6). Although many states are under federal OSHA rule, the Act does allow certain qualifying, participating states

**SCHEDULE AND REGISTRATION INSTRUCTIONS**

Fiscal Year 1998
October 1, 1997 - September 30, 1998

NOTE: This Schedule and Instructions are only for Non-OSHA Federal and Non-OSHA State Employees, County and Local Government Employees, Private Sector and Government Contractor Employees.

Federal Recycling Program
Printed on Recycled Paper

**Figure 5.5** Schedule for OSHA's Training Institute.

to develop and operate, under OSHA guidance, state job safety and health plans. States and territories having their own approved occupational safety and health programs include Alaska, Arizona, California, Connecticut, Hawaii, Indiana, Iowa, Kentucky, Maryland, Michigan, Minnesota, Nevada, New Mexico, New York, North Carolina, Oregon, Puerto Rico, South Carolina, Tennessee, Utah, Vermont, Virgin Islands, Virginia, Washington, and Wyoming. Of the 25 state plans, 23 cover both the private and public (state and local government) sectors, and 2 cover the public sector only (Connecticut and New York).

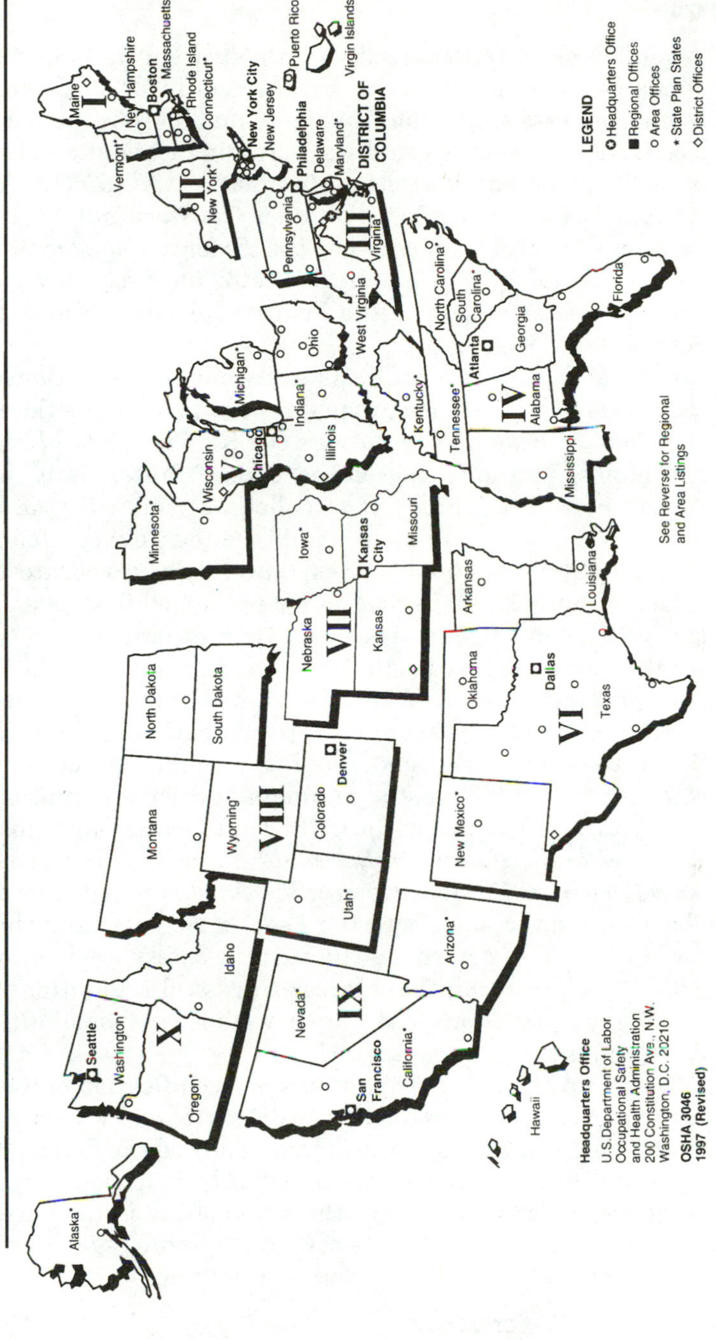


**Figure 5.6** Map of OSHA regions and offices.

## OSHA Standards

When Franklin D. Roosevelt's administration was working out a whole series of changes within the federal government in 1932, collectively called the *New Deal,* Congress started shifting more and more responsibility to federal agencies so those agencies could start creating legislation. At that time there was no central communications system, so it was difficult for the agencies that created laws and regulations to get the word out to the public in a timely manner. To resolve that issue, the *Federal Register Act* came out in 1935, which was followed by the *Administrative Procedure Act* 11 years later, in 1946. These acts enabled the Federal Register System to issue and publish government laws.

In a sense, the Federal Register System has functioned as a huge publishing house ever since. Its two major publications are the *Federal Register* and the *Code of Federal Regulations* (CFR). The *Federal Register* (see Fig. 5.7) can almost be thought of as the United States daily newsletter—although there's not much "chummy" or colloquial about it, and its coverage is quite extensive. It's published and issued every federal working day, and is distributed to libraries, law firms, corporate offices, and various information sources. Legislation is published first as a proposed rule, to give the public and businesses and other organizations time to comment on what's being proposed. The comment period throws the topic up for discussion and questioning. When the rule is finalized, it becomes a legal document a minimum of 30 days after publication in the *Federal Register.* These final rules are then rendered into numbered codes so the rules can be referred to and indexed by numbers and letters, and published in the next yearly updated edition of the CFR. That means—and this is important—*there are rules and regulations that could already have become laws that may not be reflected in the latest available hard-copy edition of the CFR.* To stay absolutely up to date, both the *Federal Register* and the CFR must be used hand in hand to determine the current safety application for any one situation. Of course, even these latest rules and regulations are further affected by the interpretations of the agency whose responsibilities are to enforce the regulations—in this case, OSHA.

The *Code of Federal Regulations* is a codification of the general and permanent rules published in the *Federal Register* by the executive departments and agencies of the federal government. The code is divided into 50 titles, which represent broad areas subject to federal regulation. Each title is divided into chapters, which usually bear the name of the issuing agency. Each chapter is further subdivided into parts covering specific regulatory areas.

## CFR Title Numbers and Descriptions

Lest you think the government—by virtue of all of the regulations and rules related to construction—is picking on home builders and related construction trades, let's review just how encompassing the *Code of Federal Regulations* is. Here are the CFR title numbers and descriptions for the whole set:

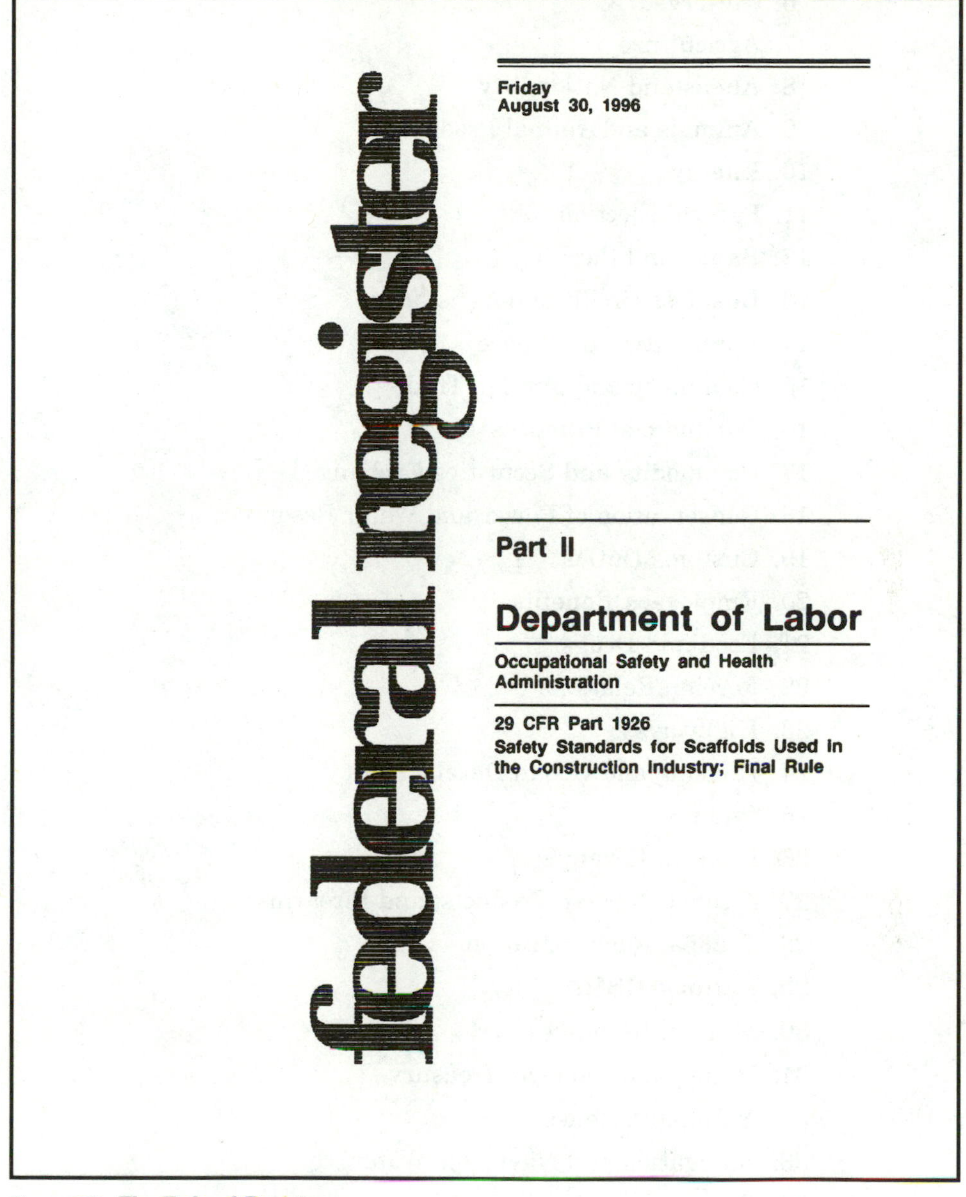
federal register

Friday
August 30, 1996

Part II

Department of Labor

Occupational Safety and Health Administration

29 CFR Part 1926
Safety Standards for Scaffolds Used in the Construction Industry; Final Rule

**Figure 5.7** The Federal Register.

1. General Provisions
2. (Reserved)
3. The President
4. Accounts
5. Administrative Personnel

6. (Reserved)
7. Agriculture
8. Aliens and Nationality
9. Animals and Animal Products
10. Energy
11. Federal Elections
12. Banks and Banking
13. Business Credit and Assistance
14. Aeronautics and Space
15. Commerce and Foreign Trade
16. Commercial Practices
17. Commodity and Securities Exchanges
18. Conservation of Power and Water Resources
19. Customs Duties
20. Employees' Benefits
21. Food and Drugs
22. Foreign Relations
23. Highways
24. Housing and Urban Development
25. Indians
26. Internal Revenue
27. Alcohol, Tobacco Products, and Firearms
28. Judicial Administration
29. **Labor—OSHA**
30. Mineral Resources
31. Money and Finance: Treasury
32. National Defense
33. Navigation and Navigable Waters
34. Education
35. Panama Canal
36. Parks, Forests, and Public Property
37. Patents, Trademarks, and Copyrights
38. Pensions, Bonuses, and Veterans' Relief
39. Postal Service
40. Protection of Environment—EPA

41. Public Contracts and Property Management
42. Public Health
43. Public Lands: Interior
44. Emergency Management and Assistance
45. Public Welfare
46. Shipping
47. Telecommunications
48. Federal Acquisition Regulations System
49. Transportation—DOT
50. Wildlife and Fisheries

To help users find the latest volumes of the CFR, the book covers change colors each year. The revision date indicates when the revision is begun, and the CFR volume may not actually be available to the public until about six months after the revision date. The titles are revised and published annually by the following schedule:

Titles 1 through 16 are revised as of January 1.

Titles 17 through 27 are revised as of April 1.

Titles 28 through 41 are revised as of July 1.

Titles 42 through 50 are revised as of October 1.

Title 29, Part 1926 contains the standards on construction.

## Index and Finding Aids

When researching by topic, the *Index and Finding Aids* found in the back of each CFR, and published annually as a separate document, are helpful. There's also a way to find up-to-date legislation or amendments on a regulation—you can consult the CFR List of Sections Affected (LSA), which is issued monthly by the Superintendent of Documents, U.S. Government Printing Office. (Request from Superintendent of Documents: Mail OP: SSOP, Washington, DC 20402-9328.)

Inquiries concerning editing procedures and reference assistance with respect to the *Code of Federal Regulations* may be addressed to the Director, Office of the Federal Register, National Archives and Records Administration, Washington, DC 20408 [phone (202) 523-3517]. Sales are handled by the Superintendent of Documents, U.S. Government Printing Office, Washington, DC 20402 [phone (202) 783-3238]. If those phone numbers have changed since this printing, contact your local OSHA office for new ones.

To look at a table of CFR titles and chapters, check near the end of any individual volume—take Title 29, Part 1926, for example (see Fig. 5.1), which happens to be the book containing the construction standards, the book that we are most concerned with.

### Title 29—Labor

| | |
|---|---|
| Subtitle A | Office of the Secretary of Labor (Parts 0–199) |
| Subtitle B | Regulations Relating to Labor |
| Chapter I | National Labor Relations Board (Parts 100–199) |
| Chapter II | Bureau of Labor-Management Relations and Cooperative Programs, Department of Labor (Parts 200–299) |
| Chapter III | National Railroad Adjustment Board (Parts 300–399) |
| Chapter IV | Office of Labor-Management Standards, Department of Labor (Parts 400–499) |
| Chapter V | Wage and Hour Division, Department of Labor (Parts 500–899) |
| Chapter IX | Construction Industry Collective Bargaining Commission (Parts 900–999) |
| Chapter X | National Mediation Board (Parts 1200–1299) |
| Chapter XII | Federal Mediation and Conciliation Service (Parts 1400–1499) |
| Chapter XIV | Equal Employment Opportunity Commission (Parts 1600–1699) |
| Chapter XVII | Occupational Safety and Health Administration, Department of Labor (Parts 1900–1999) |
| Chapter XX | Occupational Safety and Health Review Commission (Parts 2200–2499) |
| Chapter XXV | Pension and Welfare Benefits Administration, Department of Labor (Parts 2500–2599) |
| Chapter XXVI | Pension Benefit Guaranty Corporation (Parts 2600–2699) |
| Chapter XXVII | Federal Mine Safety and Health Review Commission (Parts 2700–2799) |

(Other work cited: *The Federal Register: What It Is and How to Use It,* Office of the Federal Register, National Archives and Records Administration, June 1995.)

### CFR structure

Sometimes it requires a little research to obtain information from the CFRs. A typical OSHA reference may include the title, part, section, paragraph and subparagraph numbers. For example, 29 CFR 1910.1450(a)(2)(i), reads as Title 29, CFR, Part 1910, Section 1450, Paragraph (a), Subparagraph (2)(i).

Why are the reference numbers in the CFR titles confusing? The CFR format uses a period to separate the part number from the section number, instead of a decimal system. So 29 CFR 1910.2 would precede 29 CFR 1910.15. Also, the many paragraph structural levels mix numerals and letters and include the use of lowercase *a* preceding capital *A,* which is an unusual sequence.

## What OSHA Says

OSHA's main goal is to help employers improve their safety management programs so fewer occupational injuries and illnesses are experienced.

The following paragraphs were excerpted from original nonmandatory guidelines found in the *Federal Register* 54(18):3094–3916, January 26, 1989. In essence, they review what OSHA considered to be recommended steps that companies desiring top safety management programs should follow.

### In general

Employers are advised and encouraged to institute and maintain in their establishments a program that provides adequate systematic policies, procedures, and practices to protect their employees from, and allow them to recognize, job-related safety and health hazards.

An effective program includes provisions for the systematic identification, evaluation, and prevention or control of general workplace hazards, specific job hazards, and potential hazards that may arise from foreseeable conditions.

Although compliance with the law, including specific OSHA standards, is an important objective, an effective program looks beyond specific requirements of law to address all hazards. It will seek to prevent injuries and illnesses, whether or not compliance is at issue.

The extent to which the program is described in writing is less important than how effective it is in practice. As the size of a worksite or the complexity of a hazardous operation increases, however, the need for written guidance increases to ensure clear communication of policies and priorities as well as a consistent and fair application of rules.

### Major elements

An effective occupational safety and health program will include the following four main elements: management commitment and employee involvement, worksite analysis, hazard prevention and control, and safety and health training.

### 1. Management commitment and employee involvement

The elements of management commitment and employee involvement are complimentary and form the core of any occupational safety and health program. Management's commitment provides the motivating force and the resources for organizing and controlling activities within an organization. In an effective program, management regards worker safety and health as a fundamental value of the organization and applies its commitment to safety and health protection with as much vigor as to other organizational goals.

Employee involvement provides the means by which workers develop and/or express their own commitment to safety and health protection for themselves and for their fellow workers.

In implementing a safety and health program, there are various ways to provide commitment and support by management and employees. Some recommended actions are described briefly as follows:

- State clearly a worksite policy on safety and healthful work and working conditions, so that all personnel with responsibility at the site (and personnel at other locations with responsibility for the site) fully understand the priority and importance of safety and health protection in the organization.
- Establish and communicate a clear goal for the safety and health program and define objectives for meeting that goal so that all members of the organization understand the results desired and measures planned for achieving them.
- Provide visible top management involvement in implementing the program so that all employees understand that management's commitment is serious.
- Arrange for and encourage employee involvement in the structure and operation of the program and in decisions that affect their safety and health so that they will commit their insight and energy to achieving the safety and health program's goals and objectives.
- Assign and communicate responsibility for all aspects of the program so that managers, supervisors, and employees in all parts of the organization know what performance is expected of them.
- Provide adequate authority and resources to responsible parties so that assigned responsibilities can be met.
- Hold managers, supervisors, and employees accountable for meeting their responsibilities so that essential tasks will be performed.
- Review program operations at least annually to evaluate their success in meeting goals and objectives so that deficiencies can be identified and the program and/or the objectives can be revised when they do not meet the goal of effective safety and health protection.

## 2. Worksite analysis

A practical analysis of the work environment involves a variety of worksite examinations to identify existing hazards and conditions and operations in which changes might occur to create new hazards. Unawareness of a hazard stemming from failure to examine the worksite is a sign that safety and health policies and/or practices are ineffective. Effective management activity analyzes the work and worksite to *anticipate* and prevent harmful occurrences. The following measures are recommended to identify all existing and potential hazards:

- Conduct comprehensive baseline worksite surveys for safety and health and periodic comprehensive update surveys and involve employees in this effort.
- Analyze planned and new facilities, processes, materials, and equipment.
- Perform routine job hazards analyses.
- Assess risk factors of ergonomics applications to workers' tasks.
- Conduct regular site safety and health inspections so that new or previously missed hazards and failures in hazard controls are identified.
- Provide a reliable system for employees to notify management personnel about conditions that appear hazardous and to receive timely and appropriate responses and encourage employees to use the system without fear of reprisal. This system utilizes employee insight and experience in safety and health protection and allows employee concerns to be addressed.

- Investigate accidents and "near-miss" incidents so that their causes and means of prevention can be identified.
- Analyze injury and illness trends over time so that patterns with common causes can be identified and prevented.

### 3. Hazard prevention and control

Where feasible, workplace hazards are prevented by effective design of the job site or job. Where it is not feasible to eliminate such hazards, they must be controlled to prevent unsafe and unhealthful exposure. Elimination or control must be accomplished in a timely manner once a hazard or potential hazard is recognized. Specifically, as part of the program, employers should establish procedures to correct or control present or potential hazards in a timely manner. These procedures should include measures such as the following:

- Use engineering techniques where feasible and appropriate.
- Establish, at the earliest time, safe work practices and procedures that are understood and followed by all affected parties. Understanding and compliance are a result of training, positive reinforcement, correction of unsafe performance, and if necessary, enforcement through a clearly communicated disciplinary system.
- Provide personal protective equipment when engineering controls are infeasible.
- Use administrative controls, such as reducing the duration of exposure.
- Maintain the facility and equipment to prevent equipment breakdowns.
- Plan and prepare for emergencies, and conduct training and emergency drills, as needed, to ensure that proper responses to emergencies will be "second nature" for all persons involved.
- Establish a medical program that includes first aid onsite as well as nearby physician and emergency medical care to reduce the risk of any injury or illness that occurs.

### 4. Safety and health training

Training is an essential component of an effective safety and health program. Training helps identify the safety and health responsibilities of both management and employees at the site. Training is often most effective when incorporated into other education on performance requirements and job practices. The complexity of training depends on the size and complexity of the worksite as well as the characteristics of the hazards and potential hazards at the site.

**Employee training.** Employee training programs should be designed to ensure that all employees understand and are aware of the hazards to which they may be exposed and the proper methods for avoiding such hazards.

**Supervisory training.** Supervisors should be trained to understand the key role they play in job site safety and to enable them to carry out their safety and health responsibilities effectively. Training programs for supervisors should include the following topics:

- Analyze the work under their supervision to anticipate and identify potential hazards.
- Maintain physical protection in their work areas.
- Reinforce employee training on the nature of potential hazards in their work and on needed protective measures through continual performance feedback and, if necessary, through enforcement of safe work practices.
- Understand their safety and health responsibilities.

You'll notice that many of the components mentioned by OSHA are found within the chapters of this book. It's no accident. Safety management has been evolving for years, as the importance of running injury-free operations has been rising to the top of business management priorities.

## Employer Responsibilities

OSHA, in an informative publication about itself (OSHA Publication 2056) lists the main employer responsibilities when it comes to installing and maintaining a safety management program:

- Meet your general duty responsibility to provide a workplace free from recognized hazards that are causing or are likely to cause death or serious physical harm to employees; and comply with standards, rules, and regulations issued under the Act.
- Be familiar with mandatory OSHA standards and make copies available to employees for review upon request.
- Examine workplace conditions to make sure they conform to applicable standards.
- Minimize or reduce hazards.
- Make sure employees have and use safe tools and equipment (including appropriate personal protective equipment), and that such equipment is properly maintained.
- Use color codes, posters, labels, or signs when needed to warn employees of potential hazards.
- Establish or update operating procedures and communicate them so that employees follow safety and health requirements.
- Provide medical examinations when required by OSHA standards.
- Provide training required by OSHA standards (e.g., hazard communication, lead, etc.).
- Report to the nearest OSHA office within 48 hours any fatal accident or one that results in the hospitalization of three or more employees.

- Keep OSHA-required records of work-related injuries and illnesses, and post a copy of the totals from the last page of OSHA 200 during the entire month of February each year (applicable to employers with 11 or more employees).
- Post, at a prominent location within the workplace, the OSHA poster, OSHA 2203 (see Fig. 5.4), informing employees of their rights and responsibilities. In states operating OSHA-approved job safety and health programs, the state's equivalent poster and/or OSHA 2203 may be required.
- Provide employees, former employees, and their representatives access to the Log and Summary of Occupational Injuries and Illnesses (OSHA 200) at a reasonable time and in a reasonable manner.
- Provide access to employee medical records and exposure records to employees or their authorized representatives.
- Cooperate with OSHA compliance officers by furnishing names of authorized employee representatives who may be asked to accompany the compliance officer during an inspection. If none, the compliance officer will consult with a reasonable number of employees concerning safety and health in the workplace.
- Do not discriminate against employees who properly exercise their rights under the Act.
- Post OSHA citations at or near the worksite involved. Each citation, or copy thereof, must remain posted until the violation has been abated, or for three working days, whichever is longer.
- Abate cited violations within the prescribed period.

## Working with Change

If there's one thing to remember with OSHA, it's that the organization is almost constantly in a state of change. Although many of the standards have been around for years and years, interpretations of those standards have evolved over time, and so have enforcement policies. Some of this is due to the changing of political parties and guards and climates; some is due to the personalities of the agency's directors; and some is due to the dedication, resourcefulness, and persuasiveness of individual industry lobbyists and trade organizations. Some is even due to public sentiment.

The result of these numerous interactions has been an OSHA that continues to reinvent itself. Safety and health magazines are constantly writing about the *new OSHA,* and new directions in which the agency is expected to head. At this writing, the latest new OSHA seems much more at ease with itself and with its most recent role as consultant. After years of being at odds with various trades and work groups, and working within hostile employer environments, OSHA seems well on the way to more productive times during which the agency can be viewed as more of a coach than a cop.

While some of the specific standards still seem unreasonable to individual work crafts (for example, numerous construction standards had been originally written with commercial—not residential—construction in mind), efforts have been made to accommodate those crafts which are proposing reasonable alternatives and are working closely with the agency on revising and developing new guidelines.

Part

# 2

# Setting Up an Effective Safety Management System

*Part 2 deals with the human factors of safety: It begins with the company safety policy—a policy that must actually mean something—then continues through management support and employee involvement, which are components best supported by training, incident investigations, safety audits, and peer behavior observations. There's behavior analysis: a discussion of what determines safe behavior, and why unsafe activities continue to plague our worksites. A chapter on ergonomics attempts to convince readers why it's better to adjust our worksite tasks to the capabilities of our employees, rather than forcing the workers to fit rigid job requirements. This is followed with job safety analysis of work tasks, and ends with preparations that can be made to minimize the effects of accidental worksite injuries and illnesses.*

Chapter

# 6

# Your Company Safety Policy

### Quick Scan

1. Maintain a simple, clear, but comprehensive company safety policy.
2. Train new hires on the safety policy during their safety orientation; review the safety policy with the entire workforce annually.
3. The company safety policy, by extension, should cover the safety of anyone involved with the worksite, including management, work crews, subcontractors, vendors and servicepersons, and visitors.
4. Strictly enforce all of the safety policy's components.

This chapter is a gimmee. All you have to do is come up with a simple, concise, sincere company safety policy. Write it down. Train all employees in the company on it. And then live by it.

Does your company already have a safety policy? And if so—does everyone know what it says? Does everyone understand it and agree with it? Do past practices reinforce or contradict it?

As the foundation for a healthy, effective safety management program, a company safety policy is a must. Not some elegant-sounding fancy pie-in-the-sky discourse, but a policy that can be understood, believed in, and followed by *everyone* in the company. Right about now I can hear a number of readers groaning while they're rolling their eyes. Come on, how can some grandiose policy make a difference? It will because it *does*. Companies with successful safety management programs inevitably have clear, simple safety policies that help set the proper tone for all of their employees.

Again, the safety policy has to be simply written and comprehensive—but not long-winded. In fact, it should take up no more than a single typewritten page, mostly double-spaced. A clear, effective safety policy takes only a minute or two to read. And it should be signed, personally, by the top dog—the person in charge—the owner, the CEO, the president, behind whoever's desk the buck stops at.

Unfortunately, some companies stop here, thinking that merely *having* a safety policy tucked away in some binder—ready to pull out in a pinch—is good enough. Well, it's not. Your safety policy needs to be taken off the shelf and distributed to management and employees.

The safety policy should also be reviewed with new hires at their safety orientation, and thereafter with regular employees at least annually, with discussion to ensure that no intentions of the policy are misinterpreted, and that no questions about the policy go unanswered. Subcontractors, vendors, and salespersons must also be informed of the company safety policy and of any safety procedures that may apply to their worksite visits.

What should your company's safety policy address? A number of points critical to safety management, including:

1. *Employee safety must receive a high priority among all of the other management concerns.* Your company is naturally concerned with such areas as marketing, purchasing of equipment and supplies, maintenance of vehicles and equipment, quality control, personnel issues, production records, training issues, and the like. All of these contribute to the successful operation of a home-construction firm or any contract service business. Safety, however, is also important enough to be right up there, sharing space with the top concerns. Although your company needs to make a profit to survive, a responsible operation can't allow safety to be compromised in order to do so. As seen throughout the pages of this book and others, humanitarian reasons aside, safety can have major effects on a company's bottom line, because safer worksites generally equate to higher productivity and quality.
2. *Management believes that accidental injuries are preventable, and that worksite activities can go on, day after day, week after week, month after month without accidents.* While this does not totally discount fluke incidents which may result in injuries, it does bring an attitude to the table that individuals and crews are largely in command of their own behaviors, and that an injury-free environment can be achieved. Even if injury rates exist at or above construction industry averages, the attitude that incidents are preventable must prevail, with continuous improvement the goal.
3. *The company will provide all of the support necessary to achieve an accident-free environment, including safe equipment and safe procedures.* If the safety management program at your company aggressively investigates incidents, corrects basic causes, and develops more effective procedures, ongoing safety improvements will naturally follow. But that's only part of the support required. There's a certain amount of financial assistance needed, too. This assistance should be looked at as an investment, not as an unnecessary

expense. It's providing and keeping equipment and tools in good condition. It's attention to detail. It's following up on safety concerns, and following through with corrective actions. It's talking the talk, and walking the walk.

4. *The company will supply safety training to all employees so they can work in a safe manner.* Training is a huge part of any safety management program. It shouldn't be done merely to satisfy regulatory requirements. Rather, effective training seeks to impart knowledge, understanding, and skills. To make sure that the training is really *received* by your employees, verification of understanding is necessary. Subcontractors, vendors, and servicepersons must also be informed and trained on company safety policies and procedures that may apply to their worksite visits.

5. *Company management is responsible for the development and operation of the company's safety management program.* You already know that most members of management wear quite a few different hats. They're responsible for setting safety priorities, for establishing safety inspection and feedback systems, and for providing safe environments and work systems. Ultimately, the level of safety achieved directly reflects the system that management sets up. A major part of their responsibility involves communications: making everyone in the company aware of their own safety responsibilities.

   Management involvement in the safety program is absolutely necessary, with accountability for safety performance the number-one requirement. All members of management must have the same attitude and take the same position when it comes to carrying out key safety components. Unsafe behaviors should be called out to an employee's attention wherever and whenever seen, without time lag. And only safe behaviors should be displayed by all members of management. If management representatives fail to lead by example, or let examples of unsafe behavior pass without correction, overall safety efforts will be undermined and eroded little by little. The safety program can become diluted, with the workforce (and some supervisors) developing the attitudes of apathy found in companies where desirable behaviors are not encouraged, such as "No one around here really cares about safety."

   Management has to see that safety is more than just a program by itself. Safety must instead be integrated into the company's culture. It can't stand alone. Just as the nervous system in a person's body needs the support of the other systems and organs, so does safety need to intertwine itself, from top to bottom and side to side, throughout the entire company. Remember the statement "Everyone is responsible for safety"? Well, management is responsible for *making sure* that everyone is indeed responsible for safety.

6. *All employees, hourly and salaried, must work safely and must follow regulatory safety laws and regulations plus the company's safety rules and regulations.* Everyone, from secretaries to owners, must play by the safety rules.

7. *Adherence to the company safety policy by all employees is not optional; rather, it is a condition of employment.* This may sound kind of cold and

heartless, but it's got to be said. Individuals who buck the safety system—for whatever reasons—put themselves and their fellow workers at risk. They're a bad example for new employees, and a liability to the company. If their unsafe ways can't be corrected through coaching, discipline is the next resort. If discipline doesn't work either, employees who refuse to behave safely must be replaced. Be careful that exceptions aren't made because an employee is considered to possess skills that are "too valuable" to be replaced.

Some companies maintain fancy safety binders and policies that sit on a shelf in the owner's or president's office in case OSHA or some other regulatory agency comes calling. Some companies have company safety policies that seem to have been written for only the workers, while management pays no attention to personal protection rules, worksite regulations and other important parts of the safety program. Neither will make sincere, meaningful safety blueprints for a workforce to follow.

What your operation needs is a policy that defines management's commitment to safety and outlines the expectations of everyone in the company. The policy needs to be posted out in the open, and be included in the company safety manual. It needs to be a short and concise working document that employees, subcontractors, vendors, and anyone who has any interaction with the company become thoroughly familiar with. It needs to be a policy both on paper and in action.

Although your safety policy must cover the points mentioned in this chapter, it should not be copied verbatim from another company's policy, nor should it be parroted from safety management books. Rather, your safety policy should be created with your company's own words. Yours or your management's.

Chapter

# 7

# Management Support

## Quick Scan

1. Upper management must provide resources required to maintain a safe worksite, and should make a point to periodically become involved with daily worksite safety activities to demonstrate their support.
2. Company and supervisor safety goals need to be established yearly, and performances must be measured against them.
3. Supervisors must be properly trained to run a safe worksite. They, in turn, must safety train their work crews.
4. All members of management must know and follow the company's own safety rules and procedures, and set good examples.
5. Unsafe behaviors and conditions must be corrected or isolated whenever they're found, without delay.
6. Management should encourage as much employee participation as possible.

Most management support comes from two groups of individuals in a company: *upper management* and *worksite supervisors or crew leaders.* Depending on the size of your company, upper management can mean an owner or owners, or a company president, manager, or similar position. If your company's management system is working as it should, everyone in your organization ultimately looks to members of upper management for overall leadership. However, on a day-to-day basis, worksite supervisors or leaders typically have more interaction with employees performing construction activities at the worksite.

## Upper Management

These individuals absolutely *must* provide active, ongoing leadership to enable your safety management system to succeed. Their main roles are discussed in the following:

### Developing safety goals for the company and for the supervisors

Safety goals cannot be nebulous or impossible to quantify. They've got to be described or expressed in activities, behaviors, or results that can be measured. It's absolutely worthless for a company president to establish a goal that says "During the following year, everyone in the company must become safer." For goals to be meaningful, they've got to be specific, measurable, and realistic. Lost-time injury and recordable-injury rates, while measurable, are often difficult to specifically relate to safety management activities. While they can be used in goal setting to identify and monitor general safety trends occurring in your company, they should be employed only in addition to more specifically defined goals centering around safety-related activities and behaviors. In any event, effective goals should be: (1) *specific and measurable,* and (2) *reasonable.*

**Setting specific and measurable goals.** If the activities or behaviors cannot be easily measured in some way, how can you tell if compliance, progress, or meaningful results are being achieved? They've got to be quantified in some manner to be meaningful.

*Ineffective company goal:* "All supervisors and crew leaders will receive safety training during the year." While this sounds good, it isn't specific enough. What will be the topics for training?

*Effective company goal:* "All supervisors and crew leaders will receive a minimum of additional formal training on the following topics: Supervisory Skills, Electrical Safety, Fall Protection, Scaffolds and Ladders, Excavations, and Ergonomics. This will be in addition to their participation in regular crew safety training sessions." Such a multifaceted training goal demands management attention to plan its scheduling throughout the year to budget for it and to best work the training in with regular crew safety training schedules.

*Ineffective supervisor or crew leader goal:* "Supervisors must conduct and document worksite tool-box safety meetings and prework safety discussions." That's too general. How many is enough? One a month? Two per year?

*Effective supervisor or crew leader goal:* "Supervisors must conduct and document one 10-minute tool-box safety meeting, and one 5-minute review of the morning's work at the start of each workday." That's a lot better. Supervisors will understand exactly what's expected of them, and can plan accordingly.

**Setting reasonable goals.** This is a major part of goal setting that sometimes is difficult for owners and other members of upper management to grasp. They want to start out big, and begin outlining all kinds of expectations that are way too much, way too soon. Goals developed by upper management for supervisors or crew leaders should not be decided upon and finalized without some discussion and agreement from those individuals who will be responsible for reaching those goals.

*Ineffective company goal:* The president of a company is told by its insurance carrier that if the company's total lost-time incident rate is cut in half by the end of the following year, its workers compensation insurance rates will go down by a huge amount. So the president sets a goal to "Reduce the company's current lost-time incident rate of 38.6 to 19." While such a reduction of lost-time injuries would indeed be nice to achieve, there are many other things which need to first be considered and undertaken to result in such a major rate reduction—namely, the company safety management system has to undergo major improvements to support such a drop in the injuries.

*Effective company goal:* "Enlist the aid of a reputable safety consultant to review, critique, and provide recommendations for making major step changes to the company's safety management system." Another could be for the company to become "its own consultant" by drawing information from local associations and organizations (National Association of Home Builders and related trade associations) and articles and books like this one, and reviewing, analyzing, and developing written action plans for self-improvement.

Supervisors or crew leaders, when faced with goals they feel are totally out of reach, will either flat out not cooperate, or will reluctantly go through the motions. Sometimes they will figure out ways to fool the system, such as writing and turning in "ghost" audits or safety meeting reports—making up and submitting "documentation" for audits never audited and training never done.

*Ineffective supervisor or crew leader goal:* "On a daily basis, personally inspect every power tool, electrical cord, and piece of mobile equipment used at the worksite, and document same." A supervisor couldn't possibly accomplish this at the typical worksite every day and still get the rest of the work done. His or her only alternative would be to challenge such an unrealistic expectation, or to fudge the report.

*Effective supervisor or crew leader goal:* "On a quarterly basis, see that a qualified individual (an electrician) inspects the electrical power tools, extension cords, and trouble lights, and tags them with approved tags, plus documents same." Another part of this goal would be to make sure that the supervisor had access to an electrician to perform the inspections.

### Deciding on standards of performance that upper management members and supervisors must follow

Exactly *how* the goals are achieved is important, too. Standards of performance, like goals, need to be expressed in behavioral terms as well. It's not good enough that safety meeting forms are simply dated and signed. They need to contain meaningful information—such as points covered and who attended—in legible fashion, with questions asked and answered, or asked with the information to be provided at some future date. They cannot just be completed lifelessly, to satisfy some requirement from management. The quality of the activities needed to fulfill the goals must be performed to some acceptable agreed-upon performance standards. This goes for both the company goals and the supervisor goals.

### Evaluating individual and company performances

Somebody's got to do this. If supervisors and crew leaders don't receive feedback on how they're doing—they should already know what to expect based on how well they're performing in relation to their goals and performance standards—then the entire process of setting goals and measuring results will turn into a big waste of time.

This goes for evaluating the overall company performance, too. It may help to involve some other member of the business community while reviewing how (and if) your company has improved its management of safety, for a less biased view.

### Actively supervising a company safety management improvement plan

This means making sure that a realistic, well-thought-out plan exists for making continual improvements to your company's safety management program. There's never a time when you can say enough is enough. There's always something to be improved. Regulations change. New employees are hired. New materials and construction methods are introduced. Conditions change. The market you're competing in changes. New challenges constantly arise within safety and industrial hygiene and environmental arenas. Upper management must see that a master plan for improvements is kept current and relevant to company and individual goals and performance standards.

### Updating employees on the company's overall safety goal status on a regular basis

First of all, do all employees know and understand what your company's safety goals are? They should. Along with your company safety policy, the current goals should be explained to every person who works for your company. Then, periodically throughout the year, the status of those goals should be reviewed. How else can employees feel like they're part of the safety action? If no one knows how the company is doing but a few members of upper management, how can anyone else be expected to be concerned?

### Supporting the entire safety effort

This means following company safety rules, procedures, and guidelines just like everyone else. Members of upper management must never bend or break established worksite safety rules. If the worksite is a hard-hat-only area, hard hats must be worn. If the safety goals stipulate that members of upper management should participate in audits, training sessions, and general safety meetings, then they should. Upper management *must* give the supervisors and crew leaders support at every turn. This means, at a minimum, to show up at the worksite on occasion. You'd be surprised at how many company owners or presidents get caught up in daily administration details and rarely ever visit their construction worksites, let alone participate in the safety management process.

To go beyond a minimum effort, and to effectively assist the supervisors and crew leaders, members of upper management have got to provide a frequent presence in safety activities. Since most small businesses can't afford a full-time safety manager, upper management should be there when important safety activities are taking place. An upper management trying to run safety from the office, or by memos, will likely be seen by worksite supervisors and crews as hopelessly out of touch when it comes to safety. Instead, an upper management presence is needed at the worksite where true *people interaction* can occur. To be effective, members of upper management must have their fingers on the pulse of the workplace. And they've got to be active, energetic, visible, and concerned about their employees to earn the respect of the supervisors and crews.

If upper management can't get to the worksites as much as they'd like, a partial alternative can be to ask several employees to meet at the company's main office. If those kinds of discussions go on once a week for a month or two, perhaps with one worksite crew safety representative and another worker, conventional labor-management posturing can be done away with and meaningful communications will likely take their place.

### Supplying material support

Of course, this also means providing the necessary resources: safe tools and equipment and time for planning, training, auditing, investigating incidents, carrying out corrective action plans, and discussing numerous other safety issues.

### Displaying safety leadership

Upper management can turn the safety program up to new heights just by following simple, good management principles, including the following:

**Considering themselves as managers and leaders.** Upper management should also be assertive when it comes to insisting that supervisors set and maintain good safety examples.

**Interacting with employees on an informal basis.** This doesn't necessarily mean developing fast friendships with everyone in the company, but it does mean circulating enough through the worksites to learn a little about the employees *beyond* the subjects of safety and work, so the employees know they're working with human beings, and vice versa. Talking sports or the weather or about family members, hobbies, and news items can make even members of upper management more human. A side benefit for both parties here is that everyone becomes more comfortable with each other—which helps encourage freer exchanges of ideas.

**Taking safety seriously.** This means giving more than lip service, and taking the time needed to address safety issues as they arise. It means going the extra mile to understand different viewpoints, and being reasonable and willing to compromise occasionally. It means being committed and really believing in the management of safety.

**Discussing before doing.** Management should avoid figuring things out in a vacuum. Before written procedures and rules are installed, they should be discussed with the supervisors and crews. The people doing the work should be approached first for their input. Remember employee participation? That's what's needed before a memo or procedure comes out of left field, from on management high.

## Supervisors and Crew Leaders

These individuals are relied upon as key performers in any safety management system, and they must be key performers in yours.

Sometimes, however, supervisors can't perform what is asked of them because they lack proper training. Regarding safety, many will drift from day to day, unsure of themselves, hoping that nothing goes wrong, always on the defensive. In certain companies they see themselves caught in the middle of an unfriendly situation—between upper management and work crews. Thus, the poor supervisor or crew leader absorbs complaints, ill feelings, and unreasonable requests from both sides. He or she stands alone, often indistinguishable from the workers supervised, a buffer between labor and management, subject to hindsight and guesswork.

On the other hand, properly trained and supported supervisors and crew leaders can effectively guide crews to safe and productive work habits in a positive manner.

The main roles of supervisors or crew leaders are discussed in the following:

### Participating with upper management in the development of supervisory safety goals

Again, it's necessary that supervisors have some say, and ultimately buy into their own goals. If they don't understand them, or if they don't believe the goals are reasonable, achievable, pertinent requests, the supervisors won't take them seriously. It's also important that the supervisors realize how their own goals fit into the overall company safety goals and plan.

### Maintaining OSHA and company regulation, rule, and procedure compliance

That's a mouthful, but it's true. It implies that the supervisor is aware of the various regulations and procedures applicable to his or her worksite, and also that the work crews understand them, too.

### Providing safety training for crew members

Conducting safety training for workers is a necessary part of any supervisor's responsibilities. To make an impact, the training needs to be made meaningful for use at the employees' workplace. That means taking mandatory topics and customizing them to the particular site, so workers will understand *why* it's important to use ground fault circuit interrupters, and *why* they've got to barricade wall openings, and *why* they've got to wear hard hats, and *where* they should place construction debris...and so on. Supervisors must also ensure that the training is understood, that no one misses it, and that new employees receive complete safety orientations and site-specific training before being given tasks by themselves.

### Responding to unsafe worksite behavior and conditions

The supervisor at any worksite is there representing the company, a spokesperson for the company's safety management program. He or she must recognize and react to unsafe behavior or conditions on the spot. Physical hazards must be corrected, isolated, or barricaded as soon as possible. And to walk away from an unsafe act without responding is to condone the act and encourage it to be repeated in the future. Supervisors who ignore safety violations—even when those violations do not occur on purpose—will eventually help deflate even the most aggressively planned safety management program. On the other hand, supervisors should also be quick to praise safe behavior—something that often gets lost in the shuffle of day-to-day worksite activities. Chapter 14, on reinforcing safe behavior, points out that to promote a positive worksite environment, recognition and reinforcement of *safe* behavior should be used about four times as frequently as should correcting unsafe behavior.

### Conducting incident investigations and safety audits

The importance of conducting near-miss and other safety incident investigations and auditing for safety are discussed, respectively, in Chaps. 10 and 11. Both are key activities that are needed to help recognize hazards and to help create solutions and plans to prevent specific safety incidents from happening again.

### Encouraging employee participation whenever possible

In the old days, when management wielded tremendous power over construction work crews, the workers' opinions were frequently discounted or simply ignored. Today it's plain that the old way was an extremely poor management practice. To arrive at the safest work procedures for use at the worksite, who better to participate in their development than employees who actually perform those tasks? In short, the more employee participation, the better. Autocratic, discipline-minded supervisors who won't listen to suggestions from their crew members fail to accept some of the best sources of safety assistance available. The most effective management style, now and in the future, is based on a mutual partnership in which a free, nonthreatening exchange of ideas thrives between labor and management. This applies to safety at the worksite as well. See Chap. 8 for more about employee participation.

### Communicating back and forth between upper management and the work crews

Supervisors and crew leaders play a pivotal role in such a partnership, often conveying information from upper management to the worksite, and from the worksite to upper management. To do so effectively, and to prevent the confrontational misunderstandings that so frequently used to occur between management and labor, the supervisors must understand, appreciate, and be able to communicate viewpoints on important situations back and forth between both groups.

Chapter

# 8

# Employee Involvement

## Quick Scan

1. To have an effective safety management program, almost every aspect of it needs meaningful employee involvement.
2. Try to instill in all employees what their safety responsibilities are, and make sure each member of management and of the work crews realizes that involvement with safety is a condition of employment, and is not optional.
3. Establish a position of, essentially, worksite safety representative, for individuals who volunteer or are willing when selected to assist with overall safety efforts, while performing their own work tasks.
4. Offer assistance to and support employee efforts to participate in safety and skills training beyond what is required by your annual company safety training.

When it comes to managing safety, one of the most effective ways a company can gain the cooperation of its workforce is by eliminating the perception that all safety rules and procedures are *management* driven. The way to eliminate that perception is by asking employees to help. Indeed, employee involvement has become one of the most sought-after components of today's management systems. Gone are the days when autocratic managers, superintendents, and foremen barked orders to the troops like drill sergeants, and didn't care for the opinions of the individuals doing the work. Today high-performance work teams consisting of workers and managers are supposed to perform routine and special work tasks without a lot of supervision and direction. Rather, they are now expected to actively pursue and achieve safety, quality, and productivity through various problem-solving methods, and really are expected to supervise themselves and their fellow workers by performing duties which previously were concerns only of management.

Again, it's extremely important to have employee involvement in safety at your worksites. Employees should help select personal protective equipment; should help prepare safe operating procedures; should have a say in administrative decisions; and can participate by helping with safety meetings, behavior observations and audits, and safety investigations.

By encouraging employees to participate in your safety program and reinforcing them when they do help out, everyone involved will reap huge benefits. If your company elects to do the opposite, and simply puts all kinds of safety rules and regulations into place and says "Follow them or else," your safety program will be hurting. Such a scenario elicits anything from a lukewarm attitude, to encouraging some employees—out of sheer spite—to go out of their way to violate or ignore company safety rules. On the other hand, employees who take an active role in developing procedures are likely to follow those same procedures and to see that fellow employees follow them, too.

### All Employees

All employees on a worksite should understand their own safety responsibilities. As the safety policy says, these safety responsibilities are requirements of their positions—they're not optional. Along these lines, all employees should do the following:

- Again, be held responsible for acting in a safe manner, and for helping fellow workers, visitors, and anyone else venturing onto the worksite to do the same.
- Use required personal protective equipment, and the safety equipment that the company provides, in the manner for which it's intended.
- Accept and participate in safety and skills training, adding or bringing out whatever information they can, including pertinent examples, insights, suggestions, and questions, and challenging questionable practices or decisions.
- Follow established procedures and continue to look for more effective, safer ways to perform jobs.
- Eliminate, isolate, and/or report unsafe conditions whenever and wherever they're found.
- Approach fellow workers who may be performing unsafe behaviors, encourage them to stop, and review the correct alternatives to getting the work done. This is probably the most difficult thing to ask of employees, but when it happens, when one worker takes it upon him- or herself to approach another employee about safety, it's a major safety milestone reached and a sign that the safety management program has made it to a level that not many companies ever achieve.

## Safety Representatives

A relatively recent development is establishing positions in which individuals showing a true concern for safety either volunteer or are selected (not forced) to, along with their regular work duties, become worksite safety representatives.

Safety representatives need to be aware of safety management principles and be familiar with worksite safety rules and procedures. Do they have to be the most experienced employees you have? No. Ideally, they'll be familiar enough with the job tasks to be able to field questions and answer them with some certainty for the doubting Thomases out there who will want to test the safety representatives, perhaps being a bit jealous that they themselves weren't selected. That's okay, though, because it will give the reps an opportunity to gain credibility, depending on their answers.

Safety representatives may be called *company men* at first, and may take a lot of abuse—some good natured and some not—but they should be prepared to assume some responsibility for not letting unsafe practices begin or continue.

Is there more money involved? Perhaps. Is there more training involved? Certainly. Training is critical to a safety representative's success. He or she has got to be able to lead tool-box meetings and has to know how to audit, run incident investigations, and communicate between the supervisor and the work crews. He or she must know how to tactfully resolve safety issues between supervisors and workers—or know when the necessity for going higher in the company is needed.

The safety reps cannot be afraid of conflict, because there will be differences of opinion, differences in interpreting rules and procedures. If the person is very shy and quiet, and avoids interacting with individuals he or she does not know, that's probably not an individual to select for such a position. So try to select someone who understands and *wants* the responsibility.

All kinds of people need be addressed: the high school or college hotshots, sure of their strength and athletic abilities, who may stretch the limits of danger by climbing, getting too close to hazards, and relying upon their quick reflexes to get out of trouble; to the older employees who have been doing it this way for years and have never gotten hurt—even if it was clearly a dangerous way to proceed. The safety rep has got to challenge deficiencies and not be afraid of causing or stirring up controversy on the job. It's better to have it out on a subject, then eventually figure out the proper way to do things, rather than avoid the conflict and hope that everything will work out the best way possible, with no safety or injury consequences.

One thing that safety representatives need or will find very helpful is a crash course in conflict management, because much of their time can be spent trying to convince individuals of the logic of being safe and following safe procedures. It's also a good idea to rotate the safety reps, perhaps every year or so, until all individuals who are qualified for the position have a turn at it. This will result in a well-rounded crew when it comes to safety, and will result in more knowledgeable crews. Plus, the individual crew members will be more reluctant to give the safety rep a hard time.

Employees can also participate in safety by participating in training sessions, helping to prepare and conduct meetings on topics they're familiar with or have special expertise in. For example, several employees have been sent to a hands-on training seminar and have become competent persons regarding the selection, erection, and use of scaffolding. Wouldn't it be logical to ask them to run your scaffolding safety training sessions? Same with your competent individuals on excavations and trenches. Why not ask an electrician to lead your meetings on electrical safety? Or a long-time backhoe operator to discuss how employees on the ground can best work around heavy equipment operations?

Naturally, you can't just toss these individuals into a training meeting without any preparation. Someone familiar with training methods should help them prepare for their sessions, and should arrange for whatever materials they will need.

Chapter

# 9

# Safety Training

## Quick Scan

1. There's a requirement and need in every company for safety and skills training.
2. OSHA mandates education and training for individuals in all job trades and classifications participating at construction worksites.
3. Training must include safety orientation for new hires, annual and refresher training for regular personnel, and training for all tools, equipment, methods, and materials newly introduced to the worksite.

What's the first question you're likely to hear from OSHA at the start of an employee injury investigation? I'll give you a clue. It will have to do with the injured employee. And it'll be about the training the employee received in order to do his or her work.

"Was the injured employee trained to do the job?" OSHA will want to know. And that may be followed with, "Could we please see the employee's meeting attendance records and a sample of the training materials?" Then, "And while you're at it, how about producing the rest of your training records as well."

If you haven't done your training, things will likely go downhill from there. If you haven't done your training because you "haven't had time," or your supervisors "only instructed employees as they did the work," or the regular employees "are all long-term employees who know what they're doing," or the college kids "were just summer helpers who said they had the training last year," or a similar excuse, it's not going to fly with OSHA, and it shouldn't be good enough for your company, either.

In short, you can't afford *not* to train your employees. It's the law, and it's good safety and business sense. Think of training your employees as an investment instead of an expense. Employees trained to work safely and skillfully will help your company maintain far more productive worksites than those populated by employees who are left to their own means of getting the work done. To a certain extent, skills training is also part of the safety picture. To run tools and equipment properly and to make on-the-fly decisions about various work techniques, methods, and specific situations requires familiarity with safety procedures. Don't expect employees to learn safety on the job—they need a proper grounding in safety management techniques and in safe operating procedures from the start.

## Safety Orientation Training for New Employees

Safety experts explain that those who are new on the job have a higher rate of safety-related incidents than more experienced workers. This, of course, can be attributed to ignorance of specific job hazards and/or of proper work practices—something that can be partially resolved through training.

New employee orientation and probationary periods with follow-up should be in place at your company for supervisors and hourly employees. Safety orientation training is a prime time to emphasize to new hires exactly how serious your company is about safety. It's desirable, in fact, to *start out with safety.* Put things into perspective; emphasize that nothing the employees do with the company will be more important than learning and following safe procedures on (and off) the job.

Each employee of your company needs safety orientation and training on a variety of topics, which will depend on the tasks he or she has to perform. Some of the topics are mandated by OSHA, while others just make good safety sense or are part of company safety rules and requirements. Some OSHA standards make it the employer's responsibility to limit certain job assignments to employees who are "certified," "competent," or "qualified," which means they've had special previous training, in or out of the workplace. The training should be refreshed every so often to ensure proper understanding.

Naturally, you can't train an employee in advance on everything he or she should know about the job. Some of the training comes gradually while experiencing the different facets of a job or while observing and assisting other workers. It will also depend on the level of experience the worker brings to the job. No matter what that experience level may be, the employee should be trained in the company's way of doing things. If a new employee knows from previous experience how to make a company procedure safer, that's great. On the other hand, that same employee may also have to unlearn a few bad habits or hazardous shortcuts that were encouraged or silently approved at past workplaces.

To make a lasting impression, it's a good idea to plan your safety orientations to last a minimum of one full day. First, schedule an introduction from the company owner or president, who will, by his or her very appearance, reinforce the importance of safety. The same member of upper management should then emphasize the company safety policy by discussing the role of safety throughout the daily and long-term operations. Subsequent training sessions can be completed by the company's supervisors or hourly safety representatives with whom the new employees will be working.

Another reason the safety orientation is so important is because first impressions are critical to establishing a safety mind-set. It's natural for a new employee to judge the sharpness of a company by the way he or she is received into the organization. If your company wants to attract, develop, and keep safe and alert employees, the company itself must be safe and alert. The company should be prepared to receive its new workers in a supportive way. All of the information must be organized and presented in a clear fashion, with no room for confusion or doubt when it comes to the importance of safety. Your safety policy must make it clear that working safely is neither an elective nor an option. Rather, it is a condition of employment, just like showing up every day, on time, in a mentally and physically alert condition.

If this seems like a lot of time spent on safety, put it in perspective. You already know what having incidents and injuries can cost in terms of pain, time, inconvenience, and dollars. Balance that against the time spent on a proper safety orientation, and the training becomes the best kind of preventive measure that can be taken. Not only will it help the new hires, it will further the overall reinforcement of safety throughout the company by involving the various supervisors and others who do the training.

To help construct your orientation plan, use the resources in this book and others. Hourly training should include materials on all important safety components relating to safety management and to jobs the new employees will be performing. That means reviewing the following:

- Safety management principles
- Facility safety rules
- Material handling and mobile equipment safety
- Lockout/hazardous energy control
- Hazardous material communications and handling
- Protective equipment and guards
- Tool and equipment safety
- Ergonomics
- Emergency equipment

Supervisors need the training, too. A supervisor lacking safety management and training skills will probably function at a safety efficiency level somewhere between ineffective and dangerous.

In addition to regular worksite safety topics that all employees receive, supervisors need additional training in the following areas:

- Safety management principles
- Behavior observations
- Incident investigations

- OSHA compliance
- Correction and reinforcement techniques
- Effective training methods

## Meaningful Safety Training

One of the most difficult assignments for some employers is how to arrange meaningful training for their employees. I say *meaningful* training because it's easy to toss some handouts onto a table and then read them out loud as employees follow along. In fact, some homebuilding supervisors have resorted to reading, word for word, OSHA construction regulations. If there's a quicker way to put employees to sleep, I haven't found it.

Book knowledge and theory are part of training, but only part. Hands-on skills training needs to be practiced, too. *The most effective training involves both knowledge/theory and the learning and practicing of skills.* The most effective training also seeks participation from the employees who are receiving training. By *actively participating,* they will have to think and act in response to something the trainer says or does. Training should be more than a one-way affair of just spoon-feeding information to a location that participants happen to be in. The materials must be retained or embraced by the audience. There needs to be some rational thought and some perceptible change that occurs within participants. In other words, they've got to walk away with more than they brought to the table.

Good training commands attention. Good training invites questions. "Train the trainer" is a slogan that's big in industry. Large companies realize that certain skills are needed to be a good trainer and that there's room for individual training styles. Some trainers use humor to their advantage—fine, if it's not overdone. Others attempt to use graphic what-if videos. Trainers can't totally rely on videos, though, or they risk being too general or distant from their audience. Trainees need specifics. They need customized training materials that apply to their own worksites. They need real-life examples they can relate to.

There are many ways to spice up safety training. Save and use safety-related articles from newspapers, magazines, trade newsletters, and journals. Invite guest speakers to make presentations on main topics. Manufacturers and service companies are usually very cooperative when it comes to supplying demonstrations and training on their tools, equipment, products, or services. And they work cheaply—usually for the cost of a cup of coffee and bagel. Occasionally, individuals will agree to speak on safety regarding hobbies they may have, such as hunting, boating, or woodworking.

When training, encourage trainers to avoid simply telling people what not to do and what to do. Instead, participants should be given *reasons.* Employees are much more likely to retain training if they understand and agree on why something is important to know.

## Adult Education and Training

Adults do not learn the same way that children do. Children in school are used to sitting in a classroom and receiving information and instruction from a teacher—pretty much a one-way street. Oh, they're permitted to ask questions, but that's not a major part of their learning experience. In most of the classes, children are learning something for the first time. They don't have much in the way of experiences or opinions to bring to the classroom.

Adults, on the other hand, tend to become impatient when presented with a one-way stream of information from an instructor. Adults like to share their experiences—not with a teacher who dictates what should be learned, but with someone who will guide the class and facilitate the discussions, allowing the participants to ask questions, to challenge points, to share examples, and to sort out important concepts. Supplying adults with safety training involves a lot more than just passing along information. The training leader must help the participants take an active role in the training process itself. The participants have to become interested in the material through interacting in various ways with the training leader and with the rest of the class. When employees complain about a training session being boring, they usually mean that the presentation of the material is mostly one-way. Unfortunately, the lecture method is probably the most frequently used training technique—and the least effective. Better to use part lecture and part discussion or demonstration.

When planning any training, here are some important points to consider:

1. *Identify the training needs of the employees.* What do the employees know about the topic already? What more should they know or know how to do? Try to assess their training needs so the materials are neither too basic nor too advanced for the session.

2. *Decide what the minimum training end result should be.* What do you want the participants to know or to be able to do after the training is completed? For example, you would want all participants to be able to properly danger-tag a piece of mobile equipment that's not running correctly.

3. *Plan the training session.* First, decide what method of instruction will work best. Lecture and discussion? Plus a brief demonstration with everyone doing a trial run? Outline the training session, including all the points that must be covered.

4. *Prepare the session.* Gather materials. It's best to know more than you plan to present, to have little extras available to help round out the discussions. It's not necessary to memorize presentations. They'll sound too stuffy and artificial. But be very familiar with all of the main points, so you can talk freely about them in your own words, referring to an outline, overhead transparencies, slides, flipcharts, or similar training aids. It's always advisable to tie the training into the workplace, using examples of familiar situations whenever possible.

Consider recruiting assistants from the participants, either during the training, much as a magician will occasionally ask for a volunteer from an audience, or better yet, ask for helpers in advance, so the helpers can assist in the session's planning.

5. *Decide where to have the training.* Brief training sessions, including tool-box meetings, are held right there at the worksite. More formal training should be scheduled in well-lit, quiet rooms that are large enough and sufficiently well ventilated to accommodate all of the expected participants.

6. *Arrange the room to suit adult training.* Try to avoid having the teacher in front with all the students lined up in classroom-style rows. Circular or U-shaped seating encourages equal participation and makes it less likely that individuals will drift off to sleep during the training.

7. *Train.* Coordinate. Facilitate. Present. Help draw the participants into the training. Make them feel comfortable, and ask for their opinions and comments.

8. *Verify the training.* This is often omitted by trainers because it means more work. Sometimes a lot more work. And it could mean that some of the participants are not learning what they're supposed to. That, in turn, implies potential confrontation. OSHA has been expecting and mandating that employees, when receiving training in many important topics such as hazardous energy control and hazard communications, must demonstrate their knowledge of those training materials. Written tests, demonstrations, and performance-based evaluations can be used singly or together. Written tests alone, however, are typically not the best way of verifying safety procedures requiring specific tasks.

## Keeping Training and Meeting Records

Good training concludes with some verification that the material was understood. That can be in the form of a meaningful quiz or exam, a demonstration, a bench test, an obstacle course, or a similar method. (See Fig. 9.1.)

All training session descriptions, sign-in sheets, notes, and training verification results should be recorded and kept. Recordkeeping will accomplish several worthwhile ends. First of all, it will help drive your training program. An effective training program will result in fewer injuries on the job (and generally off the job, too, because of increased overall awareness), improved interest in the work, higher morale, lower insurance rates, and reduced overall costs. Documentation also allows company management to track what training which employees have had. Furthermore, it's evidence of the company's good faith and compliance with OSHA standards that require specific types of training for particular tasks and jobs.

Verification quizzes can be prepared or purchased on most training topics. Try to mix up the questions: use true and false, fill in the blanks, multiple choice, or matching. Refer to the OSHA training requirements information later in this chapter as a resource for building training sessions. Consult your local OSHA office as well. The generic quiz will provide a sampling of verification questions. You'd do best to customize the quiz further with material particular to your operations. If you're training roofers, for example, there are loads of questions that can be asked regarding safety. Consider preparing questions from topics such as fall protection, the use of ladders, correct lifting procedures, eye protection, heat stress, cumulative trauma disorders, and safe tool use.

**WORKSITE SAFETY MEETING FORM**

**Date:** ____________________

**Attendees:** ______________________________ ______________________________

______________________________ ______________________________

______________________________ ______________________________

______________________________ ______________________________

**Follow-up from previous meetings:**

______________________________________________________________

______________________________________________________________

______________________________________________________________

**Main Topics for Discussion:**

______________________________________________________________

______________________________________________________________

______________________________________________________________

______________________________________________________________

______________________________________________________________

**Follow-up Required:**

______________________________________________________________

______________________________________________________________

______________________________________________________________

**Suggestions for future meeting topics:**

______________________________________________________________

______________________________________________________________

**Figure 9.1** Worksite Safety Meetings Form.

## Monthly Safety Meetings

Regular monthly safety meetings demonstrate the company's concern for the lives and well-being of its employees. They help build a cooperative climate, providing employees with the opportunity to contribute ideas and suggestions to improve safety, quality, productivity, and morale. The supervisor, safety representatives, and crew leaders should run the monthly safety meetings with help from guest presenters, manufacturers, and others mentioned earlier. These meetings should last from 1 to $1^1/_2$ hours, and not be rushed. Members of the support staff, office staff, and upper management should also participate in the meetings. Consider alternating the meeting leaders, giving each an advanced schedule which will allow them to plan the sessions in advance. Try to work the required topics into the monthly meetings so the meetings can be planned around safety themes. Suggested themes include the following:

Hazardous energy control/electrical safety

Power tools and equipment

Ergonomics

Fire prevention and fire extinguisher use

First aid and bloodborne pathogens

## Tool-Box Safety Meetings

Tool-box safety meetings are brief gatherings of crews or employees along with their supervisor or crew leader. They're usually held early in the morning, before the crews begin their workday or before they start some specific task so that crew members can be reminded of hazards inherent to the work at hand and can take precautions for avoiding them. Tool-box safety meetings are typically informal and brief: 10 to 20 minutes at most.

The tool-box meeting trainer or leader doesn't have to be a professional speaker or especially entertaining or clever to hold the attendees' interest because the discussions are so short. But the supervisor or leader does need to select appropriate safety topics that address the crew's current or upcoming work tasks. It does no good to speak about heat stress in February if the crews are working in upstate New York, nor does it do any good to speak about using miniloaders if there isn't one around for miles. Tool-box meetings are used by successful companies to publicize worksite hazards, assist with safe-procedure reviews, and help develop employee safety awareness. They're convenient. The employees are all there, early, with not alot on their minds.

For greatest effect, subjects that most interest the employees should be highlighted. Those topics might include accidents, inspection results, the safety program, or a particular work procedure. They may include comments on a new piece of equipment that's being purchased or rented, a new tool and its use, or an old tool and its proper use and care. Ideally, reserve a few minutes at the end of the meeting for employee participation, for their opinions, comments, and questions. Always schedule meetings at the beginning of new operations to ensure that all employees are familiar with safe job procedures and requirements of upcoming work. These meetings will save a lot of time, too, by communicating things the entire crew should be aware of, thus eliminating confusion later.

Tool-box meetings, because of their frequency, may be either instructional or motivational. The motivational meeting creates awareness and aims at worker self-protection. The instructional meeting covers a particular job task or procedure. In any event, specific aspects of the topics you'll find in the following list will lend themselves to brief tool-box safety sessions.

## Suggested safety tool-box talk topics

- Asbestos
- Back injury prevention
- Barricades
- Bloodborne pathogens
- Carpal tunnel syndrome
- CPR (cardiopulmonary resuscitation)
- Confined spaces
- Congestion and restricted movement
- Cranes, hoists, and derricks
- Drug and alcohol abuse
- Electrical safety
- Employee accountability
- EPA right-to-know regulations
- Ergonomics
- Excavations and shoring
- Eye protection: safety glasses and goggles
- Fall protection
- Fire prevention
- First-aid treatment
- Foot protection
- Forklift operation
- Grounding: electrical currents
- Hand protection and gloves
- Hand-tool use and maintenance
- Hard-hat use

- Hazard Communication Standard (HazCom)
- Hazardous materials:
  Combustible metals
  Compressed gases
  Corrosives
  Flammable and combustible liquids
  Paper, wood, straw, cloth
  Systemic poisons
  Toxins
- Hazardous waste management
- Head protection: helmets and hoods
- Hearing protection: noise exposure
- Heat stress
- Housekeeping and waste disposal
- Labeling
- Ladders
- Lifting
- Lighting: inadequate or excessive
- Lockout and tagging procedures
- Materials handling and storage
- Motor vehicle safety
- Mobile equipment
- Personal protective equipment
- Power-tool use and maintenance
- Pushing and pulling
- Radioactive materials
- Reports and recordkeeping procedures
- Respirators
- Safety program:
  Company's awards program
  Company's written program
  Safety incentive or recognition program
- Safety harnesses and lanyards
- Scaffolding
- Security: fire prevention
- Signs
- Skin protection: proper protective clothing

- Spill cleanup
- Slips, trips, and falls
- Suspended load operation
- Tie-off procedures
- Ventilation
- Workers' compensation
- Workplace violence

## What OSHA Says

On the subject of safety training, OSHA insists that employers, in all cases, supply the training their employees need to recognize and avoid hazards within the workplace.

### Construction Training Requirements, 29 CFR Part 1926

The following training requirements have been borrowed from OSHA's Construction Safety and Health Outreach Program, 1996.

Subpart C: General Safety and Health Provisions

—General Safety and Health Provisions
1926.20(b)(2) and (4)

(2) Such programs [as may be necessary to comply with this part] shall provide for frequent and regular inspections of the job sites, materials, and equipment to be made by competent persons [capable of identifying existing and predictable hazards in the surroundings or working conditions which are unsanitary, hazardous, or dangerous to employees, and who have authorization to take prompt corrective measures to eliminate them designated by the employers.]

(4) The employer shall permit only those employees qualified [one who, by possession of a recognized degree, certificate, or professional standing, or who by extensive knowledge, training, and experience, has successfully demonstrated his ability to solve or resolve problems relating to the subject matter, the work, or the project] by training or experience to operate equipment and machinery.

—Safety Training and Education
1926.21(a)

(a) General requirements. The Secretary shall, pursuant to section 107(f) of this Act, establish and supervise programs for the education and training of employers and employees in the recognition, avoidance, and prevention of unsafe conditions in employments covered by the Act.

1926.21(b)(1) through (6)(i) and (ii)

(1) The employer should avail himself of the safety and health training programs the Secretary provides.

(2) The employer shall instruct each employee in the recognition and avoidance of unsafe conditions and the regulations applicable to his work environment to control or eliminate any hazards or other exposure to illness or injury.

(3) Employees required to handle or use poisons, caustics, and other harmful substances shall be instructed regarding their safe handling and use, and be made

aware of the potential hazards, personal hygiene, and personal protective measures required.

(4) In job site areas where harmful plants or animals are present, employees who may be exposed shall be instructed regarding the potential hazards and how to avoid injury, and the first-aid procedures to be used in the event of injury.

(5) Employees required to handle or use flammable liquids, gases, or toxic materials shall be instructed in the safe handling and use of these materials and made aware of the specific requirements contained in Subparts D, F, and other applicable subparts of this part.

(6)(i) All employees required to enter into confined or enclosed spaces shall be instructed as to the nature of the hazards involved, the necessary precautions to be taken, and in the use of protective and emergency equipment required. The employer shall comply with any specific regulations that apply to work in dangerous or potentially dangerous areas.

(ii) For purposes of subdivision (i) of this subparagraph, "confined or enclosed space" means any space having a limited means of egress, which is subject to the accumulation of toxic or flammable contaminants or has an oxygen deficient atmosphere. Confined or enclosed spaces include, but are not limited to, storage tanks, process vessels, bins, boilers, ventilation or exhaust ducts, sewers, underground utility vaults, tunnels, pipelines, and open top spaces more than 4 feet in depth such as pits, tubs, vaults, and vessels.

Subpart D: Occupational Health and Environmental Controls
—Medical Services and First Aid
1926.50(c)

(c) In the absence of an infirmary, clinic, hospital, or physician that is reasonably accessible in terms of time and distance to the worksite which is available for the treatment of injured employees, a person who has a valid certificate in first-aid training from the U.S. Bureau of Mines, the American Red Cross, or equivalent training that can be verified by documentary evidence, shall be available at the worksite to render first aid.

—Ionizing Radiation
1926.53(b) [Including radioactive materials or X-rays.]
—Nonionization Radiation
1926.54(a) and (b) [Including laser equipment.]
—Gases, Vapors, Fumes, Dusts, and Mists
1926.55(b)

(b) To achieve compliance with paragraph (a) of this section, administrative or engineering controls must first be implemented whenever feasible. When such controls are not feasible to achieve full compliance, protective equipment or other protective measures shall be used to keep the exposure of employees to air contaminants within the limits prescribed in this section. Any equipment and technical measures used for this purpose must first be approved for each particular use by a competent industrial hygienist or other technically qualified person. Whenever respirators are used, their use shall comply with 1926.103.

—Asbestos
1926.58(k)(3)(i) through (iii)(A) through (E) and(4)(i) and (ii)

(i) The employer shall institute a training program for all employees exposed to airborne concentrations of asbestos, in excess of the action level and/or excursion limit and shall ensure their participation in the program.

(ii) Training shall be provided prior to or at the time of initial assignment (unless the employee has received equivalent training within the previous 12 months) and at least annually thereafter.

(iii) The training program shall be conducted in a manner that the employee is able to understand. The employer shall ensure that each such employee is informed of the following:

(A) Methods of recognizing asbestos;

(B) The health effects associated with asbestos exposure;

(C) The relationship between smoking and asbestos in producing lung cancer;

(D) The nature of operations that could result in exposure to asbestos, the importance of necessary protective controls to minimize exposure including, as applicable, engineering controls, work practices, respirators, housekeeping procedures, hygiene facilities, protective clothing, decontamination procedures, emergency procedures, and waste disposal procedures, and any necessary instruction in the use of these controls;

(E) The purpose, proper use, fitting instructions, and limitations of respirators as required by 29 CFR 1910.134;

(4) Access to training materials. (i) The employer shall make readily available to all affected employees without cost all written materials relating to the employee training program, including a copy of this regulation.

(ii) The employer shall provide to the Assistant Secretary and the Director, upon request, all information and training materials relating to the employee information and training program.

—Hazard Communication, Construction

1926.59(h)(2)(i) through (iv)

(2) Training. Employee training should include at least:

(i) Methods and observations that may be used to detect the presence or release of a hazardous chemical in the work area (such as monitoring conducted by the employer, continuous monitoring devices, appearance of odor of hazardous chemicals when being released, etc.);

(ii) The physical and health hazards of the chemicals in the work area;

(iii) The measures employees can take to protect themselves from these hazards, including specific procedures the employer has implemented to protect employees from exposure to hazardous chemicals, such as appropriate work practices, emergency procedures, and personal protective equipment to be used; and,

(iv) The details of the hazard communication program to be developed by the employer, including an explanation of the labeling system and the material safety data sheet, and how employees can obtain and use the appropriate hazard information.

—Lead in Construction

1926.62(I)(1)(i) through (iv); (2)(i) through (viii) and (3)(i) and (ii)

[This is similar to the asbestos standard.]

—Process Safety Management of Highly Hazardous Chemicals

1926.64(g)(1)(i) and (ii), and (g)(2), and (g)(3), and (h)(3)(i) through (iv), and (j)(3)

[This section includes training for situations in which highly hazardous chemicals are in use or present at the worksite, and prescribes refresher training, specific training documentation, contract-employer responsibilities, and maintenance mechanical integrity activities.]

Subpart E: Personal Protective and Life Saving Equipment

—Hearing Protection

1926.101(b)

(b) Ear protective devices inserted in the ear shall be fitted or determined individually by competent persons.

—Respiratory Protection

1926.103(c)(1)

(1) Employees required to use respiratory protective equipment approved for use in atmospheres immediately dangerous to life shall be thoroughly trained in its use. Employees required to use other types of respiratory protective equipment shall be instructed in the use and limitations of such equipment.

Subpart F: Fire Protection and Prevention

—Fire Protection

1926.150(a)(5), and (c)(1)(viii)

(viii) Portable fire extinguishers shall be inspected periodically and maintained in accordance with "Maintenance and Use of Portable Fire Extinguishers," NFPA No. 10A-1970.

From ANSI Standard 10A-1970. "The owner or occupant of a property in which fire extinguishers are located has an obligation for the care and use of these extinguishers at all times. By doing so, he is contributing to the protection of life and property. The nameplate(s) and instruction manual should be read and thoroughly understood by all persons who may be expected to use extinguishers.

"1120. To discharge this obligation he should give proper attention to the inspection, maintenance, and recharging of this fire protective equipment. He should also train his personnel in the correct use of fire extinguishers on the different types of fires which may occur on his property."

Subpart G: Signs, Signals, and Barricades

—Signaling

Subpart I: Tools—Hand and Power

—Powder-Operated Hand Tools

1926.302(e)(1) and (12)

(1) Only employees who have been trained in the operation of the particular tool in use shall be allowed to operate a powder-actuated tool.

(12) Powder-actuated tools used by employees shall meet all other applicable requirements of American National Standards Institute, A10.3-1970, "Safety Requirements for Explosive-Actuated Fastening Tools."

—Woodworking Tools

1926.304(f)

(f) Other requirements. All woodworking tools and machinery shall meet other applicable requirements of American National Standards Institute, 01.1-1961, "Safety Code for Woodworking Machinery."

From ANSI Standard 01.1-1961, Selection and Training of Operators. Before a worker is permitted to operate any woodworking machine, he shall receive instructions in the hazards of the machine and the safe method of its operation. Refer to A9.7 of the Appendix.

"A9.7 Selection and Training of Operators. Operation of Machines, Tools, and Equipment. General.

"(1) Learn the machine's applications and limitations, as well as the specific potential hazards peculiar to this machine. Follow available operating instructions and safety rules carefully.

"(2) Keep working area clean and be sure adequate lighting is available.

"(3) Do not wear loose clothing, gloves, bracelets, necklaces, or ornaments. Wear face, eye, ear, respiratory, and body protection devices, as indicated for the operation or environment.

"(4) Do not use cutting tools larger or heavier than the machine is designed to accommodate. Never operate a cutting tool at a greater speed than recommended.

"(5) Keep hands well away from saw blades and other cutting tools. Use a push stock or push block to hold or guide the work when working close to the cutting tool.

"(6) Whenever possible, use properly locked clamps, jig, or vise to hold the work.

"(7) Combs (feather boards) shall be provided for use when an applicable guard cannot be used.

"(8) Never stand directly in line with a horizontally rotating grinding tool. This is particularly true when first starting a new tool, or a new tool is initially installed on the arbor.

"(9) Be sure the power is disconnected from the machine before tools are serviced.

"(10) Never leave the machine with the power on.

"(11) Be positive that hold-downs and anti-kickback devices are positioned properly, and that the workpiece is being fed through the cutting tool in the right direction.

"(12) Do not use a dull, gummy, bent, or cracked cutting tool.

"(13) Be sure that keys and adjusting wrenches have been removed before turning power on.

"(14) Use only accessories designed for the machine.

"(15) Adjust the machine for minimum exposure of cutting tool necessary to perform the operation."

Subpart J: Welding and Cutting

—Gas Welding and Cutting

1926.350(d)(1) through (6), and (j) [Including the use of fuel gas, and reference to ANSI Z49.1-1967, "Safety in Welding and Cutting."]

—Arc Welding and Cutting

1926.351(d)(1) through (5), and 352(e) [Including operating instructions for the safe means of arc welding.]

—Fire Prevention

1926.352(e)

(e) When the welding, cutting, or heating operation is such that normal fire prevention precautions are not sufficient, additional personnel shall be assigned to guard against fire while the actual welding, cutting, or heating operation is being performed, and for a sufficient period of time after completion of the work to ensure that no possibility of fire exists. Such personnel shall be instructed as to the specific anticipated fire hazards and how the fire fighting equipment provided is to be used.

—Welding, Cutting, and Heating in Way of Preservative Coatings

1926.354(a) [Including information about performing hot works on coated surfaces.]

Subpart K: Electrical

—Ground-Fault Protection

1926.404(b)(iii)(B)

(iii)(B) The employer shall designate one or more competent persons [as defined in 1926.32(f)] to implement the program.

Subpart L: Scaffolding

—Scaffolding

1926.451, many sections. [This is a large, sprawling subpart with many training requirements associated with various types of scaffolding.]

—Guarding of Low-Pitched Roof Perimeters During the Performance of Built-Up Roofing Work

1926.500(g)(6)(i) and (ii)(a) through (f) and (iii)

(6) Training. (i) The employer shall provide a training program for all employees engaged in built-up roofing work so that they are able to recognize and deal with the hazards of falling associated with working near a roof perimeter. The employees shall also be trained in the safety procedures to be followed in order to prevent such falls. [And similar training information.]

—Fall Protection

1926.503(a)(1) and (2)(i) through (vii)

(a) Training program. (1) The employer shall provide a training program for each employee who might be exposed to fall hazards. The program shall enable each employee to recognize the hazards of falling and shall train each employee in the procedures to be followed in order to minimize these hazards.

(2) The employer shall ensure that each employee has been trained, as necessary, by a competent person qualified in the following areas:

(i) The nature of fall hazards in the work area;

(ii) The correct procedures for erecting, maintaining, disassembling, and inspecting the fall protection systems to be used;

(iii) The use and operation of guardrail systems, personal fall arrest systems, safety net systems, warning line systems, safety monitoring systems, controlled access zones, and other protection to be used;

(iv) The role of each employee in the safety monitoring system when this system is used;

(v) The limitations on the use of mechanical equipment during the performance of roofing work on low-sloped roofs.

(vi) The correct procedures for the handling and storage of equipment and materials and the erection of overhead protection; and

(vii) The standards contained in this subpart.

Subpart N: Cranes, Derricks, Hoists, Elevators, and Conveyors

—Cranes and Derricks

1926.550(a)(1), (5), and (6) [These sections address general crane and derrick training and use, and specify that a competent person must inspect all machinery and equipment prior to each use.]

—Material Hoists, Personnel Hoists, and Elevators

1926.550 and 552 [These sections include training information and requirements for qualified, competent, and/or professional personnel regarding the design and use of personnel platforms, material hoists, personnel hoists, and elevators.]

Subpart O: Motor Vehicles, Mechanized Equipment, and Marine Operations

—Material Handling Equipment

1926.602(c)(1)(vi)

(c) Lifting and hauling equipment (other than equipment covered under Subpart N of this part).

(1)(vi) All industrial trucks in use shall meet the applicable requirements of design, construction, stability, inspection, testing, maintenance, and operation, as defined in American National Standards Institute B56.1-1969, "Safety Standards for Powered Industrial Trucks."

From ANSI Standard B56.1-1969, "Operator Training. Only trained and authorized operators shall be permitted to operate a powered industrial truck. Methods shall be devised to train operators in the safe operation of powered industrial trucks. Badges or other visual indication of the operators' authorization should be displayed at all times during work period."

—Site Clearing

1926.604(a)(1)

(1) Employees engaged in site clearing shall be protected from hazards of irritant and toxic plants and suitably instructed in the first-aid treatment available.

Subpart P: Excavations

—General Protection Requirements (Excavations, Trenching, and Shoring)

1926.651(c)(1)(i)

(c) Access and egress (1) Structural ramps. (i) Structural ramps that are used solely by employees as a means of access or egress from excavations shall be designed by a competent person. Structural ramps used for access and egress of equipment shall be designed by a competent person qualified in structural design, and shall be constructed in accordance with the design.

1926.651(h)(2) and (3)

(h) Protection from hazards associated with water accumulation. (2) If water is controlled or prevented from accumulating by the use of water removal equipment, the water removal equipment and operations shall be monitored by a competent person to ensure proper operation.

(3) If excavation work interrupts the natural drainage of surface water (such as streams), diversion ditches, dikes, or other suitable means shall be used to prevent surface water from entering the excavation and to provide adequate drainage of the area adjacent to the excavation. Excavations subject to runoff from heavy rains will require an inspection by a competent person and compliance with paragraphs (h)(1) and(h)(2) of this section.

1926.651(i)(1)

(i) Stability of adjacent structures. (1) Where the stability of adjoining buildings, walls, or other structures is endangered by excavation operations, support systems such as shoring, bracing, or underpinning shall be provided to ensure the stability of such structures for the protection of employees.

1926.651(i)2)(iii)

(iii) A registered professional engineer has approved the determination that the structure is sufficiently removed from the excavation so as to be unaffected by the excavation activity; or

1926.651(i)(2)(iv)

(iv) A registered professional engineer has approved the determination that such excavation work will not pose a hazard to employees.

1926.651(k)(1) and (2)

(k) Inspections. (1) Daily inspections of excavations, the adjacent areas, and protective systems shall be made by a competent person for evidence of a situation that could result in possible cave-ins, indications of failure of protective systems, hazardous atmospheres, or other hazardous conditions. An inspection shall be conducted by the competent person prior to the start of work and as needed throughout the shift. Inspections shall also be made after every rainstorm or other hazard increasing occurrence. These inspections are only required when employee exposure can be reasonably anticipated.

(2) Where the competent person finds evidence of a situation that could result in a possible cave-in, indications of failure of protective systems, hazardous atmospheres, or other hazardous conditions, exposed employees shall be removed from the hazardous area until the necessary precautions have been taken to ensure their safety.

Subpart Q: Concrete and Masonry Construction
—Concrete and Masonry Construction
1926.701(a), and 1926.703(b)(8)(i)
(a) No construction loads shall be placed on a concrete structure or portion of a concrete structure unless the employer determines, based on information received from a person, who is qualified in structural design, that the structure or portion of the structure is capable of supporting the loads. [Also determines that shoring design and installation shall be supervised by a structural engineer].
Subpart R: Steel Erection
—Bolting, Riveting, Fitting-Up, and Plumbing-Up
1926.752(d)(4)
(4) Plumbing-up guys shall be removed only under the supervision of a competent person.
Subpart S: Underground Construction, Caissons, Cofferdams, and Compressed Air
—Underground Construction
1926.800, many sections dealing with training about the recognition and avoidance of hazards associated with underground construction.
—Compressed Air
1926.803, dealing with physical use of compressed air for breathing/decompression work.
Subpart T: Demolition
—Preparatory Operations
1926.850(a)
—Chutes
1926.852(c)
—Mechanical Demolition
1926.859(g)
(g) During demolition, continuing inspections by a competent person shall be made as the work progresses to detect hazards resulting from weakened or deteriorated floors, or walls, or loosened material. No employee shall be permitted to work where such hazards exist until they are corrected by shoring, bracing, or other effective means.
Subpart U: Blasting and Use of Explosives
—General Provisions (Blasting and Use of Explosives)
1926.900
—Blaster Qualifications 1926.901
—Surface Transportation of Explosives 1926.902
—Firing the Blast 1926.909
Subpart V: Power Transmission and Distribution
—General Requirements 1926.950
—Overhead Lines 1926.955
—Underground Lines 1926.956
—Construction in Energized Substations 1926.957
—Ladders
1926.1053(b)(15)
(15) Ladders shall be inspected by a competent person for visible defects on a periodic basis and after any occurrence that could affect their safe use.
1926.1060(a)(1)(i) through (v) and (b)
The following training provisions clarify the requirements of 1926.21(b)(2), regarding the hazards addressed in subpart X.

(a) The employer shall provide a training program for each employee using ladder and stairways, as necessary. The program shall enable each employee to recognize hazards related to ladders and stairways, and shall train each employee in the procedures to be followed to minimize these hazards.

(1) The employer shall ensure that each employee has been trained by a competent person in the following areas, as applicable:

(i) The nature of fall hazards in the work area;

(ii) The correct procedure for erecting, maintaining, and disassembling the fall protection systems to be used;

(iii) The proper construction, use, placement, and care in handling of all stairways and ladders;

(iv) The maximum intended load-carrying capacities of ladders used; and

(v) The standards contained in this subpart.

(b) Retraining shall be provided for each employee as necessary so that the employee maintains the understanding and knowledge acquired through compliance with this section.

## OSHA Training and Education

OSHA offers training courses to train and educate employees and employers in "the recognition, avoidance, and prevention of unsafe and unhealthful working conditions, and to improve the skill and knowledge levels of personnel engaged in work relating to the Occupational Safety and Health Act." Courses are presented in various OSHA-arranged locations as well as off-site at the request of interested companies and organizations. OSHA maintains facilities at:

**U.S. Department of Labor**
OSHA Training Institute
Des Plaines, Illinois
1-847-297-4913

**Georgia Tech Research Institute**
Safety, Health and Ergonomics Branch
Atlanta, Georgia
1-800-653-3629

**Great Lakes Regional OSHA Training Consortium**
(University of Minnesota, Minnesota Safety Council, University of Cincinnati)
St. Paul, Minnesota
1-800-493-2060

**Keene State College**
Manchester, New Hampshire
1-800-449-6742

**Maple Woods Community College**
Kansas City, Missouri
1-800-841-7158

**Motor City Education Center**
Eastern Michigan University/United Auto Workers
Ypsilanti, Michigan
1-800-932-8689

**The National Resource Center for OSHA Training**
Building and Construction Trades Department AFL/CIO/West Virginia University
Safety and Health Extension
Washington, DC
1-800-367-6724

**The National Safety Education Center**
Northern Illinois University, Construction Safety Council, National Safety Council
Dekalb, Illinois
1-815-753-0277

**Niagara County Community College**
Lockport, New York
1-800-280-6742

**Pacific Coast Training Center**
University of California-San Diego
LaJolla, California
1-800-358-9206

**Rocky Mountain Education Center**
Red Rocks Community College/
Trinidad State Junior College
Lakewood, Colorado
1-800-933-8394

**Southwest Education Center**
Texas Engineering Extension Service/Texas Safety Association
Arlington, Texas
1-800-723-3811

**University of Washington**
Department of Environmental Health
Seattle, Washington
1-800-326-7568

The OSHA Training Institute Education Centers are available to conduct courses at sites other than their campuses. Contact the closest education center for information on fees, registration procedures, course availability, hotels, and transportation.

## Other Training Resources

- The National Safety Council offers a large array of safety training classes and sessions.
- Many universities offer safety programs.
- Trade associations have training seminars and sessions at annual membership meetings and conventions.

- Homebuilders' associations put on seminars.
- Manufacturers of tools, equipment, and personal protective equipment may offer safety training.
- Companies that market safety literature and audiovisuals, including videotapes and supporting books, booklets, and other written materials, are good resources.

Chapter

# 10

# Incident Investigations

## Quick Scan

1. All safety incidents involving actual or potential workplace injuries, illnesses, or property damage must be investigated as soon as possible.
2. Each investigation must include fact gathering and analyzing that lead to the basic cause(s) of the incident, and recommendations for immediate and long-term corrective actions.
3. Preprinted incident investigation forms, witness interview forms, and corrective action forms, as appropriate, should be used for each incident investigation.
4. All worksite employees should receive training on conducting and participating in effective incident investigations, and on why such investigations benefit everyone in the company.
5. Incident-investigation cases should not be considered finished until all items requiring corrective action have been completed.
6. Data from incident investigations should be used to help identify and correct management and safety program deficiencies, and to help train employees regarding safety.

When a worksite incident involving injury or illness occurs, the first priorities, of course, are to treat the injured worker and to ensure that no one else is harmed while doing so. When a safety incident occurs without injury or illness—which is another way of referring to a near miss—the emphasis must first be to prevent anyone from being affected if it happens again.

For our purposes, an *incident* is an unwanted or unexpected event or exposure that results, or has the potential for resulting, in some kind of loss. *Loss* is meant as an outcome of what's commonly called an accident, further defined as personal injury, property damage, or materials damage. But it can also be an illness that has developed over a long time span. Simply put, incident investigations are needed to prevent the same or similar undesirable incidents or exposures from happening again. Consider what unfolds in the airline industry when a manufacturing or design flaw causes a plane to fail—at times disastrously. A detailed investigation is undertaken to discover what happened, so that the root cause of the incident can be communicated and corrected before the same thing happens again. Usually, due to the incident investigation and subsequent corrective action, the entire industry benefits. On a local level, you too can learn from incident investigations at your worksites. Although everyone hopes that such learning incidents involve no personal injury or loss of life, members of the home building industry and associated trades are not always that lucky. Thus, incident investigations are used not only to identify and correct immediate causes associated with the specific incident, but more important, to uncover and eliminate *basic causes*. Recognizing and eliminating the basic causes of incidents is management's responsibility. In every case, it is management who must undertake the prevention of unsafe behaviors or conditions that result in worksite near misses or accidental injuries and illnesses.

For example, an employee is nailing temporary flooring to a deck. An hour into the job he misstrikes a nail with his hammer. The nail glances off the hammerhead against a metal tool box and ricochets into the corner of his eye, causing damage which luckily is not permanent but which keeps him off work for a week and a half. The immediate cause for that incident is that *the employee wasn't wearing proper eye protection while using a striking tool.* That cause can be corrected simply, by the employee wearing safety glasses with side shields whenever nailing with a hammer. The basic causes, though, address several larger concerns:

1. The first basic cause addresses *the company's overall inadequate policy on personal protective equipment: there wasn't one.* Employees were told to be careful, and employee selection and use of personal protective equipment was optional. In short, they were allowed to do what they wanted. And no personal protective equipment was made available by the company.

2. The second basic cause reflects *a deficiency in training and knowledge about personal protective equipment in general.* The injured employee should have known enough to have worn safety glasses while nailing and should have been aware of what could happen without the eyewear—to the point of not using a hammer without it.

From incidents that are investigated, data can be compiled and reviewed to reveal trends that will show where to concentrate safety planning and training efforts.

## Why Safety Incidents Occur

Over the years, different theories have surfaced on why incidents occur in the first place. We've already discussed the relationship between amounts of unsafe behavior and the frequency of safety incidents. For each incident, though, early theories pointed to a very specific, immediate cause as reason enough—a single precipitating act that triggered a particular unfavorable result. Those early theories were later refined to suggest that *more than one cause* was to blame for each incident: some immediate cause(s) and some basic cause(s) combined to elicit each incident. The single-act theory proved deficient in its design because after the immediate cause of an incident was identified and corrected, nothing went on to address the basic causes, which permitted the unsafe behavior or immediate cause to recur, and presto—the same incident would happen again.

Individual immediate causes are directly associated with specific incidents. Unfortunately, they frequently are only symptoms of the basic causes—the underlying reasons in the system which let conditions exist in which the incident can occur. Basic causes are just that: basic deficiencies in management that must be addressed if true improvement is to take place.

## Why Investigate Safety Incidents

It's important for everyone to understand that the purpose of an incident investigation is not to yell at or discipline or even to find fault with employees. The reason for conducting investigations is to learn both *what* happened and *why* it happened, so that solutions can be arrived at that will prevent such an incident from happening again. That's pretty simple. It's critical that employees know why investigations are thus stressed, because if they expect discipline to result from each investigation, what kind of cooperation do you think you'll get? Employees who are fearful that blame will be assessed will be less cooperative and may even cover up key information. They have to be comfortable with the investigation process, and they should have a certain amount of training and experience on exactly what is expected for a typical incident investigation. Again, the main goals should be to find factual and objective data, to accurately determine the incident's causes, and to develop and map out appropriate solutions for preventing future incidents. On a more fundamental level, incident investigations can be used to help evaluate the effectiveness of the company's overall safety management program, and to reveal where additional efforts are needed for improvement.

## What Incidents Should Be Investigated

Some company owners or managers believe that only incidents that result in recordable injuries or illnesses or considerable damages should be investigated and documented. That's a poor idea. By leaving out near misses, first-aid incidents, and property-damage cases, the middle part of our safety tree model won't get the consideration it deserves. To gain the full benefits that incident investigations provide, *all* incidents should be included. Remember that the reason a near miss, a first-aid incident, or property damage does not result in a serious injury is due to pure chance. If a partial square of roofing shingles falls from a roof, the incident should be investigated whether or not anyone gets hurt. And if someone receives a shock while working around a temporary electric service line, the incident must be investigated and corrective action taken to prevent a similar recurrence. If ignored, repeats of either instance could involve a serious injury in the future. In short, all safety incidents should be investigated: near misses, first-aid incidents, no-lost-time recordable injuries and illnesses, lost-time recordable injuries and illnesses, motor vehicle accidents, and, of course, fatalities.

Although all incidents, even near misses, should be investigated, some are so cut-and-dried that they demand little more than a simple documentation of what occurred by completing just the bare bones of your standard incident investigation report. As the severity levels of incidents and near misses increase, additional investigative resources should be brought into play. For example, you and one employee could probably handle all of the investigation required for an incident involving the employee, a butt knife, and a lacerated thumb. But you're going to want more help on gathering data relating to an incident involving an employee who broke several vertebrae in his neck in a fall from a scaffold.

## Who Should Conduct Incident Investigations

Ideally, an incident investigation should be conducted by the injured employee's supervisor. But if the crew has a looser organization and there is no direct foreman or supervisor on the site, then a crew leader or ranking individual should start the process. For serious incidents, if a foreman, manager, or owner is within a reasonable driving or transportation distance, one of them should make himself or herself available.

What are serious incidents? You or your company should have at least loose guidelines regarding what's to be considered a serious incident. Many companies include in their definition events or exposures which result in any of the following events:

- Fatalities
- Serious neurological and/or spinal damage
- Loss of consciousness
- Amputation of a limb or digit
- Loss of or impairment to vision
- Second- or third-degree burns over 25 percent or more of the body
- Medical treatment or hospitalization of three or more employees
- Motor vehicle incidents
- More than one recordable injury or illness from the same incident

Certainly, it pays to have the most senior or highest member of management at the scene whenever possible—as long as that individual understands that his or her reason for being there is not to point fingers and find fault, but to conduct an investigation to reveal the immediate and basic causes, and then to arrive at corrective actions. Such a policy will help demonstrate the company's commitment to investigating incidents, while making sure that someone in charge has firsthand knowledge of the situation.

Irrespective of who leads the subsequent investigation, the ranking member of the on-site crew must make an immediate assessment and capture the earliest details possible, even while others are en route. This person can record the reactions and statements of observers before important details become forgotten or confused. As time passes, there's an increasing likelihood that individuals as well as equipment and other important pieces of evidence may be moved. Also, in many cases, the on-site supervisor typically knows more about the operation of the crew than anyone else.

It makes sense to supply training to individual crews so they can investigate incidents in your absence. Depending on how many employees you have in a crew, and how many crews you run simultaneously, another option is to have several employees trained to work together at the site to record and investigate incidents. The more employees you have that are familiar with how to proceed, the greater the likelihood will be of getting accurate and meaningful reports.

## Investigations

Learning the 10 steps to a thorough investigation will acquaint us with the basic incident investigation process. These key steps are:

1. Preparing to conduct investigations
2. Training the employees who will investigate
3. Gathering incident-related facts
4. Analyzing the facts
5. Developing conclusions
6. Writing the incident-investigation report
7. Recommending immediate and long-term corrective actions
8. Communicating what's in the report
9. Initiating corrective actions
10. Following up on recommendations

### Preparing to conduct investigations

The time to prepare your company for doing incident investigations is long before those necessary skills will be called for. Arrangements for investigating all categories of safety incidents, from near misses to lost-workday cases, should be planned for in advance. Reacting to an emergency with no idea of what to do leaves everything up to chance, and the situation will not likely go as well as it could.

As mentioned in the introduction to Sec. 2, you'll need a plan to ensure the availability of first-aid procedures and emergency services. These plans should be worked out in advance. First, what is the means of communication you want your employees to use? A cellular phone is ideal. A radio is another efficient way—provided that someone is listening at the other end. Otherwise your employees will have to use a nearby phone in the neighborhood, walking or driving to reach it, as necessary. Again, contacts with emergency responders and fire departments should be made in advance, to learn what those professionals recommend and what their expected response times will be. Here's also where it pays to have someone on the crew trained in first aid, first-responder techniques, and incipient (small, early stage) fire-fighting skills. Practically every community has low-cost programs that will train interested parties in CPR, first-aid procedures, and emergency response and incipient-fire skills.

The other part of the preliminary plan is the means to contact additional individuals who will help conduct the investigation, if needed. Naturally, the level of complexity demanded by any particular incident investigation increases with the severity of incident. You're not going to call for the cavalry if someone receives a minor cut or abrasion, but if someone falls from a second floor and can't move his legs, whoever is in charge of the company should be informed right away.

The gathering of information relating to safety incidents will be enhanced if a pad of worksite injury/illness report forms is already on hand and available for immediate use by on-site employees. The simplest forms (see Fig. 10.1) include requests for information about the injured employee, the incident, witnesses, subcontractors, and other incident-related data. More comprehensive forms (see Figs. 10.2, 10.3, and 10.4) feature additional sections that are filled in as more information is gathered. Some companies use a simple incident report form in the field, then attach it to a more inclusive form later—especially for incidents involving serious injuries or costly damages. You can make up your own company incident-investigation report forms, or, if you'd rather, use OSHA's reporting form No. 101, designed to elaborate on recordable injuries and illnesses required to be entered on OSHA 200 logs. Additional forms that can be helpful include witness-interview checklists (see Figs. 10.5 and 10.6), and corrective action follow-up forms (see Fig. 10.7).

### Training the employees who will investigate

In addition to learning how to investigate incidents and how to use the related incident report forms, everyone who may be called to a worksite incident aftermath should also receive *scene-safe* training. Scene-safe training relates to the necessary precautions to take and hazards to look for when responding to accident situations. In the manufacturing industry, an accident situation is referred to as one kind of upset condition that occurs during out-of-the-ordinary events, that tends to put individuals in hurried, stressful states. An upset condition like that found during a major safety incident where injury has occurred can tax responding employees' good judgment and pose serious risks that at the time may not be recognized as such. Scene-safe training teaches investigators not to

**INCIDENT INVESTIGATION REPORT** **Case No.** ___________

**BASIC CASE INFORMATION**

Incident Date ______________________ Diagnosis Date ______________

Time of Incident ______________________________________________

Location ______________________________________________

Task employee was performing at time of incident ______________________

______________________________________________

Employee's Date of Hire ______________________________________________

Investigation participants ______________________________________________

______________________________________________

______________________________________________

Date and time the investigation began ______________________________

Employee's occupation at time of incident ____________________________

Employee's supervisor at time of incident ____________________________

**MEDICAL INFORMATION**

Specific Treatment Provided ______________________________________________

______________________________________________

Treatment Date ______________ Treatment Time ______________________

Prescription Medicine Dispensed ______________________________________

Initial Treatment Given by ______________________________________________

Health-Care Professional Name and Address ____________________________

Health-Care Facility Name and Address ______________________________

Health-Care Facility Phone Number ______________________________________

**Figure 10.1** Incident Investigation Report Form.

rush onto the scene of an accident without first giving it a quick but comprehensive visual assessment. Is high-voltage electricity present? Is a dangerous tool or piece of equipment still running or energized? Is it safe to enter the site?

As part of their training, employees should learn how various pieces of equipment—including cameras, video equipment, tape recorders, measuring tapes, and sketchbooks—can be used to help document incidents.

**WORKSITE INJURY/ILLNESS REPORT**

Name of Injured ______________________________

Social Security Number ______________________________

Incident Date ______________ Incident Time ______________

Injured Employee's Occupation ______________________________

Describe Injury/Body Part(s) Affected ______________________________

______________________________

Exact Location of Incident ______________________________

______________________________

Describe How the Incident Occurred & What Employee Was Doing ______________

______________________________

______________________________

______________________________

______________________________

Equipment/Tool/Material/Machinery Involved ______________________________

______________________________

Immediate Corrective Action to Prevent Repeat of Incident ______________

______________________________

Witnesses:

Name ______________ Address ______________ Phone# ________

Name ______________ Address ______________ Phone# ________

Subcontractor Involved? ______ Name & Address ______________________

Sub. Employee Name ______________ Address ______________ Phone# ______

Sub. Employee Name ______________ Address ______________ Phone# ______

Prepared By ______________ Title ______________ Date ________

Approved By ______________ Title ______________ Date ________

**Figure 10.2** Worksite Injury Report, 1st Page.

**INCIDENT DESCRIPTION**

**Write a detailed narrative of what happened. Include all relevant facts dealing with individuals involved, exact locations, tools, equipment, materials, processes, procedures, and personal protection devices.**

______________________________________________________________________________

______________________________________________________________________________

______________________________________________________________________________

______________________________________________________________________________

______________________________________________________________________________

______________________________________________________________________________

______________________________________________________________________________

______________________________________________________________________________

______________________________________________________________________________

**INCIDENT CAUSE(S)**

**Why did this incident occur? Identify the immediate and basic cause(s) which led to this incident.**

______________________________________________________________________________

______________________________________________________________________________

______________________________________________________________________________

______________________________________________________________________________

**CORRECTIVE ACTION**

| Corrective Action(s) | Person Responsible | Date Complete |
|---|---|---|
| | | |
| | | |
| | | |
| | | |
| | | |
| | | |

**Figure 10.3** Worksite Injury Report, 2nd Page.

**SUPPLEMENTAL INFORMATION**

**Figure 10.4** Worksite Injury Report, 3rd Page.

**WITNESS INTERVIEW FORM**

**Incident Date** ______________________________

**Name(s) of Injured Person(s)**

______________________________

______________________________

______________________________

**Date and Time of Interview** ______________________________

**WITNESS INFORMATION:**

**Witness Name** ______________________________

**Witness SS Number** ______________________________

**Witness Address** ______________________________

______________________________

______________________________

**Witness Phone Number** ______________________________

**Witness Employer Name, Address, and Phone Number**

______________________________

______________________________

______________________________

**Figure 10.5** Witness Interview Form, 1st Page.

**SUPPLEMENTAL INFORMATION**

Figure 10.6 Witness Interview Form, 2nd Page.

## CORRECTIVE ACTION FOLLOW-UP FORM

**Name of Injured** ______________________________ **Incident Date** __________

**Brief Incident Description** ______________________________

______________________________

**Investigation Findings** ______________________________

______________________________

______________________________

______________________________

______________________________

______________________________

______________________________

| Corrective Action(s) | Individual Responsible | Date Completed |
|---|---|---|
| | | |
| | | |
| | | |
| | | |
| | | |
| | | |
| | | |
| | | |

**Comments**

______________________________

______________________________

______________________________

**Figure 10.7** Corrective Action Follow-up Form.

Employees also should be aware that certain reporting regulations may kick in when an incident occurs; OSHA logs could be required to be filled out and local OSHA offices called within prescribed time periods. Insurance carriers should be notified of serious incident investigations as quickly as possible. To take the element of guesswork out of the picture, it's best to have a procedure for investigating safety incidents. Write down exactly how employees should proceed, then use sample incidents as examples to train with. The plans should be as simple as the incident allows. Overcomplicated investigative efforts may, in fact, lead to distorted results or discourage employees from participating at all.

### Gathering incident-related facts

The gathering of facts should begin as soon as possible after the incident occurs. The longer the delay, the greater the likelihood of error. Time, as the saying goes, forgets all. Accurate incident investigations begin as soon after the incident time as possible, when witnesses are nearby, memories are still fresh, and the scene is still intact, with evidence not yet moved or tampered with. Witnesses should be interviewed separately, if possible, so they won't influence each other, inadvertently or otherwise. It's generally best to conduct the interviews at or near the place where the incident occurred so the witnesses can show or point out what they're describing. Try to get the full names and social security numbers of witnesses so as to leave no doubt as to who they are. Make sure their names are spelled correctly. The investigators should ask the witnesses to write a description of what they saw, and sign it, and should then get them to help fill out the Witness Interview Form (see Figs. 10.5 and 10.6).

Seek pertinent details about what, why, who, when, and how the incident occurred. If possible, take photos or make a sketch of what happened. Don't worry about the quality of the artwork. Note anything that might have had a bearing on the incident, such as tools or materials involved, environmental conditions such as rain, high temperatures, low lighting levels, or even distracting noise. Determine whether measurements must be taken (such as the size of an opening, the weight of a piece of plywood, or the depth and width of an excavation), and if so, arrange for them to be made. Do you need a drill motor's operating manual? Or a roofing chemical's material safety data sheet? An electric box wiring diagram? Although gathering this kind of information may seem like overkill while it's being collected, it may come in handy later—especially if a serious injury or illness develops from a seemingly minor condition, and litigation eventually comes into the picture.

The incident investigators must not confuse gathering facts with assessing blame. Assessing blame may address the immediate cause of the accident, but it won't help solve the long-term problem.

### Analyzing the facts

After all available facts have been assembled, you should begin to review them with a discerning eye. What about the witnesses? Are they saying the same things? Is there any conflicting information? Do the facts support the incident results? Are there any reasons to doubt what the witnesses have reported? Have the witnesses given facts—or opinions? Have all the facts (even those that may have been discovered at a later time) been written down and attached to the main incident report?

### Developing conclusions

You're familiar with the term "jumping to conclusions." In a serious incident investigation, you may be pressured one way or another to arrive at a preconceived conclusion. In the interest of truth, investigators must resist any such temptation. Certainly, as information is gathered, it's natural for people to begin to draw conclusions about what happened and why. From the answers to an incident investigation's questions, immediate causes and basic root causes of the incident can be posed.

Conclusions must be based on facts, not opinions or speculation. For some individuals, a potentially tricky part of developing conclusions is determining whether (1) unsafe conditions or (2) unsafe behavior played the main role in causing the incident. As we discussed earlier, the odds heavily favor a combination of unsafe behaviors rather than unsafe conditions—but the conclusion must be borne out by the facts. What's needed is a combination of immediate and basic causes. That's why an attribution that an incident was caused by carelessness, or not paying attention is practically worthless. You've got to attempt to understand why the employee was allowed to be careless, and, if he or she *is* routinely behaving in a careless manner, *are others,* too? *Does the employee's carelessness represent a standard among the company's employees that is not being recognized by management?*

In past years, carelessness was often cited as a chief cause for accidents. Little follow up was done except to discipline the employee or to tell him or her to be more careful in the future. Today, instead of blaming the employee, a more effective approach is to identify the immediate cause(s) (which *could* include the fact that the employee was totally in the wrong) and the meaningful basic cause(s) that helped trigger the incident, so those causes can be counteracted before the same incident repeats itself.

As you recall from earlier in this chapter, immediate causes are associated with the specific incident. Typically, immediate causes can focus on:

- Working without personal protective equipment
- Electrical or power-tool failures
- Lack of specific procedures to follow
- Poor housekeeping or materials storage
- Lack of safety equipment, guards, or barricades
- Taking shortcuts in the handling of materials
- Improper lifting techniques

Basic causes imply overall worksite safety management deficiencies. Numerous safety management experts have agreed that there are fewer than a dozen categories of basic causes—of which one or more contribute to practically every safety-related incident that occurs in all kinds of industries. The basic-cause categories most frequently associated with safety incidents experienced in the construction industry include deficiencies in the following areas:

- Knowledge and training
- Supervision
- Procedures and practices
- Personal protective equipment
- Employee hiring standards
- Inspection and maintenance programs

### Writing the incident investigation report

You may be able to obtain investigation report forms from your insurance carrier or your local home builder's association, or you can copy the one in this book. You may choose to use any of them as an aid in customizing your company's own report form. Again, having preprinted forms will ensure that all of the important details are covered during an incident investigation. Just using plain paper for a report can easily result in a collection of reports of uneven quality, where important facts are inadvertently left out.

A good report is concise and clear, leaving no room for opinion or conjecture. The language is simple and direct; flowery or difficult writing need not apply. If your report is a comprehensive one and covers all of the bases, it may be used to substitute for recordable-injury or illness reports on the OSHA No. 101 form, state-plan OSHA incident-reporting forms, or many state worker's compensation incident-investigation reports as well.

The typical company incident investigation report includes the following sections:

#### Basic case information.

- A case number, with chronological numbers the best option, such as #01-98, #02-98, and so on. This number should appear on one of the top corners of the report.
- The date (day, month, and year) on which the incident occurred. If the employee cannot pin down the date specifically (or if the incident involved an illness), then use the date it was reported or diagnosed.
- The time of day (or night) when the incident occurred.
- The location of the incident. The more specific this information is, the more useful it will be.
- The task or activity the employee was performing when the incident occurred.
- The employee's date of hire. Although the employee's name has already been recorded on the worksite injury/illness report, it's a good idea to write it here, too. And it's not enough to use a nickname or first name. Record the employee's correctly spelled name.
- The names of the individuals participating in the investigation.
- The date and time the investigation began.
- The name of the employee's supervisor.

**Medical information.** This, of course, may be left blank when reporting on a near miss or property-damage incident. Otherwise, include whatever injury or illness that was treated or diagnosed. Identify the body part(s) affected. Note the date and time the treatment was given, and the name and credentials of the persons giving the treatment.

**Incident description.** Prepare a concise, clear descriptive narrative of what happened, including the facts of the case such as who, where, when, and what happened. What were the conditions of tools and equipment that were involved? Were there any special weather or environmental factors, such as loud noise, high temperature, poor lighting, wind, rain, or ice? Were any eyewitnesses available for interviews? Was personal protective equipment worn or available?

**Incident cause(s).** Why did the incident happen? Was unsafe behavior involved? Was there an unsafe condition? Identify the immediate cause(s), and the root cause(s).

**Corrective action.** Describe the corrective action decided upon for each of the basic causes that led to the incident. Name the people who will be responsible for carrying them out or seeing that they're carried out, and state when corrective actions are to be completed.

**Supplemental information.** This part is for witness statements, drawings, photos, and additional supporting details, such as original write-ups.

*Note:* For motor vehicle incident reports involving noncompany personnel, rely on the official police report for the investigation findings section, and involve your insurance carrier as quickly as possible.

## Recommending immediate and long-term corrective action

This is one of the most important parts of the report. It looks toward to the future instead of just documenting the past. Immediate corrective actions are usually taken before the written report is completed, and include such measures as tagging out a defective tool, barricading a physical hazard, or repairing a vehicle's faulty brakes. Taking care of the basic causes is usually not so simple, and often requires the input of several or more individuals. Corrective action plans should note the individuals responsible for completing the corrective actions along with specific target dates for each recommendation. The sample corrective action follow-up form (see Fig. 10.7) should be filed in an active incident-report file, and tracked until all of its corrective actions have been completed. The incident investigators must ensure that every cause identified in the report is addressed by a corrective action. Each recommendation should cover just one item, spelling out precisely what should be done to correct the situation. A report may present a list of several recommendations, each separately stated, along with specific required actions.

### Communicating what's in the report

After making sure that the report is as complete and accurate as possible, share any pertinent information (but *not* personal and confidential data such as a worker's social security number) with all employees whose jobs could be affected by the incident's circumstances. Use information from the incident investigations to help train new employees and as refresher training for long-term employees. Major injuries involving noncompany personnel should be reported to your insurance company and legal representative as soon as possible. Incidents resulting in fatalities or in the hospitalization or medical treatment of three or more employees must be reported to OSHA (within eight hours, as of this writing).

### Correcting the situation

The report can be effectively prepared, and an appropriate corrective action mapped out—but unless the process is followed through to completion, both the report and the investigation become academic, all for naught. The stuff has to get done. Company management is ultimately responsible for seeing it through. An accountability system has to be established. Far too often, once it becomes apparent that an injury is not serious, all meaningful corrective action stops. But we must not forget that the severity of an incident is largely a matter of luck, and a minor outcome can easily take a turn for the worse if the given situation repeats itself.

### Following up on recommendations

You need to follow up. If *you* don't double-check, who will? It's a good idea to maintain a tickler file or other simple yet formal tracking mechanism to ensure that recommendations are completed by specific target dates. Be certain to document any corrective actions that are taken. Refer to the corrective action follow-up form (see Fig. 10.7) until every action item has been completed. Typically, there won't be many things to do for a single incident; home builders and members of related trades generally do not face the complexities that are encountered in major manufacturing or process-oriented industries.

## Incident Analysis and Data Collection

Regular collective analysis of injuries and illnesses is needed to identify trends or patterns. What is the history of accidents at your job sites? Are there any discernible patterns to them? You need to look at near misses and first-aid incidents in addition to any recordable cases you may have experienced, because only the total picture will provide you with an accurate backdrop for identifying problems and trends. Pay particular attention to the following factors:

- Injury types
- Body parts injured

- Injury and illness frequencies
- Costs of injuries and illnesses
- Incident breakdowns, by crews
- Employees who have numerous or repeat-incidents
- Basic causes of incidents
- Tools, equipment or materials involved
- Day of week and time of incidents

### Overcoming employee resistance to incident investigations

Some employees, no matter how well you explain the necessity of doing accurate incident investigations, will just not get it. Or will refuse to get it. They'll have a litany of excuses:

- "ll get disciplined."
- "I'm not a squealer."
- "It takes too much time and effort."
- "There's no need because no one was hurt."
- "I can't stop working or I won't finish my job."
- "Let the pencil pushers do it."
- "I'm terrible with paperwork."
- "We'll ruin our safety record."
- "We might lose our monthly safety prize."

If any fears behind those or similar excuses have merit, you'd better change gears and put them to rest, or your company's employees won't exactly be pushing to the front of the line when it comes to investigating incidents. Certainly, for incident investigation reporting to yield meaningful results, the employees will have to become integral parts of the process. Only then can incident investigations, reports, and corrective actions help to identify and correct weaknesses in the company's overall safety management program.

### Recordkeeping guidelines

Unless specified by other laws or rules, incident investigation reports for recordable injuries and illnesses, motor vehicle incidents, and other serious incidents should be retained for five or six years. Those involving lesser degrees of severity, such as first-aid incidents and near misses, should be kept for three to four years.

Before we're off to a different chapter, be aware that even when all employees are trained on the importance of reporting incidents (including near misses), they still typically report less than actual totals. With OSHA recordables, the estimated reporting rate is between 70 and 85 percent of those incidents that actually occur. With first-aid incidents, depending on the employees (some are more macho than others and prefer to suffer in silence or simply ignore an injury), the reporting rate has been estimated to be as low as 10 percent. And for near misses, the rate dwindles even more.

Unfortunately, this underreporting means that valuable information and decision-making data are lost. It also underlines why it is so important to take advantage of every safety incident by making sure an effective investigation is carried out, and a report with corrective action is completed.

Although at first it may seem difficult to find anything positive in the fact that you've got to investigate safety incidents, consider that by examining near misses and first-aid incidents you may be able to identify problems, patterns, and trends that if allowed to go unrecognized can lead to more serious and severe incidents, injuries, and illnesses. By following through investigations and completing their subsequent reports with corrective action plans, you can make changes or corrections that will effectively prevent many workplace injuries, illnesses, and damages.

Chapter

# 11

# Safety Audits

## Quick Scan

1. Safety audits are planned observations of worksite activities that record both safe and at-risk behaviors and conditions.
2. It's beneficial to custom-prepare safety audit forms for use with your own worksite operations.
3. All members of the company, including management, should be trained and expected to participate in safety audits—both as auditors and individuals audited.
4. Safety audits should never be used to discipline employees.
5. Safety audits should play an important role in helping to monitor the effectiveness of your company's safety management program and helping to determine where additional emphasis is needed.

Safety auditing is a simple technique that takes a concentrated look at worksite activities and conditions and makes particular note of the safe and at-risk behaviors and situations present: a kind of "safety snapshot" of the worksite that can be used for a variety of purposes.

## Why Audit?

Numerous reasons exist for developing and maintaining an auditing program at any company, including yours. They include the following:

1. *In a most general sense, to see how well the employees are doing safety-wise.* If you always wait until an incident occurs, you'll consistently be finding out what happened *after the fact.* By auditing current job behaviors and conditions, you'll be looking at the *now,* when safe behavior can be identified

and reinforced and at-risk behavior can be corrected and prevented from repeating itself in the future.

2. *To see if your crew leaders, foremen, or supervisors are managing safety at the worksite.* If audits turn up rampant unsafe behaviors and conditions, it becomes plain to see that the supervisors are not doing their jobs when it comes to safety. *Why* they're not doing an adequate job is another story, but first of all you need to know that risky behaviors are present.

3. *To provide a method of increasing safety awareness.* Safety auditing is a process that is out in the open, performed in front of workers and supervisors alike. If an audit is going on around you, it's usually very obvious.

4. *To help train workers and supervisors.* Auditing cannot be done without proper training. Those doing the auditing, as well as those being audited, need to know what the process is for and about. Hopefully, all employees will have turns both at being audited and at auditing their coworkers as the weeks and months of auditing roll by. Remember, to be adequately prepared for auditing, employees need to know what to look for—what distinguishes safe from unsafe behaviors and conditions. Part of sharpening a crew's auditing skills requires the additional understanding of work procedures, equipment use, ergonomic principles, and many other job- and task-related factors.

5. *To measure the effects of skill and safety training.* Let's say a new nailing gun, along with a unique method of operation, is introduced to the crew. Later, an audit of employees using the gun will verify if the introductory training was clear and thorough enough or if additional training is necessary. Another topic could be the importance of wearing safety glasses with side shields. If only 60 percent of crew members are observed wearing the proper eye protection at a jobsite, it's obvious that something—perhaps the understanding of why it's so important to wear protection—is lacking, and therefore additional training and emphasis are needed. Training programs can be evaluated and made stronger for future use based on the results of safety audits.

6. *To help identify and correct at-risk behavior, providing immediate feedback to prevent possible incidents from happening.* Keeping to the general guidelines of behavior observations, auditors are instructed to immediately bring unsafe behaviors that could result in injury to the attention of the person being audited. There should be no pussyfooting around. The sooner an individual is tactfully corrected of at-risk behavior, the less likely it is that he or she will be injured. That's a direct and immediate benefit of auditing.

7. *To help determine where additional safety efforts are needed and to provide opportunities for improvement.* The safety audits, once written up, provide an accurate record for follow-up. They can help determine what additional training is needed, who needs the training, if additional equipment or tools should be obtained, and where resources should be allocated in the company's safety management plan.

8. *To yield audit results for supplying tool-box safety meetings with discussion topics and feedback.* Written safety audits can be quickly reviewed at tool-box meetings. Audit results can be to the point and can usually be related to the day's activities—especially if the audits were completed in the same area.

9. *To identify and reinforce positive, safe behavior encountered on the job.* The behavioral observation process recommends that four or five times as much effort should be spent reinforcing positive, safe behavior than correcting negative or unsafe behavior. Safety auditing is a perfect mechanism for identifying employees who are doing things right or taking extra safety steps and for complementing their thoughtfulness.

10. *To reinforce management's support for safety at the worksite.* Auditing takes time and effort. Time and effort spent on auditing will be recognized by employees as something valued by management—or it wouldn't be done. It helps reinforce the safety policy and management's commitment to it.

11. *To help maintain a certain level of safety performance.* Auditing will at the very least help maintain the level of safety your company has already achieved. And as the quality of audits improve (that is, as they get tougher and tougher by holding employees to ever-more-stringent safety standards), the auditing process will play a role in the continuous improvement desired.

## How to Arrange Effective Auditing at Your Worksite

Effective auditing does not come naturally to most individuals. Untrained auditors tend to dwell mainly on identifying negative or unsafe behavior and will rarely point out positive activities for reinforcement. Untrained auditors may also improperly use information gained during an audit to embarrass or to initiate discipline. In short, there's more to auditing than just running around with pencil and paper in hand. Effective auditing requires an understanding of people and procedures that can only come through training and practice. Here are the most important training steps you need to review with employees you select to audit:

1. *Explain the theory and purpose of auditing to everyone involved.* Remember that over 90 percent of workplace injuries can be traced to unsafe work practices and behaviors. Although auditing for conditions is better than nothing, the most effective audits will be ones focused on actual work in process. To understand the theory and purpose of auditing, the basics of safety management, including the safety tree model, should be briefly reviewed.

2. *Auditors must plan to spend a certain amount of time on the audit.* To allow enough time for the observations and brief discussions along the way, 20 to 30 minutes is usually adequate. (It could be more or less depending on what's being audited.) Then another few moments are needed for recording the findings. Although a clipboard with the audit form could be carried by the auditor during the observations, that may make employees too uncomfortable. Seasoned auditors often find it easier to make mental notes during the observations, then jot down the points afterward, when they are by themselves.

3. *Audits should be scheduled in advance, with enough leeway to catch work in process on the day the auditing takes place.* Sometimes it's okay to mention an audit in advance to employees who will be audited, but don't do it all the time. Mix in a few surprise audits.

4. *Vary the participants.* Crew leaders, supervisors, safety representatives, and workers doing the audits should whenever possible recruit a fellow employee to help with the observing. Try eventually getting everyone accustomed to both observing and being observed.

5. *Auditors should not concentrate on anything but safety during the audit.* It's a mistake to try to combine safety auditing with passing out work instructions for the morning, for example, or with making quality inspections. Auditors need to just observe and talk about safety during the audit, or the benefits and effectiveness will be diluted.

6. *Auditors must prepare for an audit by making sure they know the correct and safe procedures for whichever operations are going to be observed.* At times this may require reviewing a piece of equipment's operating instructions, a safe job analysis, or a construction method or technique.

7. *Auditors should be reminded to look for positions and actions of employees instead of just conditions.* Although auditing efforts should concentrate mainly on behavior, conditions that could present an immediate or serious safety hazard cannot be allowed to continue and should be stopped or taken care of on the spot.

8. *Unsafe behavior or conditions that could lead to an injury should be addressed immediately or physically identified/barricaded/stopped.* Allowing an unsafe situation to continue will completely undermine the purpose of auditing, which is to identify and correct unsafe behavior and to highlight and reinforce safe behavior throughout the worksite.

9. *Auditors must give feedback on positive behaviors observed.* Unless something is way out of whack, the opportunity to compliment and reinforce safe behavior should exist at least four, five, or more times as often as unsafe behavior is noticed.

10. *Sometimes it's counterproductive for auditors to carry a large clipboard with a pad of $8^1/_2$- × 11-inch audit forms.* It can have an unsettling effect on the employees being audited—so much so that the employees may stop performing their normal work, and thus a true picture of their activities will not be observed. It's often better for the auditors to make mental notes of what's observed, or, if they must, to carry a small card or notebook and pen so they can unobtrusively jot down reminders during or immediately following the audit.

11. *Auditors must avoid making accusations and finger-pointing.* Care and tact should be used or employees will not cooperate with the auditing process. The intent here is not to find fault with and publicly embarrass employees. Employees should be trained not to argue or cause a scene while performing a safety audit. Auditing at your worksite should eventually be associated much more with reinforcing positive, safe behaviors rather than with faultfinding and correction.

12. *Remind auditors to write up the audit when finished with the observations.* A written record is important to ensure that meaningful information is not forgotten or lost. This will enable corrective actions to be planned and completed and allow trends (safe and otherwise) to be identified. Many audits are ineffective because auditors do not document them. They *intend* to, but they may get inter-

rupted by someone or decide to take care of it later in the day. One interruption leads to another, and soon the day is gone—and so is the audit. The sooner the audit form is completed after the audit, the more accurate and comprehensive it's likely to be.

13. *Employees will naturally be curious about what their worksite audits reveal.* See that those results are reviewed at tool-box and other safety meetings. Share the information with all interested and affected parties.

14. *Expect increased quality of audits from your employees as the audits continue.* As procedures and conditions improve, make sure the audit expectations are gradually increased as well. Something that may remain unmentioned in an early audit should be highlighted in the future, as the auditing focuses increasingly on the finer points of safe and unsafe behavior. Continual improvements will naturally occur as auditing skills sharpen and workplace behavior becomes safer and safer.

## Activities to Audit

Because about 85 percent of all accidental injuries result from unsafe behavior, safety audits should usually be undertaken when activities are taking place at the worksite. What should you audit? Consider typical and nontypical activities of employees and subcontractors. Here are some examples:

- Delivery and unloading of building materials
- Mobile equipment operation
- Hand- and power-tool usage
- Painting walls and ceilings
- High work, ladder use
- Lifting and carrying tasks
- Landscaping tasks
- Performing maintenance on mobile equipment
- Laying brick and concrete block
- Changing a vehicle's flat tire
- Pouring concrete or laying asphalt
- Carpentry work
- Welding, burning, grinding, or other hot jobs
- Drywall or plaster operations

If a safe job analysis (SJA) has already been prepared on the task or activity, it should be reviewed by the auditors beforehand as a basis for the audit. Do the observed behaviors match those that the SJA requires? Or do they differ? Do unsafe behaviors occur instead? If the activity being audited is frequently performed, consider completing a SJA (see Chap. 16) for future reference.

## Audit Forms for Your Worksites

Make a few dozen copies of the blank Worksite Safety Audit Form (see Fig. 11.1). You may revise the form to better suit your worksites, if desired, and have copies of them—complete with your logo—printed by a local printer and made into pads. Distribute the pads among your supervisors or crews.

### Worksite audit instructions

When ready for an audit, select another member of the worksite crew to help you audit, and remind him or her how the auditing process works. Have a pad of your company safety audit forms ready. Review the forms, but typically refrain from carrying a clipboard with the audit sheet during the actual observations. Employees may feel too self-conscious with someone writing comments as the work progresses, even to a point where the activity being observed may be artificially changed *just while the audit is taking place,* thus giving the auditors false data.

### During the audit

There are at least four categories of behavior and conditions you should be looking at during a worksite safety audit, including *personal protective equipment, tools and other equipment, procedures,* and *housekeeping.*

#### Personal protective equipment (PPE)

- Is the proper PPE being worn? For example, if an employee is operating a loud, gasoline-powered chain saw, is he or she wearing safety goggles, hearing protection, vibration-dampening gloves, sturdy work shoes, chaps?
- Is the PPE being worn properly? Is it in good condition? Are employees wearing safety glasses with side shields? If grinding or sawing, are they wearing goggles or face shields? If operating loud equipment, is hearing protection being worn? What about a hard hat? If handling rough materials such as lumber or concrete block, are gloves being worn? What about safety shoes or boots, respiratory protection, or fall protection?

#### Tools and equipment

- Are tools adequate for the job?
- Are they being used properly? For example, using the back of a pipe wrench to hammer a nail is improper tool usage.
- Could the safety of a particular job improve if different tools were used?
- Is equipment being operated with safety guards in place?

**WORKSITE SAFETY AUDIT FORM**

Auditors: ______________________________

______________________________

Jobsite/tasks Inspected: ______________________________

______________________________

______________________________ Date: ______________

TYPES OF OBSERVATIONS

Personal Protective Equipment:

--Eyes and Face
--Ears
--Head
--Hands and Arms
--Feet and Legs
--Respiratory Protection
--Fall Protection

Tools and Equipment:

--Right for Job?
--Used Correctly?
--In Safe Condition?
--Warning Signs?
--Mobile Equipment Operated Safely?

Procedures:

--Safe Procedures Known?
--Safe Procedures Been Written Down?
--Safe Procedures Being Followed?
--Body Positioning OK?
--Reactions to auditor?

Housekeeping:

--Slipping/Tripping Hazards Removed?
--Tools and Mtrls. Kept Orderly?
--Waste Properly Disposed of?

Safety Observations

______________________________

______________________________

______________________________

______________________________

______________________________

_____ Follow-Up Required: ______________________________

______________________________

**Figure 11.1** Worksite Safety Audit Form.

- Is it being operated as designed?
- Is it being maintained in safe condition? Are ladders free from defects?
- Did mobile equipment operators do a safety check on their machines before starting work?

#### Procedures

- Have the people doing the activity been properly trained?
- Is equipment locked out and brought to a state of zero energy before being worked on?
- Do procedures exist for the tasks, and does everyone know and use them?
- Are employees working in safe positions, or are they contorting their bodies, reaching too far, standing between or beneath objects that could shift or fall.
- The "reaction to auditor" question refers to behavior that auditors sometimes come across when observing work crews, including adjusting personal protective equipment (dropping glasses down into position from a forehead), changing from an unsafe position to a safe stance, or stopping some other unsafe practice.

#### Housekeeping

- Are materials and tools kept in an orderly fashion? Are tools lying around, or are they kept in holders or boxes when not in use? Are extension cords causing tripping hazards?
- Are spills cleaned up immediately?
- Is trash disposed of in a safe manner?
- Is waste accumulating throughout the worksite? Is it being dropped into the excavation around the foundation or collecting along the basement or interior walls?

#### Additional pointers

- If you see an unsafe action or condition, address it right away, and eliminate the unsafe behavior or hazard.
- Reinforce safe behavior by recognizing and praising it.
- If necessary, take notes during the audit casually on small cards, then fill out the audit form afterward.

### Filling out the audit form

- See Fig. 11.2. Write your names on the line that says "auditors."
- Fill out the jobsite or activity inspected and the date.

- Jot down the most important safe and unsafe behaviors and conditions you notice under the "Safety Observations" section.
- Note if there's follow-up attention needed on the last two lines, and use the back of the sheet if necessary.
- Keep the completed audits in a binder and review them periodically. Use them as topics of discussion at tool-box safety meetings. When discussing audit information, it's recommended that you don't specifically point out who was performing unsafe behaviors. Some of the completed audits can be used to start Safe Job Analyses on activities that are repeated with any regularity.

**WORKSITE SAFETY AUDIT FORM**

Auditors: Phil
Steve R.

Jobsite/tasks Inspected: New Construction – Canterbury Drive

Date: April 2nd 1998

TYPES OF OBSERVATIONS

**Personal Protective Equipment:**

X --Eyes and Face
X --Ears
--Head
--Hands and Arms
--Feet and Legs
--Respiratory Protection
X --Fall Protection

**Tools and Equipment:**

--Right for Job?
--Used Correctly?
--In Safe Condition?
X -Warning Signs?
--Mobile Equipment Operated Safely?

**Procedures:**

--Safe Procedures Known?
--Safe Procedures Been Written Down?
--Safe Procedures Being Followed?
--Body Positioning OK?
X -Reactions to auditor?

**Housekeeping:**

--Slipping/Tripping Hazards Removed?
--Tools and Mtrls. Kept Orderly?
--Waste Properly Disposed of?

**Safety Observations**

Worker jackhammering sidewalk blocks – no hearing protection
Two men on roof working on dormer – both had harnesses/lines on
Backhoe back-up alarm not working. College helper forgot his
safety glasses (we got him an extra pair from the trailer).
Rear foundation trench has 1½ feet of water.

X Follow-Up Required: Backhoe Alarm needs repaired
Rear foundation trench – water needs pumped out

**Figure 11.2** Sample of a Completed Audit Form.

Chapter

# 12

# General Safety Rules

## Quick Scan

1. Prepare a set of general safety rules—those key guidelines to be followed by all company employees.
2. Review the general safety rules at all new-hire safety orientations and during annual refresher training for all company employees.

Every company has boilerplate rules of one kind or another. They are so obvious that they practically go without saying, but say them you must. This chapter provides a sampling of the rules and guidelines that are common to the operation of many companies. If your company doesn't already have one version or another, it will be a good idea to create your own list of important general safety rules that can be taught to newly hired employees at their safety orientations, and reviewed yearly with all other personnel.

## Recommended Basic Rules

Here are some assorted practices, procedures, and topics that other companies have included in their general safety rules:

1. *Unsafe behavior.* Make it a habit to warn fellow employees or others who are in danger at the worksite. This includes warning employees who may be taking chances with tools, with equipment, with material handling, or with construction methods and procedures.

2. *Unsafe conditions.* Correct or barricade and isolate, if possible, unsafe conditions as they are discovered, or report them to your supervisor.

3. *Near misses and injuries.* Report *all* near misses and injuries to your supervisor immediately.

4. *Personal protection.* Hard hats, eye protection, and safety shoes or boots must be worn at the worksite at all times, except in designated places outside the construction area. Additional safety gear, including hearing protection, fall protection, gloves, respirators, and other devices are also required as needed.

5. *Suitable clothing.* Neat and clean work clothing—meaning at least long pants and sleeved shirts—must be worn at all times. Never wear loose-fitting clothing that could be caught in equipment or machinery, and also avoid wearing jewelry at the worksite.

6. *Horseplay.* Practical jokes or constant teasing and badgering of employees cannot be tolerated. If allowed to persist, such behavior can get carried away—easily leading not only to serious injury, but also to human-resource or personnel problems that lead further to lawsuits and other complications.

7. *Theft.* The stealing of time, materials, equipment, or the personal possessions of other employees cannot be tolerated.

8. *Firearms and weapons.* Firearms and other weapons are forbidden at the worksite.

9. *Fighting.* Workplace violence has escalated into a major concern during the past decade. Avoid confrontations, and report suspect behavior to your supervisor or crew leader. Walk away from any situation that appears to be leading to a confrontation.

10. *Drugs and other illegal substances.* Reporting for work under the influence of drugs or alcohol is strictly prohibited.

Your company has a right to develop, maintain, and enforce a policy regarding substance abuse. It's a proven fact that drug abuse in the worksite creates problems for everyone involved. Drug abusers bring many ills into play, including absenteeism, tardiness, lack of concentration, poor judgment, and mood swings and erratic behavior.

New-hire, random drug, and postincident drug testing can all contribute to improving safety at the worksite by protecting nonusing employees from those who do, by removing the fear of people who must work near others who do, and by eliminating the associated risks and costs. In short, you have the right to know that company employees are free from the influence of illegal substances in their bodies.

Approach your local manufacturers' association about setting up a program patterned after others that may be currently in use. Of course, any effective program entails strict confidentiality, specimen integrity procedures, and administrative procedures that respect employee rights. Many programs have a certain leeway for a step toward rehabilitation following a positive test result, by which the employee can seek professional assessment of his or her problem, and accept rehabilitation through an intervention program.

In the interest of ensuring a healthy and productive workforce, consider contacting a company specializing in a full range of employee health-screening services. There are major companies in the marketplace that can customize a program that includes everything from substance-abuse testing and background checks to the complete management of the program. Optional services include:

- Alcohol testing: evidential breath alcohol testing devices, urine, blood, and on-site saliva collection and testing.
- Drug screening: preemployment, random, follow-up and emergency services; specimen collection for urine, blood, hair, and saliva; on-site testing, plus on-site-approved testing devices.
- Background checks: employment and reference verification; education and licensing verification; motor vehicle and criminal record searches.
- Wellness programs: cholesterol and health risk assessments.

11. *If you don't know, ask.* When in doubt about the safe or correct way to perform a job, get instructions from a supervisor or knowledgeable employee *before* attempting the task.

12. *Seatbelts.* The use of seatbelts is required in company cars, trucks, lift trucks, front-end loaders, and other moving equipment.

13. *Riding outside vehicle cabs.* No employees may ride on a running board or hang onto the outside of trucks, front-end loaders, backhoes, bulldozers, tractors, skid-steer loaders, or other mobile equipment while that equipment is running or in motion. Never ride or stand in or on a loader or backhoe bucket to perform work.

14. *Unauthorized operation of equipment.* Never attempt to operate company equipment or vehicles unless trained in their use and authorized to do so by your supervisor.

15. *Suspended loads.* Never walk or stand under or next to suspended loads. Stand clear of materials being unloaded from open-sided flatbed trailers.

16. *Equipment guards.* Do not remove protective guards while tools, machinery, or mobile equipment are in operation. Never complete an equipment maintenance or repair job without replacing guards that had to be removed for servicing, cleaning, or repairs. Report damaged or missing guards to your supervisor immediately.

17. *Electrical equipment* All portable electric tools and equipment must be properly grounded.

18. *Ladders.* All ladders must be inspected before every use.

19. *Scaffolding and rigging.* All scaffolding and rigging work must be supervised and inspected by experienced, competent individuals in accordance with approved standards and regulations.

20. *Digging, drilling, and cutting.* Never dig or begin an excavation without permission from your supervisor or crew leader and acknowledgement that the area has been reviewed for the presence of electrical, natural gas, and other underground utilities. Never drill or cut into floors or walls unless similarly authorized.

21. *Excavations and trenches.* All excavating and trench digging must be supervised and inspected by experienced, competent individuals in accordance with approved standards and regulations.

22. *Practice good housekeeping.* When discarding boards, always remove or bend down protruding nails. Keep work areas clean as the day progresses. Do not allow debris to be strewn about work areas, or to collect on floors or in excavations or basements.

23. *Reinforcement rods.* Exposed, vertical reinforcement rods and similar protuberances must be shielded with large enough covers to prevent employees or visitors who may accidentally fall onto them from impaling or injuring themselves.

24. *Lifting.* When lifting a heavy object, keep your back straight and as nearly upright as possible, and lift with your legs instead of your back. If in doubt regarding your ability to lift an object, get help.

Chapter

# 13

# Correcting At-Risk Behavior

## Quick Scan

1. Unsafe behaviors must be corrected whenever and wherever they're seen.
2. Supervisors *must not fail* to recognize, interrupt, and correct at-risk behaviors whenever and wherever they're encountered throughout their worksites.
3. Correcting unsafe behavior requires six relatively simple steps be performed by the individual doing the correcting:
   - Interrupt the behavior right away.
   - Put yourself in the correct frame of mind (no anger).
   - Calmly describe the unsafe behavior.
   - Review why the unsafe behavior is dangerous.
   - Describe the safe behavior that should be used.
   - Supply follow-up positive reinforcement for the safe behavior.
4. Work toward establishing a safety program where work-crew employees observe each other, help correct unsafe behaviors, and assist with reinforcing safe behaviors at the worksite.
5. In general, avoid relying on discipline alone for correcting unsafe behavior. It's not conducive to developing cooperation between management and labor, and in turn will produce at best mediocre results.

As discussed in Chap. 2 on behavior, worksite injuries and illnesses can be greatly reduced if employees don't perform unsafe acts. How much simpler can it get? A strong safety orientation program, coupled with continuous examples of safe behaviors and correction of unsafe behavior by coworkers is the absolute best way to avoid worksite injuries and illnesses.

Again, why don't employees follow safe procedures? A number of possibilities exist:

- They don't know the safe ways to do things.
- The employee cannot perform the safe methods due to physical or other restrictions.
- The standard procedure is incorrect and needs to be changed.
- The employee has simply decided not to perform the task safely for a variety of reasons, including expediency, laziness, daring, and/or contrariness.

Since numerous kinds of unsafe behaviors can occur at the typical worksite, it makes sense to classify them in order to understand them, so that corrective actions can be taken. The first classification is by *company position:* either *management* or *labor.* Granted, sometimes there's a fine line between the two, but as a rule, if an individual supervises and is responsible for the behavior of others, he or she can be considered management.

### Correcting Management At-Risk Behaviors

At-risk behaviors by management can mean two things: failing to supervise others within proper safety management standards or performing at-risk behaviors while doing job tasks at the worksite. Of the two, the latter behaviors are simpler to correct—by consistently applying a combination of negative consequences and positive reinforcements for each respective unsafe or safe behavior, much the same as at-risk and safe behaviors by work-crew members are handled. But if you've got supervisors or crew leaders who, for whatever reason, haven't bought into your safety management program and who display frequent at-risk behaviors, immediate action must be taken. Supervisors who fail to recognize or, worse yet, recognize but fail to correct unsafe behaviors at the worksite must be dealt with quickly or your program will falter. Unfortunately, that's a relatively common situation.

Indeed, one of the most difficult challenges for supervisors, foremen, and crew leaders is to correct inappropriate, incorrect, and at-risk behaviors of the employees they supervise. Why? Because to supervisors, that implies confrontation. Confrontation could easily lead to the supervisor's embarrassment. Additional factors may also come into play. Perhaps the individual who needs correcting is a friend or relative of the supervisor. Maybe the supervisor has seen the employee occasionally perform the unsafe behavior before, with no accompanying negative results. Perhaps the supervisor has decided to give more emphasis to production—getting work accomplished—than to safety. There are many reasons that a supervisor will not attempt to correct someone who is performing unsafe acts, but none of them should be allowed to continue at your worksites.

Corrective actions for a supervisor who permits at-risk behaviors to occur include additional training in basic management principles. Maybe the supervisor is not fully aware of how serious such a situation is. Just because individuals have been working in construction for many years and intimately know all of the job skills and tasks doesn't mean they automatically can become good supervisors or leaders. It takes considerable communications skills, an understanding of how to motivate workers, and specific management techniques for someone to develop into a sharp supervisor. It's smart for a company to invest in its supervisors, and if you are one, invest in yourself. Numerous seminars, classes, and programs are offered by reputable schools and organizations all across the country. Only through training and experience will supervisors develop the skills and confidence needed to confront and correct unsafe behavior.

Supervisors who fail to respond to additional training, reason, or discipline (including the possibility of reduced wages) are probably not worth keeping at the worksite or (unless they can function effectively and safely on a work crew) on your company payroll. In other words, supervisors *must* manage a safe worksite, or they shouldn't continue as supervisors.

## Correcting Worker At-Risk Behaviors

Essentially, there are two ways to correct worker at-risk behavior:

1. *Correction that comes from crew leaders, foremen, supervisors, or company management.* This has been the traditional way. While it's still an important method of correcting behavior used at many companies, it's not the way of the future.
2. *Correction that comes from an employee's coworkers or peers.* This is the correction technique most desired now and for the future: workers understanding and following safe procedures on their own, keeping an eye out for each other, participating in practically every aspect of the company's safety management program. To arrive at this level of correction, however, requires considerable planning and follow-through. Education and training will have to be done on a regular basis, and an employee observation process will have to be set up and be accepted by the workers. Unlike large-scale observation processes undertaken and set up by large manufacturing facilities, a construction worksite can be so done on a slightly less formal basis. For more on the employee observation process, see the next chapter, which discusses reinforcing safe behavior.

## Correcting Unsafe Behavior at the Worksite

Correction, as described in Chap. 2, is the opposite of reinforcement. To be sure, correction should not be used either at the same time or immediately before or after positive reinforcement, because mixed signals will confuse employees. For most individuals, correction implies some form of discipline or confrontation, which is historically what has dominated workplaces since the Industrial Revolution. Such scenarios typically play out like this: The people in charge see an employee doing something wrong or, in this case, unsafe, and the violator is corrected and disciplined so that, theoretically, he or she is afraid to perform the unsafe behavior again. There are hard feelings involved, often disagreements, and frequently misunderstandings and polarizing of attitudes—labor versus management—to the point where the actual thing that caused the discipline is forgotten, and the confrontation and discipline become the issue, taking on a life of their own.

A major misconception here is that correction always implies confrontation—in the negative sense of the word. It doesn't need to be that way, though, if a few simple steps are followed. Correcting unsafe behavior at the worksite can usually be accomplished through six steps, all of which can be taught to supervisors and members of the work crews. When you observe an at-risk act, do the following:

1. *Interrupt the unsafe activity as soon as it's encountered.* Don't wait. Hesitation can be perceived as uncertainty. Unsafe behavior needs to be corrected whenever and wherever it raises its ugly head. If a violating employee is not taken to task, he or she will consider the unsafe behavior acceptable, and *wham,* that's reinforcement...a license to do it again. Also, any negative consequences justifiably resulting from the unsafe behavior should occur as soon as possible, so the employee will make the association and retain the lesson.

2. *Keep your cool.* Never let anger show when correcting employees. Even if you're at the tail end of being upset with something else, take a few deep breaths and realize that if you let your anger show, confrontation will likely take over, with the employee not listening to reason and becoming extremely defensive instead. Using an angry tone to correct someone tends to bring the discussions to a personal level, which won't be conducive to getting your point across.

3. *Describe the unsafe behavior in no uncertain terms.* Don't beat around the bush. Tell the violating employee exactly what was observed and what needs to be corrected. It's not good enough to say "Gee, you weren't being very safe over there with your front-end loader," when you really mean "You shouldn't have lifted Tom up to the roof of the garage in your machine's bucket." Be specific, not vague. *If you're not specific enough, the employee may not realize what he or she did wrong.* The employee needs a definite understanding of exactly what unsafe behavior has occurred.

4. *If you know why the unsafe behavior shouldn't be used, say so.* State the facts. Discuss the behavior, *not* the individual. Avoid saying things like, "You can't keep doing this," or, "I've seen you do this before, and..." Don't become sidetracked into the emotional side of things. Avoid bringing up an employee's attitude, appearance, lifestyle beliefs, or previous behavior. Most people will become defensive when openly faced with something they have done wrong, and they may try to deny it or justify why it was necessary—even when they realize the behavior was incorrect. Rationally review what could happen when the job is done unsafely and why selecting an unsafe method of performing the work is too risky. Facts, on the other hand, are indisputable, even to an employee expressing displeasure when he or she is being corrected.

5. *Describe the safe behavior which should have been used.* It's not enough to merely tell someone they're doing something wrong. You need to point out the correct way or ways of doing the work to ensure employee understanding. Maybe the employee doesn't know how to accomplish the work safely. Perhaps he or she has never been trained. In an instructive manner, provide the safe

way to perform the behavior. A demonstration of the safe behavior may be helpful, followed by having the employee repeat the behavior so you can verify his or her understanding.

6. *Reinforce the safe behavior shortly afterward.* Once you see that the employee understands how to correctly do the work and realizes why the previously attempted unsafe behavior was dangerous, you can let the employee get back to work. After a while, make a special effort to return and observe the employee performing the work. If he or she is still doing it correctly, be supportive, supplying a healthy dose of positive reinforcement. Afterward—days, weeks, or even months later—whenever the situation presents itself, remind the employee that doing the work safely has really been paying off.

Remember that effective use of positive reinforcement is a coaching skill. Just as a coach continually works with an athlete to develop a certain skill or form, so must a supervisor observe, correct, or remind his team members.

Note, if supervisors are a little weak in any one of the six steps, it's likely to be with the last one. Anything requiring follow-up needs conscious effort to remember and plan for.

## Discipline

Discipline has long been considered by typical members of upper management as something to be used to prevent employees from behaving in a way that's undesirable to the company. In short, sometimes discipline is necessary—but not as often as you might think. It must never be used by itself for correcting unsafe behavior. While it may make the supervisors and managers feel that they're accomplishing something, it doesn't work very well on the employees whose behaviors are supposed to be changed.

Effective discipline, like correction, is best applied in a structured manner. When issuing discipline for unsafe behavior, consider the following:

- The discipline should occur as soon as possible after the violation. Remember the sooner-or-later aspect of consequences from Chap. 2?
- Tell the violating employee specifically what he or she did wrong. Don't leave it up to conjecture.
- Let the employee know why the violation is dangerous or wrong, inform him or her of the action that will follow.
- Don't bring personalities into the picture. You're concentrating on the behavior, the facts of the situation, not the person's character traits, background, beliefs, or appearances. Keep calm and collected during the discussion.
- Indicate that you expect the employee not to repeat the unsafe behavior in the future, and that you'll be available to help, if needed.
- Don't hold a grudge or nag about the violation. Instead, expect the best from similar situations in the future.

Chapter

# 14

# Reinforcing Safe Behavior

## Quick Scan

1. Establishing and maintaining a safe worksite takes frequent and consistent applications of positive reinforcement.
2. Overall, strive to use four times more positive reinforcement than negative (i.e., disciplining employees or correcting at-risk behavior).
3. There are two types of positive reinforcement: tangible and intangible.
4. Tangible rewards, such as money, airplane tickets, jackets, televisions, and similar desirable items, won't inspire a consistent pattern of safe behaviors.
5. Intangible rewards are the true motivators of employees. They include recognition and praise from supervisors and fellow employees, a sense of achievement, job satisfaction, and similar personal reinforcement.
6. Do not mix correcting at-risk behavior with positive reinforcement.
7. Tangible reinforcements can be applied concurrently with intangible rewards as incentives or recognition in well-planned safety programs.

This chapter complements the previous one. But instead of being about correcting at-risk behavior, it concentrates on why and how safe behavior should be encouraged through reinforcement. It's important to note that reinforcing safe behavior at the worksite (to keep in line with expert recommendations) should be done *four times as often* as correcting unsafe behavior. That means a supervisor who spends an hour a week correcting employees who are performing at-risk behavior should also be spending four hours encouraging and reinforcing safe behavior with those same employees. That 1 to 4 ratio tends to produce the most bang for the buck, safetywise, at the worksite, and will ensure continual improvement.

Reinforcement, in the scheme of behavioral theory and practice, essentially means a positive consequence: something that increases the probability of a safe behavior continuing in the future. Statistics show that reinforcement works. Employees *need* to be reinforced to keep doing what they should be doing. If positive reinforcement for safe behavior isn't given frequently by supervisors and fellow employees, unsafe behaviors will gradually creep into the worksite.

An employee taking a shortcut, even though it's risky, will perceive the shortcut as positive reinforcement if nothing bad happens—that is, if no negative consequence results from the shortcut. If such a "positive" reinforcement for the employee can be achieved from taking a shortcut on a job, and if nothing else contrary intervenes or follows, that shortcut will likely become the standard operating practice from then on. For instance, say an employee needs to install a skylight in an entrance foyer in a new home that's under construction. He sets up an 8-foot stepladder and starts working. An hour into the job he needs to reach even higher, so he steps up and stands on the very top of the ladder, balancing himself by leaning against the wall, in clear violation of what he's been told. To get a longer ladder would mean trooping out through the snow and mud a half block away, unstrapping the ladder from the truck's side rack, cleaning off the ice, and setting it up inside. It would take an hour, he figures. So he continues working on the same ladder and finishes the skylight 45 minutes before quitting time. No supervisor is around to check on him, and his helper (a college kid working part-time) doesn't know better. Thus, using the too-short stepladder becomes an acceptable practice, and one reinforced positively to those two employees because it requires less time and effort.

To recall the discussion about reinforcement in Chap. 2, some sort of consequence always follows a behavior. If an employee perceives the consequence as negative, it's less likely that the behavior will be repeated in the future. If the employee considers the consequence as desirable, it's likely that the behavior will persist.

Now all we need to do is identify how to arrange for employees to receive positive consequences, which equate to positive reinforcement, by following safe behavior. Reinforcement can be tangible, intangible, or a combination of both.

## Tangible Reinforcement

Tangible reinforcement is some kind of reward that can be seen and owned or experienced. It usually has an inherent value that anyone could appreciate. There are two schools of thought about tangible reinforcement, which can consist of any items or activities perceived by the employees as desirable. Proponents suggest that such material rewards, when tied to desirable safety behaviors or given in recognition of some achievement, will cause employees to consciously strive toward those rewards by working safely. Further, when employees perform safe behavior in the hope of achieving some goal that will trigger the tangible reward, proponents of these programs say those safe behaviors are likely to develop into habits for the employees involved.

Companies have used items such as jackets, coats, sweaters, knives, and catalog gifts as rewards in safety incentive programs. Companies have also used cash, tool kits, restaurant gift certificates, extra days of vacation, airplane tickets, color televisions, and...well, you name it, they've been used by someone as safety rewards.

The problem with tangible reinforcements, or rewards, say individuals who don't believe in their use, is that they're typically given for reaching a particular goal rather than for (and therefore associated with) a desired and continuing specific safe behavior. Critics of tangible rewards doubt that long-term good-behavior habits are formed during goal-specific "contests." Tangible reinforcers can become a negative factor if employees start repeating a behavior in order to win a color television or VCR. After the goal is reached, then what? Why would further attention or action go on? What then will keep reinforcing the desired behavior? Yet another goal/reward? And another? How many jackets or tool kits or VCRs does an employee need? Because behavior is continuous, shouldn't the reinforcement of safe behavior be frequent and ongoing as well?

## Intangible Reinforcement

Intangible reinforcement is typically something that cannot be seen or owned, something appreciated only by those who receive it (or, if it can be seen, has little innate value to anyone other than those who receive it, such as a note of appreciation). Sometimes antecedents are mistaken for reinforcement. Telling an employee to wear safety-toed work shoes instead of sneakers because you don't want him to smash his toes is not reinforcement. Telling that employee to wear safety shoes is an antecedent—something he reacts to by changing into his safety shoes. Reinforcement would be if you saw the same employee wearing his safety shoes and you showed him an article you cut out of the paper that describes how a worker in another city lost four toes in a crushing accident on the job. Antecedents certainly have certain effects on behavior, but they won't encourage consistent, lasting behavioral changes by themselves.

Perhaps the best way to describe intangible reinforcements is to list some examples:

- Praise from a supervisor or another employee
- Asking for an employee's opinion or suggestion
- Allowing an employee leeway in problem solving with a difficult worksite task
- Thanking an employee for some aspect of a job done safely
- Asking an employee to participate in preparing a safe job analysis
- Giving an employee a helper
- Including a write-up about the employee in the company newsletter
- Permitting an employee to select his or her job assignment
- Not assigning undesirable work tasks to an employee
- Increasing an employee's responsibility

- Asking the employee for assistance in safety training, auditing, incident investigations, personal protective equipment selection, equipment purchases, and similar safety- and job-related tasks
- Encouraging an employee to take an active role in observing and correcting others who are not working safely
- At a company safety function, recognition of an employee as a model of a safe worker

## How to Reinforce Safe Behavior

The first group of individuals learning to correctly reinforce safe behavior at the worksite should be the company's management. Of prime concern are supervisors. Supervisors are the ones who will make things happen where the work is getting done. Before the supervisors can get started, upper management needs to supply them with the proper training, which can include materials from the chapters on behavior management in this book and from similar books and articles in safety magazines and trade journals, possibly along with sessions put on by proponents of behavioral management in your area. Upper management needs the same training as well. They've got to understand and support the application of safe behavior reinforcement.

Although giving positive reinforcement for safe behavior doesn't have to be formal or extensive, using as many of the following guidelines as possible will help.

### Intangible reinforcement guidelines

**Be immediate.** Remember Pavlov and his dog? The dog grew to associate the sound of a bell with food (or something similar). The dog began salivating and anticipating a culinary reward as soon as it heard the signal. One reason it made such a strong association was the immediacy with which the food reward was supplied. A similar concept holds true at the worksite, where, to be effective, reinforcement should take place immediately after a safe behavior occurs. "Hey," an employee thinks, "it was nice that the foreman appreciated how I danger-tagged the rip saw after Frank pulled it through the closed glass slider and stripped the cord's insulation." The employee must be able to associate a specific behavior with the reinforcement. He tagged out the tool. The foreman saw it and complimented the action right away. If too much time passes before safe behavior is reinforced, the circumstances may change, leading to an incorrect association—for example, with some other behavior having no positive relation to the job. Delayed or late reinforcement is a poor motivator. If the foreman who saw the employee tag out the rip saw waited until the end of the day—five hours after the tag-out—to say, "By the way, I'm glad you tagged out the saw this morning," it would still have some positive effect, but not as much. The immediacy would have been lost.

**Be specific.** Being as specific as possible about the behavior being reinforced will greatly increase the reinforcement's effectiveness. Tell someone exactly what caught your eye, exactly what they did that you liked. Avoid general remarks such as "Good job," or "You're doing great." Instead, dwell on the specific behavior. For example, "Jim, I just saw you tag out the rip saw. It wasn't your job, and you didn't have to do it. But you recognized a situation that could easily have shocked Frank—especially while he was leaning against that steel door frame. You explained what had to be done, and you did it. I really appreciate the effort, and so should Frank."

What's the likelihood that Jim will repeat such a behavior in the future? Or that Frank will learn a lesson, too? Being specific with positive reinforcement makes a lasting impression on all participants.

**Have a personalized delivery.** The supervisor or fellow employee reacting to a situation and providing positive reinforcement to another worker should do so spontaneously, in his or her own style. And be personal about it. Far better to say something such as, "I just saw you barricading the area beneath the scaffold out back, and that's a really good idea. I'd sure hate to get hit with a brick from up there," than to say, "That's playing by the company's rules." The more personal and colloquial the message, the better.

**Be sincere.** If positive reinforcement is delivered in a cold, calculating way and said mechanically, it won't ring true and sincere. Show genuine concern for the coworkers' safety. Lukewarm, insincere attempts at reinforcement will result in employees who will be suspicious of all compliments, coaching, and other reinforcement in the future. No matter how good an actor or actress a person is, it's almost impossible to fake sincerity. If a supervisor or worker has a regard for the safety of his or her fellow workers, it's bound to show, and that will go a long way toward making positive reinforcement meaningful.

**Provide frequent feedback.** Remember the desired ratio: reinforce positive behavior four times as often as you correct unsafe behavior. That translates to four instances of positive reinforcement per each instance of punishment or negative reinforcement. Supervisors are usually far more comfortable confronting unsafe behavior than reinforcing safe behavior. After all, they think, employees should work safely because it's part of the job. They get paid for working safely, don't they? Supervisors are busy people. Their jobs involve planning the work, observing, directing, correcting, and reinforcing their work crews. How, then, can supervisors possibly devote the amount of attention suggested here on positive reinforcement?

All positive reinforcement doesn't have to consist of lengthy encounters. In fact, short contacts are often best. A smile, an approving nod or look, brief thanks or praise—all help to reinforce safe behaviors at the worksite.

**Never combine discipline with positive reinforcement.** "You did a fine job installing that hand basin yesterday, but the tub you put in this morning looks awfully scratched up. Can't you be more careful next time?"

If you were the plumber on the receiving end of that feedback, which part would you tend to remember? People are usually a lot more sensitive to criticism, and tend to think about it and resent it for quite a while. Praise given too close to criticism will be lost in the fray.

**Practice.** Effective reinforcement requires conscious effort and frequent practice. Praising and encouraging coworkers, especially among peers, does not come naturally. So effective reinforcement requires practice. Traditionally, criticism and correction have been far more popular with supervisors, so practice is important if new positive reinforcement is to replace old ways. Employees, too, may eventually realize that they can apply the same principles with their supervisors—they can effectively change supervisory behavior through similar methods of positive reinforcement. Supervisors are only human. The more that some of their actions are appreciated, the more often they'll likely repeat them.

### Combining tangible and intangible reinforcement

Look back at the list of intangible reinforcers. They're ego-builders. Compliments and appreciation everyone loves. And better yet, they're always available at the tip of a supervisor's or fellow employee's tongue. They can be used frequently. They're available at the right price. To compare intangible rewards with tangible ones, time and attention from supervisors and coworkers win hands down when supporting and encouraging safe behavior at the worksite. Consider a child in second grade, bringing home her report card. How does she feel when her teacher, parents, or grandparents praise her for an excellent performance at school? Which will she remember, her parents' words of appreciation and encouragement, or the dollar apiece she collected for each A? Which will help support an environment that encourages her to study in the future?

Is there any place, then, for tangible reinforcements? Yes. They can be used effectively when combined with intangible rewards. Tangible reinforcers can help spice up a safety program. They can help promote safety themes and call attention to short-term goals and programs. They can help make a safety incentive program more interesting. However, apply the following guidelines:

- Use tangible reinforcers only occasionally.
- The perceived value of the tangible reinforcer should reflect the importance of the behavior being recognized.
- Distribute tangible reinforcements only after repeated occurrences of safe behavior.
- Do not offer tangible rewards before the safe behaviors occur. Tangible items offered before behavior occurs are bribes, not reinforcers.

- Small, clever items, often imprinted with the company name or logo and a safety theme, work well when presented with a related safety message. Usually, they're items that are handy to have around work or home, but which the employees don't usually buy for themselves. They also increase reinforcing value through family involvement. Let employee safety representatives or the safer employees help select the items. These tangibles also help to get the employees' families involved in and thinking about safety.

### Safety incentive programs

These are also known as safety motivation programs, safety recognition programs, safety contests, and safety award programs. As long as they're understood and properly planned, they can complement intangible reinforcements while helping to reduce worksite injuries. Above all, don't just purchase items and hand them out without rhyme or reason in the name of safety. Employees won't understand what it's for. They might like it, but any opportunity to remind and strengthen employee knowledge and awareness about areas of safety you'd like to see improved will be wasted with the give-stuff-away approach.

The more effective safety incentive programs incorporate the following components:

- *Upper management support.* Think of expenses for safety incentive programs not as a cost, but as an investment. Upper management should play a prominent role in the program, with promotion and participation.
- *Incentives.* Choose incentives that employees will *like.* Encourage employee participation in the selection.
- *Program rules.* Keep things simple. Complicated rules discourage understanding and participation.
- *Program promotion.* The program should be promoted at and by all levels in the company, from upper management through the supervisors and workers. Small, imprinted items work well for promotion.
- *Communication of safety information.* Periodic letters about important safety topics that are key to the worksites mesh nicely with safety incentive programs. In addition to conveying information on safety issues, they can also help discuss the incentive program itself. And when sent to the employee's home (possibly along with an inexpensive but carefully chosen imprinted item), it's a way of involving families as well.
- *Motivators.* Again, this is the intangible side of reinforcement. It's recognition. It's a sense of achievement, of contributing, of participating, and of being appreciated. It's this *in addition to* the gift items and other material rewards.
- *Employee participation.* Without being embraced by the lion's share of employees, any company safety incentive program is doomed to mediocrity or failure. Enthusiastic employee participation is a must. That participation will come if the program is reasonably planned and if the communications

have been set up to make sure everyone understands the rules of the program before it begins. Make sure that everyone is kept informed *as the program develops*—preferably weekly—and that everyone knows how they can impact or affect the outcome.

- *Communication of program feedback.* There needs to be a method that tracks the program's details so the employees are supplied with continuous updates and feedback. Otherwise they'll lose interest.
- *Recognition.* Arrange for award ceremonies. Celebrate. This should be a festive occasion. Even if the program wasn't 100 percent successful (meaning that everyone didn't qualify for a major reward), there can be small giveaways, and presentations, and safety-related materials to discuss. It's a great opportunity to create and reinforce a higher level of safety awareness throughout the entire workforce. Just the fact that the company went to the time and effort to hold the ceremony will stand out in most employees' minds.

For safety incentive or recognition programs to be effective, all of the preceding components must be present in a particular environment created for the employees. That environment should promote employee discussions about the safety campaign or contest, sharing ideas, and asking questions. Communications must come in a planned manner to keep things flowing and should include weekly or monthly feedback of those being recognized and/or receiving awards. The work crews in some companies lend themselves nicely to forming teams, which can add yet another level of interest to the program.

Along the way, a safety incentive or recognition program can also contain programs within the program. For example, you could reward individuals for the following behaviors:

- Wearing their personal protective equipment properly
- Displaying good housekeeping patterns
- Using the right techniques while operating tools and equipment
- Using fall protection when necessary
- Reporting a hazardous situation or near miss

# Chapter 15

# Ergonomics

## Quick Scan

1. Ergonomics is the science of fitting the job to the employee (rather than fitting the employee to the job).
2. Ergonomics risk factors to look for within individual jobs or tasks include *strength or force, length of exertion time, frequency of exertion, recovery time, posture, environmental factors,* and *job-related factors.*
3. Educate and train supervisors and other employees in the basic principles of ergonomics.
4. Repetitive-motion injuries and illnesses or cumulative trauma disorders are among the chief targets of ergonomics prevention techniques.
5. An ergonomics problem-solving model includes:

- Identifying problem jobs or tasks
- Completing an ergonomics job analysis
- Identifying possible solutions
- Evaluating the solutions
- Implement

Maybe you've already heard the term "ergonomics." If so, you might associate it with assembly-line work, where employees repeat the same tasks hour after hour. It may surprise you to learn that ergonomics plays a role in practically every industry, including numerous trades involved with construction.

Ergonomics has gradually evolved, from the time the word was first coined by a Polish scientist more than 120 years ago, to mean the science of adjusting or modifying the work to fit the employee. The word "ergonomics" hails from two

Greek words: *ergos,* meaning "work," and *nomos,* meaning "law"; put them together and you've got "the laws of work." From a practical point of view, it means fitting the job to the employee, rather than fitting the employee to the job. This has long been important to manufacturers of products such as aircraft and automobiles, and to industries such as meat packing and food processing, where the physical and psychological stresses of manufacturing tasks became major issues. Professional engineers, psychologists, anthropologists, and physiologists were called upon to share their expertise on what could be done to identify worker physical capabilities so that equipment could be developed and procedures planned to enable employees to work to their maximum capabilities without taking undue risks or injuring themselves.

Today, ergonomics is considerably more than a set of laws or principles. The application of ergonomics can mean much more than just preventing injuries and illnesses: an effective ergonomics program can improve productivity and quality, reduce waste, and make individual tasks physically less demanding.

The building blocks of ergonomics include:

- Lifting training
- Strength capabilities
- Hand-grip techniques
- Personal protective equipment
- Specialized tools
- Duration of tasks
- Frequency of tasks
- Posture
- Body positioning
- Repetitive motions
- Body mechanics
- Working surfaces
- Job complexity
- Equipment controls
- Environmental temperature
- Environmental lighting
- Vibration
- Environmental noise
- Employee involvement

One goal of any effective ergonomics program is to identify which of the above-mentioned factors may tax employee abilities or limitations to a point where the chance of error or injury becomes greater than it should be. Much of the end result of any such program will depend upon employee involvement. A typical ergonomics evaluation starts by identifying risks, by reviewing past injuries. Then jobs or tasks where employees have received injuries related to possible ergonomic deficiencies will take priority for analysis. The evaluation process involves observing the job and measuring or analyzing such elements as weight, distance, exertion time, body position, and data from additional job factors. Critical to this information is input from the employees performing the work. Once solutions have been decided upon and implemented, follow up needs to be done to confirm their effectiveness. Not all solutions will be successful, so modifications may have to be considered after a trial period.

## Ergonomic Risk Factors

Ergonomics looks at various task or job factors. These include:

**Strength or force requirements.** Force is a requirement determined by the job. Driving a nail into a particular piece of lumber takes a certain minimum amount of force. Strength is an employee's capacity to apply force. It takes a certain minimum amount of strength to swing a hammer that drives a nail. An employee's physical strength varies with the height of the work surface and its distance from the body. We have optimum strength when working reasonably close to the body and between the knee and hip levels. An employee standing up and driving nails down into a horizontal neck-high surface will not be working at his or her maximum strength. As a rule, men are stronger than women, and strength for both sexes generally—but not always—decreases with age. As with most components of ergonomics, strength is idiosyncratic, varying with different employees.

A worker's ability to grip an object decreases as tool widths vary from the optimal grip span (between about $1\frac{1}{4}$ to 2 inches). This also depends on the size of the employee's hand. Gloves can reduce a worker's effective strength by 30 to 60 percent. Although gloves help protect the skin from abrasions or injury, most of them have the detrimental, fatigue-inducing effect of causing the hands to squeeze tighter around a tool handle or object being grasped.

**Length of exertion time.** Naturally, tasks requiring exertion for longer periods of time require greater strength.

As the duration of the task increases, recovery time for the task must also be lengthened. Tables for exact recommended recovery rates are available, but they can also be approximated simply by paying attention to how tired employees get while performing repetitive tasks, or by just asking the employee for feedback as the job progresses.

**Frequency of exertion.** The potential for fatigue increases as the number of times the task is performed increases. Repetitive motions (doing the same task over and over) result in continuous exertion of specific muscle groups, leading to fatigue. Fatigue causes strain or injury, or slows a person's ability to react to hazards.

**Recovery time.** Recovery time is the time period between muscle exertions that allows the muscles to rest. As force, duration, and frequency of the task are increased, recovery time also must be increased. Failure to provide adequate recovery time results in more and more fatigue. But not allowing enough recovery time is an administrative decision that can be easily changed. When repetitive tasks must be accomplished, supervisors can alternate the employees doing the job, or arrange for other tasks to be interspersed between the repetitive ones. Extensive periods of intense repetitive tasks should neither be scheduled nor encouraged.

**Posture.** Posture refers to the position of the body. Working in an awkward posture can result in overexertion and muscle strains. Maintaining the same posture, such as sitting or standing in the same place for a long period of time, can also result in fatigue. At the worksite, this has often created a difficult situation: either perform a task relatively quickly in an awkward posture, or take a relatively long time to prepare work platforms, ladder setups, temporary stair installations, or similar arrangements first. The problem is that the time to prepare such access takes longer than does performing the work once those access aids are in place.

**Environmental factors.** Environmental conditions can have dramatic effects on both physical and mental performance. Excessive heat or cold, humidity, noise, vibration, pollutants, and inadequate lighting can all lead to increased errors. Although such conditions may not cause immediate problems, they can have a dangerous cumulative effect on employees who are susceptible to individual environmental factors. (Refer to Chap. 36 regarding working in hot and cold temperatures.).

**Job-related factors**

*Comfort.* A reasonable level of job comfort is desirable, with as few distractions as possible. This is often difficult to control at the worksite, but keeping tools and equipment in good condition can make a big difference.

*Task complexity.* Some employees like complex work tasks, while others are far more comfortable with simple ones. Supervisors should be aware of individual preferences when assigning work.

**Controls and display compatibility.** Having clear controls on equipment and tools, with no confusing or unmarked control arrangements, is important to an efficient operation. Because pieces of heavy equipment such as backhoes, front-end loaders, and dump trucks require large capital expenses, sometimes a construction company will rely on older models. The controls of those older models are frequently worn to a point where the operational markings are no longer legible. This deficiency can create special hazards for individuals who occasionally must run the equipment when regular operators are absent.

## Identifying Ergonomic Risks

Educate employees in basic ergonomic principles, so that supervisors and employees can audit jobs and identify employees who are:

- Performing repetitive activities. In other words, doing the same motion over and over. The longer the same motion must be maintained, the more likelihood of injury.
- Bending over for more than 6 seconds continuously because of extended reaches or vision requirements, or bending repeatedly at a rate of more than 10 times a minute.
- Working above shoulder height or below knee height for more than 10 seconds continuously.
- Working for long periods with the body turned or twisted to one side, or with weight unevenly distributed on one leg.
- Working with the head tilted forward or tilted to one side for long periods, or with the head back so as to see through bifocals. Individual situations must be considered.
- Displaying shoulder and neck tension instead of relaxation. Tight-looking shoulders pulled up toward the head are a sign of tension, while loose-moving shoulders that hang down indicate a state of relaxation.
- Using equipment that fails to support the feet on the floor or on a foot rail when seated, or that provides inadequate clearance for stretching the legs forward. This could occur when operating heavy equipment or other vehicles.
- Displaying signs of static muscle loading, from such activities as using one hand to hold a part or piece of material or performing intense visual work for extended periods. Symptoms include shaking/rubbing a limb, squinting, or rubbing the eyes and head.
- Using hand tools that aren't right for the job, or that don't fit the hands. Using a very wide (more than 3-inch) or very narrow (less than $1^1/_4$-inch) grip to hold a tool or object can quickly lead to problems. So can using pinch grips to exert force or lay down materials.
- Using strong wrist angles (flexion, extension, twisting to the sides of the hand) or rotating the wrist or forearm while applying heavy forces.
- Operating vibrating power tools or equipment for extended periods.

What does OSHA say? It recognizes ergonomic issues and over the years has used the General Duty Clause to issue citations to companies who choose to ignore or violate ergonomics principles. OSHA wants companies to train workers in problem jobs, along with their supervisors; then the employer needs to evaluate the effectiveness of the training. OSHA has been formulating an ergonomics standard for a number of years. OSHA asks companies to direct particular attention to the following kinds of employee risk factors:

- Repetition of the same motion or motion pattern every few seconds for more than two, three, or four hours at a time.
- Fixed or awkward postures for more than two, three, or four hours.
- Use of vibrating or impact tools or equipment for more than two, three, or four hours during a work shift.
- Forceful hand exertions for more than two, three, or four hours at a time.
- Unassisted frequent or heavy lifting for more than one or two hours.

Additional information regarding OSHA standards can be found (as of April 1997) on OSHA's web page on ergonomics, located at http://www.osha.gov/ergo. Visitors to the ergonomics page will typically find directions for setting up effective worksite ergonomics programs, useful extracts from OSHA publications, details about OSHA special-emphasis initiatives to avoid repetitive-stress injuries in such environments as meatpacking plants and nursing homes, listings of international ergonomics standards, and links to technical OSHA information.

How can ergonomics fit into your program? OSHA recommends at least four parts to a good ergonomics program:

1. Worksite analysis
2. Hazard prevention and control
3. Medical management
4. Training and education

Much of the program will involve education and training: your company's supervisors and employees should learn to recognize potential problems early on so that they can be corrected before injuries or illnesses develop. Most problems can be dealt with quickly and inexpensively at the worksite. Such preventative solutions are particularly attractive, especially when you compare their low costs with the high costs—in pain, inconvenience, and dollars—of injuries or illnesses. Applying ergonomics at the worksite will reap gains in safety, productivity, and in many cases, morale.

It's also important to be aware of employee physical limitations. Administrative decisions can play a role in reducing risks. Can someone who is nailing shingles be moved to a different task for a while? How long has an employee been handling bricks? In truth, you can get far, far into ergonomics—into strength capabilities, the effectiveness of certain grip techniques, the hazards associated with repetitive motions, recommended rest periods between movements, work heights, twisting and lifting, and many more ergonomics components, both obvious and subtle.

## Repetitive-Motion Injuries and Illnesses

Repetitive-motion injuries and illnesses (RMI) and cumulative trauma disorders (CTDs) are sometimes difficult to recognize because their "cause and effect" relationship may not be clearly defined or observed. A puncture wound, a laceration, a burn, or a sudden ankle sprain all have obvious causes. A repetitive-motion injury or illness may not have such easily discerned origins.

In most cases, overexertions are the likely causes for repetitive-motion or cumulative damage injuries or illnesses, and damage that usually occurs is "musculoskeletal"—to muscles, ligaments, tendons, nerves, bones, and/or joints. It's not so much due to a single instance as it is to cumulative damages, when employees repeatedly attempt to do more work than they normally should or can handle.

Cumulative trauma disorders or repetitive-motion injuries include:

**Carpal tunnel syndrome.** This painful wrist and hand condition is caused by an overextension or twisting of the wrist, especially under force. It affects the median nerve, which runs through a channel in the wrist called the carpal tunnel. If inflammation or swelling occurs, compression of the ligaments can cause uncomfortable burning, itching, prickling, or tingling feelings of the wrist or first three fingers and thumb. Severe conditions can result in atrophy of the thumb with an overall weakness of the hand.

**Tendinitis.** This inflammation of a tendon results from using the wrist or shoulder too much or in ways that are demanding beyond their normal capabilities. If no relief is given, the tendon fibers can even fray or tear.

**Reynaud's syndrome.** Also known as "white finger," this condition is caused by repeated, prolonged exposure to vibration, when the skin and muscles don't get enough oxygen from the blood. Symptoms include tingling, numbness, loss of finger dexterity, pain, and fingers turning white and losing feeling. This can result from frequent and prolonged use of grinders, pneumatic tools, chain saws, and other equipment that vibrates.

The longer that cumulative trauma disorders are ignored, the worse they'll become. If an employee keeps using the damaged parts, the condition may eventually progress too far to be cured. One goal of education is to encourage employees to report symptoms early, such as pain or aching, numbness or tingling, burning, swelling, stiffness, and weakness, especially around the joints in the wrist, hands, and arms.

## Ergonomic Problem-Solving Model

Here's a basic ergonomics model you can use when analyzing jobs or tasks at your worksites:

**Step 1: Identify problem jobs or tasks.** Identifying problem jobs or tasks requires a complete incident analysis. Use the company's incident investigation backlog to screen incidents for ergonomic problems. Review individual incident reports to prioritize jobs for analysis, assigning the highest priorities to the jobs or tasks causing the highest frequency of injuries.

**Step 2: Complete an ergonomic job analysis.** If videotaping the task will make studying the activities involved easier, do so. Identify risk factors influencing the task. Provide specific measurements of risk factors, frequencies, recovery times, weights, distances and similar factors. Identify body parts affected by the risk factors.

**Step 3: Identify possible solutions.**

Reduce force requirements of the task.

Reduce the frequency of the task.

Reduce the duration of the task.

Provide required recovery time.

Eliminate unnatural or awkward postures. Have employees work with their palms down (instead of face up) as much as possible. And work with straight wrists instead of bent wrists. Perform tasks with two hands instead of one whenever possible.

When feasible, use a power grip technique (see Fig. 15.1) instead of a finger pinch grip (see Fig. 15.2) or a palm pinch grip (see Fig. 15.3). The power grip spreads your grasping force out over the whole hand and fingers, while a pinch grip requires too much squeezing power against too few fingers. Don't apply too much pressure to a tool's grip with the center of the palm; use the surrounding parts of the hands that have more padding instead.

Use hand-friendly tools. New ones are entering the market every day. New designs feature padded handles and textured grips. Textured grips are preferable to the old types that had precut or molded finger grooves—the grooves

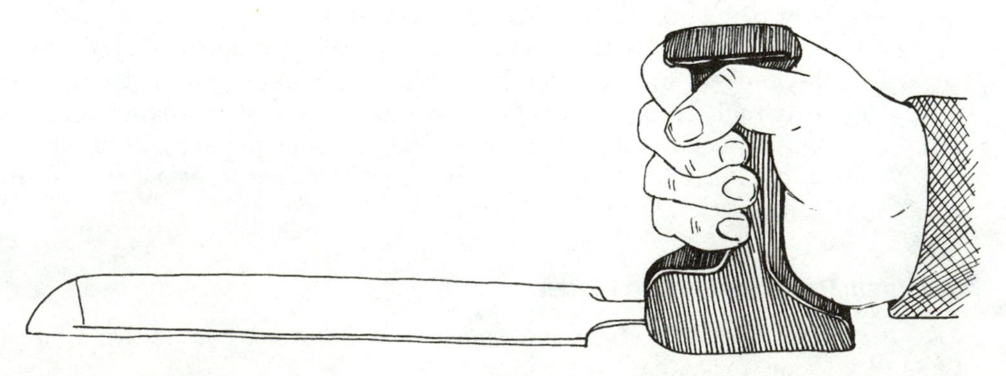

**Figure 15.1** Power grip.

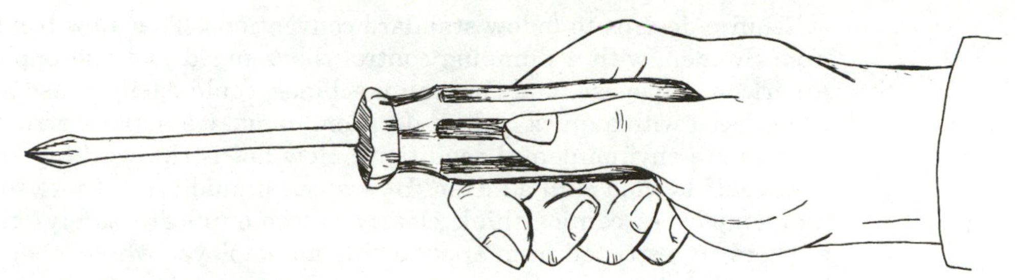

**Figure 15.2** Finger Pinch Grip.

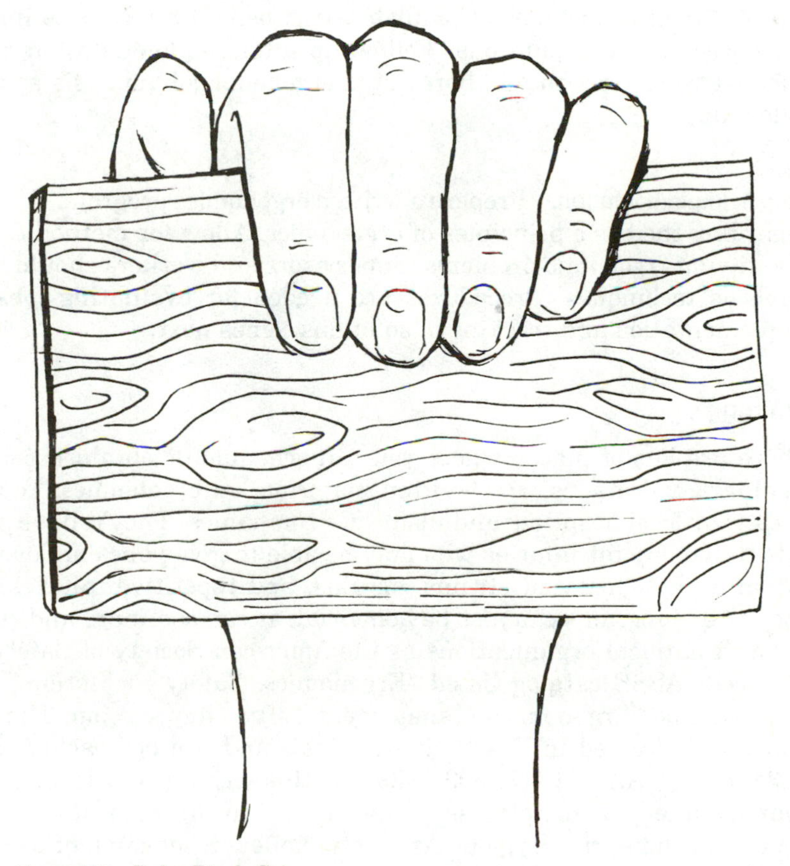

**Figure 15.3** Palm Pinch Grip.

were too large for some employees and too small for others. Many newer tools have larger triggers, distributing the load to more than one finger. Other new tools are being designed for use with two hands, or even with an overhead suspension system that carries their weight while in action.

Modify work stations to accommodate reaches, clearances, and heights for employees. Try to arrange work heights at about waist height. Especially avoid conditions that require working six or more inches below the waist.

Require designs to follow standard conventions. That new front-end loader from Sweden, with a dumping control that's piped just the opposite of most American, Japanese, and German machines, could easily cause an employee, when faced with a quick control decision, to make a serious error.

Evaluate environmental conditions. How hot is the working environment? How cold? Is wind chill a factor? How about humidity? Is there so much noise that employees cannot think clearly or communicate safely? Should other workers be restricted from approaching an employee who is operating a jackhammer?

**Step 4: Evaluate solutions.** Complete a cost/benefit analysis to implement the most cost-effective solutions. Follow up after implementation to determine effectiveness. Perform an abbreviated ergonomic job analysis as a part of the follow-up.

**Step 5: Implementation.** Prepare a written ergonomics program. Train employees regarding the basic principles of ergonomics. Allow for methodical analysis of identifying ergonomic problems. Supervisors and workers should be trained in analysis techniques. Procedures are needed for evaluating jobs and tasks. Implementation and tracking of solutions comes next.

## Ergonomics Training

There are lots of places where you can schedule or obtain excellent classes, seminars, videotapes, and written materials on ergonomics. You could start with your local hospitals and insurance companies. They'll have specialists in soft-tissue or joint injuries who may be able to give your employees a presentation on how to avoid strains, sprains, and repetitive injuries. University-sponsored programs can also be beneficial, as can seminars and classes put on by such national organizations as the American Society of Safety Engineers. (A recent ASSE catalog listed "Ergonomics, Safety in Motion," a three-day seminar, and "Ergonomics Management," a two-day session. The ASSE headquarters is located in Des Plaines, Illinois and can be reached at (847) 768-3434, or by FAX at (847) 699-2929.) At this writing, OSHA offers a four-day course called "Principles of Ergonomics" in its training schedule. The American Industrial Hygiene Association offers a selection of books and other publications on ergonomics, and many major publishers (including the publisher of this book) feature helpful volumes on all aspects of ergonomics. Magazines and newsletters on ergonomics are also available; these include *Workplace Ergonomics,* a quarterly published by the Stevens Publishing Corporation.

Manufacturers have also become more ergonomically conscious, and are coming out with new tools, equipment, and other products that are much more ergonomically friendly than their regular tool and equipment counterparts.

Chapter

# 16

# Safe Job Analyses

## Quick Scan

1. A safe job analysis (SJA) is simply a detailed safe operating procedure that breaks down a particular job (like operating a chain saw) into simple steps or tasks, then analyzes the potential hazards of each step, and explains how those hazards can be avoided.
2. SJAs should be written for jobs at the worksite that put your employees at risk for injury.
3. Employees who perform the job for which an SJA is being written should take part in the preparation of that particular SJA.
4. Whenever possible, use existing manufacturer operating procedures to help prepare SJAs for tools, equipment, and vehicles.
5. Completed SJAs should become resources for employee training, and worksite audits.

A safe job analysis or SJA, also sometimes referred to within manufacturing and industrial environments as a behavioral job analyses (BJA), is simply a set of written procedures or guidelines on how to safely complete a particular job. The theory behind an SJA is: why not write down the safest way to perform a particular job so that safe way can be shared with others, and used again and again both as an orientation training aid and as a refresher to individuals who may only infrequently do the job?

SJAs recognize that most jobs can be broken down into a series of consecutive individual tasks or steps, and that each of those steps may involve specific hazards. SJAs take those hazards into account as they appear in the job, by performing job hazard analyses and by prescribing safe behaviors to counter the hazards so those individual steps of a job can each be completed safely.

To customize this process to your company's worksites, individual tasks must be evaluated for hazards and risk levels.

## Selecting Jobs for Analysis

Safe job analyses can be performed for all tasks or jobs at the worksite, whether the jobs involve routine or special tasks. To help determine which jobs should be analyzed first, consider or review your company's safety incident investigations or reports from the previous few years. It should become apparent that SJAs should be completed first for jobs having the highest rates of disabling injuries and incidents. Jobs where serious near misses have occurred should also be high on to-do lists, as should tasks and jobs incorporating recent changes.

Completed SJAs make ideal resources for training new employees and cross-training other workers. They also provide documented tasks that employees can use for performing safety audits, complete with handy observation checklists.

## Involving Your Employees

The preparation of an SJA is a good opportunity for employee involvement; who knows the job better than the individuals who actually perform the work? Revisions to SJAs and other work procedures, if needed, can be based upon incident investigations and additional observations and employee discussions.

Examples of tasks that can benefit from SJAs include:

- Operating powder-actuated tools
- Operating a chain saw
- Using an open-end wrench
- Changing a backhoe tire
- Excavating a trench
- Working on a roof
- Supplying brick to a scaffold
- Felling a large tree
- Working with a temporary power supply
- Pouring concrete

Before an employee is asked to help, the supervisor or someone fairly knowledgeable about the job should prepare a general outline checklist for the job or task being studied. Next, he or she should enlist the assistance of employees who routinely perform the to-be-studied work as part of their jobs. The supervisor should follow with a discussion about how *employee performance* won't be the subject of the observations, but that the actual procedures they're following—the various steps of their jobs—are what's being looked at. For SJAs to be effective, employees must be involved in this early part, and then in all remaining phases of the analysis, from listing specific job steps to discussing potential hazards to arriving at recommended solutions. Similar input can be sought from others who have performed the jobs, too. Depending on what's being looked at, here are some sample questions to ask employees while collecting pertinent information:

- Do employees wear proper personal protective equipment for the jobs they're performing?
- Are the tools needed for the job—including hand tools, machines, and equipment—available and in good condition, or are they in need of repair?
- Is there typically excessive noise or similar distractions in the area due to tool usage or the operation of nearby compressors or heavy mobile equipment?
- Are trucks and mobile equipment properly outfitted with back-up alarms, horns, overhead guards, recently serviced brakes, and other functional safety equipment?
- Is ventilation in the work area adequate?
- Is lighting in the work area adequate?
- Are materials or equipment too heavy for manual lifting?
- Have employees who perform this job complained of headaches, breathing problems, dizziness, sprains, strains, or other soreness?
- Are all employees who operate heavy mobile equipment and power tools properly trained in their use and authorized to do so? Are recently hired individuals learning on the job in a haphazard way?

## Performing a Safe Job Analysis

Once a working checklist of tasks has been prepared, and one or several employees have been enlisted to help, someone knowledgeable about SJA preparation must review the basic steps for preparing a standard safe job analysis with the supervisor and employee(s).

### Instruction checklist for filling out a safe job analysis form

1. *Prepare the forms.* Make copies of Fig. 16.1 to use as worksheets and for final versions of the completed SJAs. For a sample completed form, see Fig. 16.2.
2. *Get the participants together.* Assemble the individuals who will be performing and studying the job being discussed. It's important to include at least several of the employees who are normally responsible for doing the job.
3. *Write a short title for the job.*
4. *List personal protective equipment needed.* Determine what personal protective equipment the job requires.
5 *List tools and equipment needed.* What tools or other equipment, if any, are needed to perform the job safely?

**JOB HAZARD ANALYSIS FORM**

JOB TITLE: DATE OF ANALYSIS:

JOB LOCATION:

| STEP | HAZARD | CAUSE | PREVENTIVE MEASURE |
|---|---|---|---|
| | | | |

**Figure 16.1** OSHA Blank Safe Job Analysis Form (Job Hazard Analysis).

**SAMPLE JOB HAZARD ANALYSIS**
**GRINDING CASTINGS**

| STEP | HAZARD | CAUSE | PREVENTIVE MEASURE |
|---|---|---|---|
| 1. Reach into right box and select casting | Strike hand on wheel | Box is located beneath wheel | Relocate box to side of wheel |
| | Tear hand on corner of casters | Corners of casters are sharp | Require wearing of leather gloves |
| 2. Grasp casting, lift and position | Strain shoulder/elbow by lifting with elbow extended | Box too low | Place box on pallet |
| | Drop casting on toe during positioning | Slips from hand | Require wearing of safety shoes |
| 3. Push casting against wheel and grind burr | Strike hand against wheel | Wheel guard is too small | Provide larger guard |
| | Wheel explodes | Incorrect wheel Installed | Check rpm rating of wheel |
| | | Cracked wheel | Inspect wheel for cracks |
| | Flying sparks/chips | Wheel friction with caster | Require wearing of eye goggles |
| | Respirable dust | Dust from caster metal and wheel material | Provide local exhaust system |
| | Sleeves caught in machinery | Loose sleeves | Provide bands to retain sleeves |
| 4. Place finished casting into box | Strike hand on castings | Buildup of completed stock | Remove completed stock routinely |

**Figure 16.2** OSHA Sample Job Hazard Analysis.

6. *Observe the job.* Either observe someone else while they perform the job, or think through the complete job, using memory recall. It's generally most effective to combine actual job performance, with additional input from memory. It also helps to have manufacturer operating instructions on hand, if they're available. Some individuals find that videotaping a particular job can provide an exact source of data that can conveniently be replayed, studied, and discussed later.
7. *Break the job down into individual tasks, or task steps.* Each job can typically be broken down into individual tasks, or task steps. Identify major tasks, and number and list the tasks or task steps as they occur, in chronological order as the job is typically worked, in the first column on the safe job analysis worksheet. Each task or step must be an observable behavior.
8. *Identify hazards.* Identify possible hazards associated with each task step, listing them in the middle column of the safe job analysis worksheet. SJA preparers need to think about what events could lead to an injury or illness for each hazard identified. A first draft should include as many as can be thought of. You can edit out inappropriate ones later. It's important that specific information be provided, with short, active statements to describe the hazards.

   Typical questions that should be asked when thinking about hazards include:

   - Is hazardous energy control a part of the job? Are lockout procedures being used for equipment deactivation during maintenance procedures?
   - Are dusts or chemicals involved?

- Noise or heat?
- Sharp surfaces?
- Falling or flying materials?
- Pinch points?
- Possibility of crushing injuries?
- Awkward work positions?
- Heavy lifting
- Repetitive motions?

9. *Counteract each potential hazard with safe behaviors.* In the third column of the safe job analysis worksheet, specify the required safe behaviors. Specific safe behaviors must correspond to identified hazards; in other words, for each hazard there should be a corresponding safe behavior that will avoid that hazard. Again, use concise descriptions of tools and equipment, and precise behavioral terms to describe actions or events. Avoid writing general statements like "Use caution," or "Pay attention."

## Revising and Updating Safe Job Analyses

Safe job analyses must be kept current. That means someone in the company should be reviewing and updating them periodically. Even if no changes occur to a job, that doesn't mean that better methods of doing a task can't be arrived at later. Maybe combining several steps or changing a sequence would make a task safer and more effective. If an incident or injury occurs with a specific job, the SJA should be reviewed immediately to see if a job procedure should be changed. And if an injury results from someone *not following* the SJA, that should be reviewed with all employees performing the job.

Naturally, whenever an SJA is changed, those changes, along with the entire SJA, should be discussed with all employees affected by the changes, at dedicated training sessions.

You might be tempted to ask the question, "Hasn't this all been done before?" And if so, why reinvent the wheel? Why reinvent a safe job analysis for working from a scaffold, if all you've got to do is duplicate copies of a scaffolding manufacturer's manual and pass them out? Too, why not just go out and find safe work procedures for excavating a trench for a footer, or running a radial saw?

That certainly could be done, but it's not recommended here. If you want to compare your SJAs to those prepared elsewhere, fine. But an important part of preparing SJAs *is the process itself*—getting employees to think about individual steps in tasks and jobs, and encouraging them to consider corresponding potential hazards. Recognizing that something could actually go wrong during a job is one giant step for employees who may have worked for years without giving safety its due.

For more information on safe job analysis, review OSHA booklet #3071 (see Fig. 16.3), available through your local OSHA office or from the U.S. Government Printing Office.

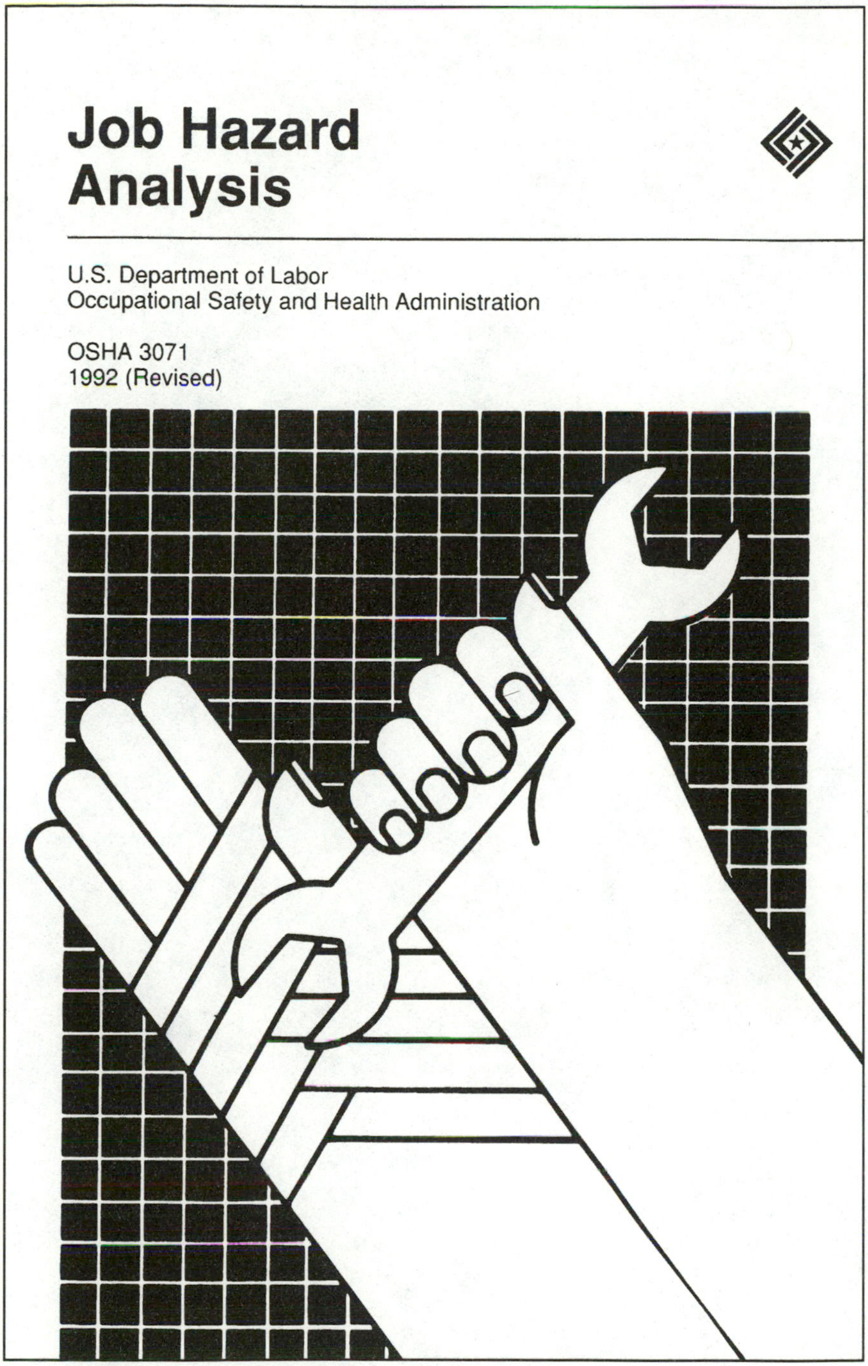

# Job Hazard Analysis

U.S. Department of Labor
Occupational Safety and Health Administration

OSHA 3071
1992 (Revised)

**Figure 16.3** OSHA Booklet #3071.

Chapter

# 17

# First-Aid/Responder Arrangements

## Quick Scan

1. Have a written plan for worksite first-aid and emergency medical services.
2. Train all employees on how to access first-aid and emergency medical services.
3. Arrange for emergency communications at the worksite, and ensure that all employees know how to use them.
4. Provide first-aid kits and related equipment at the worksite and with selected vehicles and pieces of heavy equipment.
5. Encourage and support formal first-aid training, cardiopulmonary resuscitation (CPR) instruction, and first-responder training for employees. Provide ongoing refresher training and support.
6. Train all employees in blood-borne pathogen hazards, precautions, and related safe practices.

This is one chapter that can be considered helpful *after* an injury-producing accident or worksite illness occurs. This material won't help to prevent an incident, but it *will* help to prevent the incident from becoming worse after it occurs. As such, minimizing the effects of an incident should be an important part of your safety management system.

## Being Prepared

Safety preparedness is a matter dictated by both common sense and law. You've got to assure the availability of first-aid services on the jobsite, and you've got to have a trained person available if medical facilities are not located nearby. First-aid supplies recommended and approved by a physician or medical supply company must be available, and emergency phone numbers must be conspicuously posted, and employees informed of their location.

A wide variety of incidents could occur on a construction worksite. Examples include:

- Unconsciousness
- Bleeding
- Stopped breathing
- Electric shock
- Head injury
- Heart arrest
- Choking
- Bone injury
- Eye injury
- Poisoning
- Burns
- Fever/seizures
- Bites/stings
- Heat-caused illness

## What OSHA Says

> 1926.50 Medical services and first-aid (a) The employer shall insure the availability of medical personnel for advice and consultation on matters of occupational health.

A company doctor should be selected who is familiar with the work tasks and the worksite hazards encountered by your employees. In addition to providing medical treatment, the physician can advise you regarding the suitability of first-aid supplies and related equipment.

> (b) Provisions shall be made prior to commencement of the project for prompt medical attention in case of serious injury.

This regulation means that arrangements must be made in advance of any emergency. The confusion that often ensues following an emergency situation often makes things worse. There has to be a medical plan in place. Who will provide first-aid treatment? What first-aid supplies should be on hand? How will emergency medical services be notified?

> (c) In the absence of an infirmary, clinic, hospital, or physician, that is reasonably accessible in terms of time and distance to the worksite, which is available for the treatment of injured employees, a person who has a valid certificate in first-aid training from the U.S. Bureau of Mines, the American Red Cross, or equivalent

training that can be verified by documentary evidence, shall be available at the worksite to render first aid.

It certainly makes sense to ensure that medical help is available *before* that help is actually needed. Many companies depend on the quick response of local first-aid providers through either volunteer or paid fire departments, ambulance services, or even police or other law enforcement agencies. And if such arrangements are feasible for a particular worksite, that's fine. The problem arises in some locations where those response times are not immediate or reasonable because of the distance between the worksite and emergency care. Some companies are fortunate enough to have employees who are also members of volunteer fire departments. Many volunteers, in turn, are trained first-responders or emergency medical technicians.

While having employees trained in first aid is definitely an asset in providing certain levels of care before the emergency medical system kicks in, their training in patient assessment and certain types of techniques is limited. First-responders, in contrast, are highly trained individuals possessing the skills necessary to begin assessing and caring for patients at the worksite or scene of an injury or illness. The whole point for such training is that a sudden, severe accidental injury could cause a death to occur before a patient can be taken to the hospital, or in some cases, even before ambulance emergency medical technicians (EMTs) can arrive. Patients can need immediate care before the EMTs reach the scene. The care that first-aid givers and first-responders can provide may reduce suffering, prevent additional injuries, and even save a life at your worksite. Hundreds of thousands of individuals have been trained as emergency responders and are providing this important care within the emergency medical system across the country. Many are members of law enforcement agencies or fire departments, while others are private citizens or employees like yours, who are willing to be trained and involved with providing medical care.

How quick is a *reasonable* response time? That can vary, depending on the type of injury. Many single or compound fractures can wait 10 or 15 minutes or longer for a satisfactory response. But consider an injury involving severe bleeding or an interruption in breathing, and reasonable response time shrinks down to about four minutes. If professional medical attention isn't available within those four minutes, then someone trained in first aid must be available at the site.

Progressive contractor companies arrange for the availability of both services for on-site first-aid response: they train employees in first aid, have first-aid resources on hand, and make advance arrangements with local emergency providers.

(d)(1) First-aid supplies approved by the consulting physician shall be easily accessible when required.

This doesn't mean that the first-aid supplies are at the main office, 20 miles away. Nor in the back of a foreman's pickup truck while that foreman is away from the site delivering supplies. It means that the supplies are readily available on site—and that everyone knows where they are.

> (d) (2) The first-aid kit shall consist of materials approved by the consulting physician in a weatherproof container with individual sealed packages for each type of item. The contents of the first-aid kit shall be checked by the employer before being sent out on each job and at least weekly on each job to ensure that the expended items are replaced.

First-aid kits are a must. You can make up your own, or you can purchase complete kits (see Fig. 17.1). Numerous styles of kits are available through local industrial supply or medical equipment shops. The size and breadth of your kit or kits will depend somewhat on your circumstances. Are three employees working at the jobsite, or twelve? The whole point of a first-aid kit is to have the materials available—now. You don't want someone driving 20 minutes to a pharmacy for aspirin, or for tweezers to remove a painful sliver. A typical contractor's first-aid kit may be carried in a sturdy metal or plastic case, and should include items such as:

1 pair tweezers

1 pair scissors

4 extra-long bandages

1 bottle eyewash , 2 oz

1 bottle peroxide, 4 oz

1 roll of adhesive tape, $^1/_2$ in $\times$ $2^1/_2$ yd

4 nonstick adhesive pads, 2 in $\times$ 3 in

6 Dristan tablets

6 Advil Cold and Sinus tablets

1 bottle Advil

1 bottle Tylenol Extra Strength

1 bottle Bayer aspirin

8 throat lozenges

2 elastic bandages, 2 in $\times$ 5 yd

1 triangular bandage

1 box assorted (36) elastic strip bandages

2 instant cold packs

8 antiseptic wipes

1 CPR mouth shield

2 pair disposable latex gloves

2 eye pads

1 roll 2-in stretch gauze

1 small bottle SPF 15 sunscreen

1 tube or container first-aid cream

1 first-aid instruction booklet

6 antacid tablets

6 Alka-Seltzer tablets

Larger or smaller kits are also available. You may want a large kit for every worksite, and smaller, personal kits to be carried in all company vehicles. In any case, training should be provided on how to use the contents of each kit, and what the limitations are.

Recently there have been discussions about the liability of providing over-the-counter medicines to employees at the worksite. Check with your local OSHA office and trade association for further developments.

> (e) Proper equipment for prompt transportation of the injured person to a physician or hospital, or a communication system for contacting necessary ambulance service, shall be provided.

By the very nature of construction worksites, there are usually plenty of vehicles available for transportation.

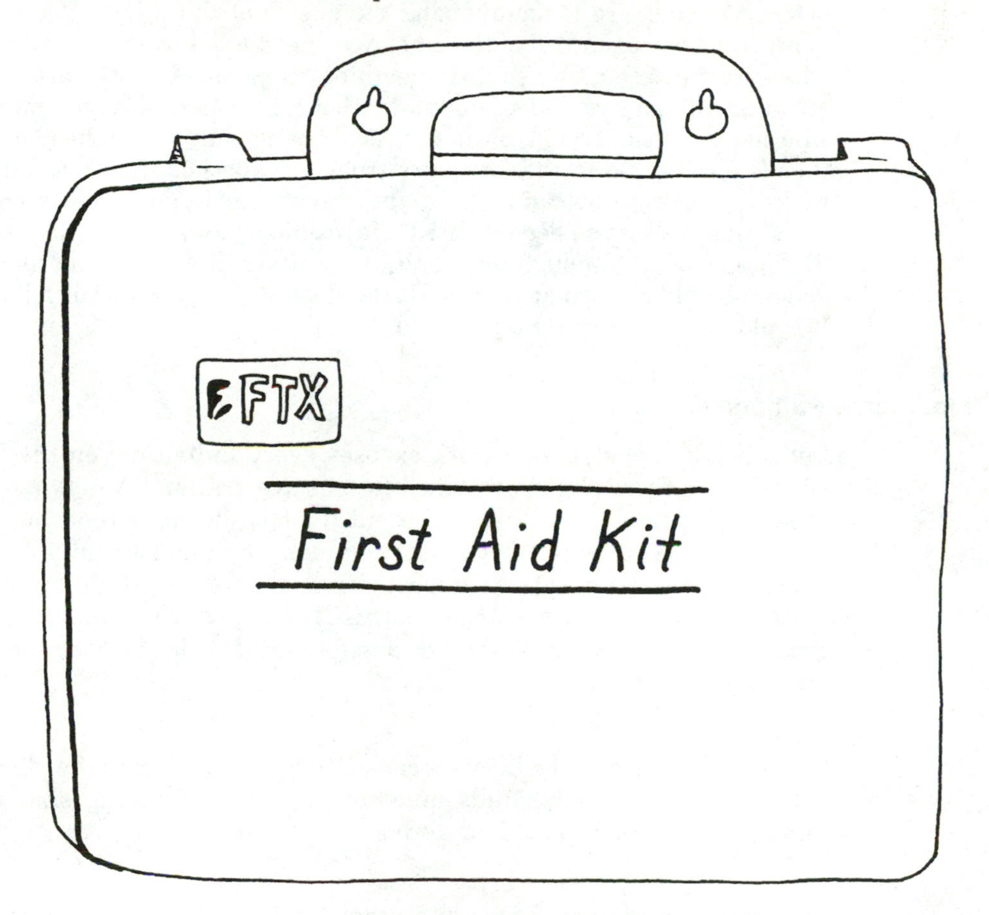


**Figure 17.1** First-Aid Kit.

> (f) The telephone numbers of the physicians, hospitals, or ambulances shall be conspicuously posted.

This regulation addresses the means of communication available at the worksite. Is the site in the immediate vicinity of emergency facilities, or within a residential neighborhood where arrangements can be made for placing emergency calls? Other communication possibilities include cellular phones and two-way radio systems. In any event, the proper phone numbers must be posted and available for immediate use in case of an emergency.

> (g) Where the eyes or body of any person may be exposed to injurious corrosive materials, suitable facilities for quick drenching or flushing of the eyes and body shall be provided within the work area for immediate emergency use.

Portable eyewash units are available for worksite setup. (See Chap. 19, on Eye, Face, and Respiratory Protection.)

In any case, the main idea is to consider the various possibilities in advance. What should your employees do in the event case of a serious or life-threatening injury? The sooner that medical assistance can reach the injured employee, the better the odds are for minimizing the effects of the injury. Plans have to be made in advance, and all of the employees need to be trained on what to do and what to expect. Don't wait until something happens. Make the arrangements in advance. Employees have to know about the available communications—whether they're expected to run to a neighbor's house to use the phone, whether there's a CB radio in the foreman's truck, or whether there's a cellular phone with the worksite first-aid kit. Do they know which employees are certified in CPR? Has the system been audited? Have employees asked "What would you do if..." questions? Although everyone hopes never to face a true medical emergency, it could happen at your worksite. It could happen to your fellow workers. It could happen to you. Be prepared.

## Blood-borne Pathogens

Let's face it. Construction work exposes every individual on the site to possible injury, from slight scratches to massive trauma. And when blood and other bodily fluids flow, risks inherent to blood-borne pathogens occur.

Blood-borne pathogens include the human immunodeficiency virus, or HIV, which leads to acquired immunodeficiency syndrome (AIDS). And there's the more common liver-destroying hepatitis-B virus, which is sometimes fatal and greatly increases the risk of other possibly fatal liver diseases such as cancer and cirrhosis.

**Exposure.** Exposure to blood-borne pathogens can occur by direct contact between infected body fluids and broken skin—including skin that's only minutely scratched, chafed, or abraded.

**Universal precautions.** Blood and other bodily fluids need to be treated like the hazardous materials they could very well be. How would you protect yourself against a strong acid or caustic fluid? Would you allow it to get onto your exposed skin, or splash into your eyes? Of course not. Neither must blood or other potentially infectious fluids come in contact with your bare skin. That means that a barrier must be placed between an injured person's blood and the aid-provider's skin, to prevent contact. Emergency responders use thin medical nitrile or latex gloves, and so can individuals trained in first aid. Others must use common sense, and if gloves are not available to keep the blood from the helper's skin, something else may suffice—such as a rolled-up towel or jacket, raincoat, or similar material. Blood-borne pathogen medical kits are also available, complete with gloves, CPR microshields, face mask with eye shield, antiseptic swipes, disinfectants, and related supplies (see Fig. 17.2).

**Figure 17.2** Blood-borne Pathogens Kit.

Part

# 3

# Personal Protective Equipment

*What are you doing right now? At this very moment, you've started to read this paragraph, using your eyes. It's a proven fact that the typical person's eyes provide about 80 percent of all the knowledge that he or she will ever know. They also control more than 80 percent of a person's actions. Your eyes—anyone's eyes—even in this modern age, are irreplaceable.*

*Yet each year about 1 million Americans suffer eye injuries at work, and hundreds of thousands more injure their eyes at home or at play. Most eye injuries are caused by flying or falling objects. Daily construction activities, rotating machinery, power and hand tools, chemicals, dust, and pieces of raw materials can all cause serious eye injuries.*

*On the other hand, consider that over 90 percent of eye injuries are preventable. The best way to prevent them is to use safety eyewear whenever you're in a potentially dangerous area, or whenever you're performing a task in which something could fly, fall, blow, drip, splash, leak, or strike at your eyes. Chapter 19 deals with eye and face protection.*

*The same theory holds true for other parts of the body exposed to hazards. Your hearing organs, for instance. More than 20 million Americans are exposed to hazardous noise on or off the job. Those hazards can be avoided by wearing proper hearing protection.*

*People's hands, too, have historically taken a beating. That's why we stress the importance of wearing gloves whenever there's a chance of sustaining slivers, cuts, abrasions, or burns.*

*OSHA's personal protective equipment (PPE) standards are mostly performance oriented; they don't supply specific guidelines on how a personal protective equipment hazard assessment should be conducted. But OSHA* will *hold the employer accountable if things go wrong and people get injured. Companies must ensure the quality of the hazard assessment, plus the adequacy of the PPE used by employees.*

*OSHA guidelines—specifically 29 CFR Part 1910.132, Subparts (d)(l)(i) through (iii)—require that employers perform worksite hazard assessments to determine if hazards that require the use of PPE are present or likely to be present. If hazards do exist, the employer must select and ensure use of appropriate personal protective equipment, and communicate those selection decisions to affected employees. OSHA Appendices A and B provide additional guidance for employers on how to perform a hazard assessment. OSHA also requires—in 29 CFR Part 1910.132(d)(2)—that a certified written record detailing that the hazard assessment was conducted be kept by the employer. This form does not have to detail the methodology used when performing the hazard assessment, but it must record the date of the hazard assessment, the workplace evaluated, and the name of the individual performing the evaluation. The form must be readily identifiable.*

*Personal protective equipment only works if it fits, if it is in good condition, and if it is worn properly. OSHA forbids the use of damaged or defective PPE, and so should you. This matter is addressed in 29 CFR part 1910.132(e). OSHA does not define any methods for ensuring that PPE use is adequate. One recommended method for identifying defective PPE is having employees conduct frequent visual inspections before, during, and after use. Employers should ensure that defective equipment is replaced immediately and train their workers accordingly.*

*Comprehensive training requirements are addressed in 29 CFR Part 1910.132, Subpart (f)(1) through (3). In these sections, OSHA mandates that the employer train each employee who uses PPE. OSHA details its expectations of the contents of the training in (f)(1)(i) through (iv). Requirements include provisions that employees receive instruction on the reasons why personal protective equipment is necessary, the types to use, and how to properly don, remove, adjust, and wear them. In addition, employees must receive training on the proper care, maintenance, useful life, limitations, and disposal of PPE. An employer must ensure that each employee who has received training understands the content of the training before using the PPE. This is mandated by 29 CFR Part 1910.132(f)(2).*

*OSHA also details the circumstances that require retraining. This is addressed in 29 CFR Part 1910.132, Subparts(f)(3)(i) through (iii). One such situation where retraining becomes mandatory is when the employer believes that an affected employee does not have the requisite knowledge, understanding, or skills when using the PPE. OSHA does not define lack of requisite knowledge. A suggested method of compliance would be to conduct frequent audits and inspections of the employees using the PPE to ensure that it is being used properly. If an accident or incident occurs that results in an injury or illness of an employee, retraining probably would be necessary. Other situations defined by OSHA when retraining would be required include changes in work practices and/or changes in the type of PPE to be used.*

*Documentation of training and retraining is required by the standard in 29 CFR Part 1910.132(f)(4). This subparagraph requires that companies prepare a written certificate that documents the name of each employee trained and when the training occurred. The form must clearly indicate that it is the written certification of training.*

*Employees who do not wear PPE not only are jeopardizing their own health; they're putting a crimp in your company's safety program as well. They're bad examples, and they'll ultimately be expensive examples—to themselves, to their families, and to your company.*

*Developing a PPE program begins with determining which exposures are present in the worksite. The supervisors, foremen, or lead persons know best what's encountered on the job. They're ideal sources of information when doing a hazard assessment. A hazard assessment may sound intimidating to those who have never partaken in one, but it is simply a survey of worksite risks and of what PPE is available to counteract those risks. A walk-through assessment works best. Even if you've been involved with home-building all your life, a walk-through is important. Try to look at the worksite with a fresh view. You may find:*

- Harmful dust
- Sharp objects
- Rolling or pinching objects
- Electrical hazards
- Sources of motion that could cause a collision
- High or low temperatures
- Harmful chemicals
- Light radiation
- Danger from falling or dropping objects

*Personal protective equipment has limits, of course. It's effective for reducing risks, but it should never be used in the place of safe procedures. A pair of gloves or safety glasses, for example, should never be the only control measure between an employee and an observed risk. They're really a last resort, a last line of defense—something to guard your eyes or hands if something goes wrong.*

If employees are trained to recognize situations which call for PPE on the job, this will likely extend to their off-the-job activities as well. Soon they'll be using PPE in home basement projects, while landscaping, and even during recreational activities.

Some employees get it through their heads that the only time to wear personal protective equipment is when a boss is around. The rest of the time, they won't bother. Many companies that have never had PPE programs, upon starting one, find a few individuals who have all of the excuses ready:

"I never got hurt before. Why do I need safety glasses now?"

"I'm tough. A few bumps, bruises, and scratches won't hurt me."

"They're too hot."

"I can't bend down with the helmet on."

"The glasses bother my peripheral vision."

"They just get in the way."

"I can't hear with them on."

*For those individuals, read Chap. 13, on correcting at-risk behavior, and then apply the applicable information.*

# Chapter 18

# Head Protection

## Quick Scan

1. The main goal of protective helmets or hard hats is to minimize the rate and level at which impact forces are transmitted to the brain, neck, and spine.
2. A secondary goal of helmets is to partially protect the head, face, and neck from electrical current and from other environmental hazards such as sun rays, rain, snow, wind, and extreme temperatures.
3. There are two basic types of hard hats: Type I and Type II. Type I hard hats have a full brim not less than $1\frac{1}{4}$ inches wide around the entire helmet. Type II hard hats include helmets without brims but having a bill or peak in the front to help protect the eyes and face.
4. There are four classes of helmets:

   Class A—For general service and protection against impact hazards. Some defense against low-voltage electrical current.

   Class B—For general service and protection against impact hazards, plus some defense against high-voltage electrical current.

   Class C—Some impact protection, but no voltage protection. Usually made of aluminum.

   Class D—Impact protection, fire resistant, and will not conduct electricity. Made chiefly for firefighters.

4. Hard hats consist of a shell and a suspension system. The shell actually bends or flexes under impact, resulting in some absorption of energy. By stretching slightly and distributing the force of impact equally to the wearer's head, the suspension system then absorbs impact energy that isn't handled by the shell's denting and flexing.

5. Train employees on helmet use, inspection, and care.
6. Insist that employees, subcontractors, and visitors present during precompletion stages or on-site construction activities wear Type I or Type II, Class A or Class B helmets.

I can remember attending a college picnic near State College, Pennsylvania, back in the early seventies, when a number of states were passing mandatory helmet laws for motorcyclists. We could see (and hear) from far into the valley the approach of a lone Triumph Bonneville motorcycle with two riders. Both driver and passenger wore what first appeared to be bright silver helmets that shone in the midmorning sun. But when they pulled onto the picnic grounds, it became obvious that they were wearing sheets of tinfoil that they had pressed around their heads—giving each the appearance of wearing a helmet, yet affording the riders absolutely none of the protection that a real helmet provides.

Unfortunately, numerous home-construction crew members treat head protection in a similar cavalier manner, electing to work without the benefits offered by helmets. Individuals who believe that wearing hard hats is overkill may not fully understand how fragile a human skull can be, nor how serious daily construction risks are. Nothing is ever going to happen to *their* heads, they think.

## Protecting the Head, Neck, and Spine

Everyone knows that along with the spinal cord, the brain is critical to the central nervous system. The brain, of course, is naturally protected somewhat by a person's skull. The skull, however, is not as durable as most people think. Collisions with stationary objects or with flying or falling tools and materials can not only inflict scalp cuts and other soft-tissue damages, but can cause fractures of the cranium (upper and side parts of the skull) and face, plus direct and indirect injuries to the brain. Indirect injuries include concussions and contusions or bruises to the brain caused by jarring impact forces, when the bones of the cranium are neither cracked nor broken, but the brain is shaken up considerably.

The main goal of protective helmets or hard hats is to minimize the rate and level at which impact forces are transmitted to the brain, neck, and spine. Consider participants in sports such as football, baseball, hockey, bicycling, motorcycling, wrestling, skydiving, white-water rafting, kayaking, and race-car driving. They all recognize the importance of wearing protective helmets and know what can happen to individuals who fail to do so. The same situation applies to members of the home construction industry. Home builders who have received head injuries in the past, or who have seen fellow workers get injured, have no problem understanding why wearing protective helmets is the smart thing to do.

## What OSHA Says About Helmets

Construction work is...well, construction. There are plenty of opportunities for head injuries. OSHA requires, in 29 CFR 1926.100, appropriate head protection to be worn "where there is a danger of head injury from impact, or from falling or flying objects, or from electrical shock or burns." Although specific situations are not covered by OSHA, the implication is that due to the very nature of construction methods and conditions, an unprotected head is at considerable risk of injury from striking or being hit by objects, or from electrical hazards, throughout the construction process, from initial lot preparation and excavation to completing the roof, walls, and interior. There are too many different hazards to keep track of—some of which crop up only temporarily—while a house is being put together. Individuals from numerous trades come and go, the quarters may be tight, and the potential for falls great, plus new fixtures, sharp corners, and similar items are being added daily that could easily win in a random encounter with a worker's skull.

OSHA further states that the employer is responsible for determining which employees are at risk for head injuries and which ones must wear effective head protection. In our case, effective head protection means helmets.

## What To Do

Require *all* persons entering the jobsite to wear appropriate helmets or hard hats. And the helmets you select might as well be Type II, Class B models, which also provide some protection from high-voltage electricity. Why should *everyone* wear helmets? you may ask. Surely there are some tasks and some areas at the jobsite having little or no risk of head injury. Consider those exceptions carefully. If you use a case-by-case analysis of job tasks for determining who must wear hard hats, you'll find many gray areas where employees will opt not to use helmets when they should otherwise be protected, such as when walking through hazardous parts of the site en route to a "safe" area. Start making exceptions for employees, and soon you'll be arguing with workers about "walking through a certain part of the site to get to a work area," or "because I didn't think I'd be working there, I didn't carry my helmet today," and so on. Better to just insist that hard hats be worn on the jobsite. It's a lot simpler and safer to have an across-the-board rule that everyone—employees, contractors, realtors, potential buyers, regulatory officials, and yourself—entering and/or working at an active, under-construction site wear a helmet. Period. After a while the wearing of a helmet will become so natural that individuals will feel uncomfortable without them, or they won't even realize they have them on. Notify subcontractors and others of your helmet rule in advance, and/or have extra helmets available at the site. Post "Hard-Hat Area" signs to remind employees and to notify subcontractors and visitors of the rule.

## Enforcing Helmet Rules

Just the fact that you provide your workers with helmets, and instruct employees on when to wear them, is not good enough. You've also got to enforce the rule. If you don't make it happen, and a head- or neck-related accidental injury occurs, you could be found at fault for allowing such a lax environment to develop and go unchecked. Even if you've dispensed discipline in the past to employees for not wearing helmets, that's still not good enough. Your employees *must* wear helmets on the job. Management has got to make that happen. This concept holds true for other protective measures as well. OSHA considers a company's management responsible for making the company's employees follow safe procedures. Even if an employee supplies his or her own helmet, and it's the correct type and style, the employer must still make sure that the helmet is in good condition, correctly worn, and properly maintained.

## Helmet Standards, Types, and Classes

Standards recognized by OSHA for protective helmets purchased after July 5, 1994 can be found in *American National Standards Institute Personnel Protection—Protective Headwear for Industrial Workers—Requirements,* which is ANSI Z89.1-1986. For helmets purchased before that date, check with *American National Standards Institute Requirements for Industrial Head Protection* (ANSI Z89.1-1969), and *Requirements for Industrial Protective Helmets for Electrical Workers* (ANSI Z89.2-1971).

On all helmets, be sure to look for where the manufacturer's name and the ANSI designation and class are marked on the inner shell.

### Helmet types

There are two basic types of protective helmets we're interested in:

Type 1 helmets (see Fig. 18.1) include those with a full brim not less that $1^1/_4$ inches wide.

Type II helmets (see Fig. 18.2) include those without brims but having a bill or peak in the front to help protect the eyes and face. This is the type of helmet generally preferred throughout the construction industry.

### Helmet classes

Beyond types, helmets are further described as belonging to one of four classes or groups:

Class A helmets are manufactured for general service and protection against impact hazards, and they also offer some defense against low-voltage electrical current (tested at 2,200 volts phase-to-ground). They're water-resistant and slow to burn when exposed to fire.

Class B helmets combine impact-resistance with some high-voltage protection (tested at 22,000 volts phase-to-ground). The voltages the helmets are tested at, however, do not mean that an employee would be safe when encountering electricity at those levels. The helmet is intended to provide protection against accidental (not intentional) contact with electricity. The maximum voltage that a hard hat will protect against depends on numerous factors, starting with the

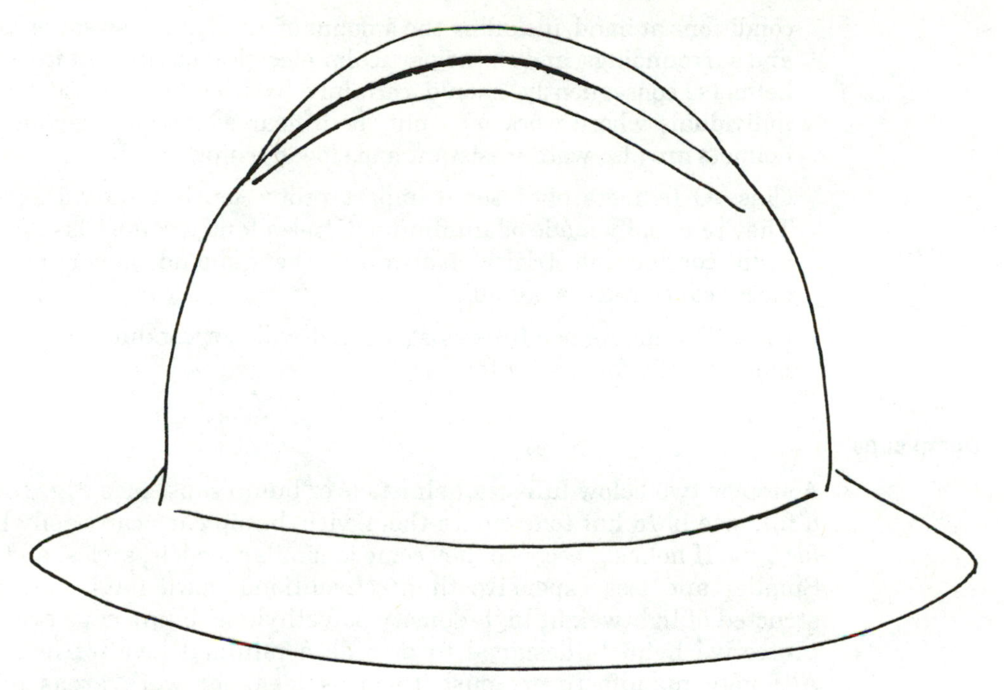

**Figure 18.1** Type I Hard Hat.

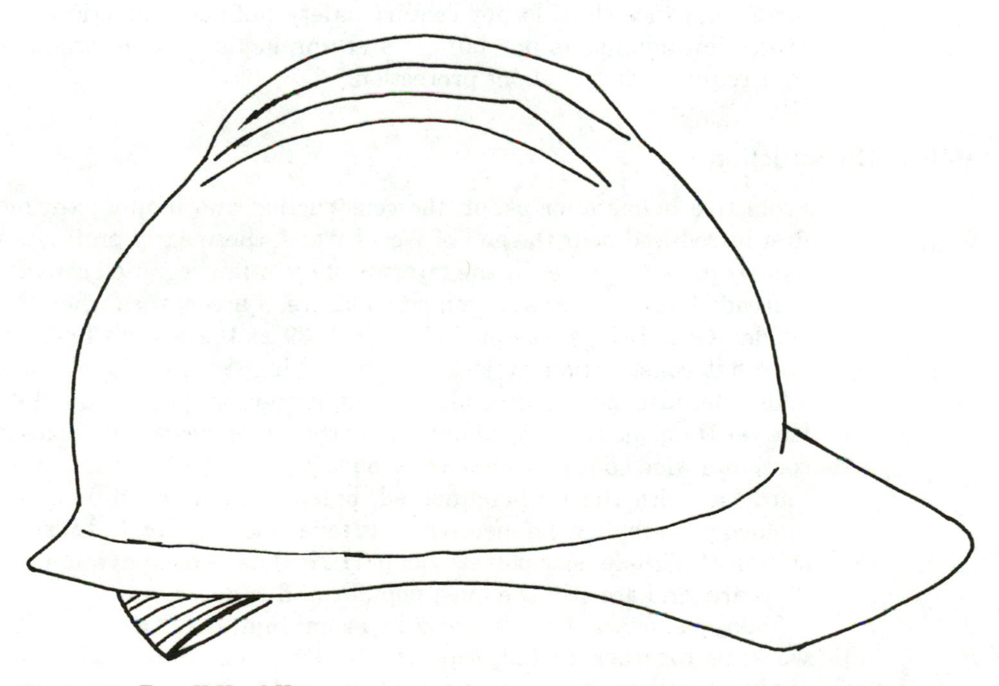

**Figure 18.2** Type II Hard Hat.

conditions at hand, including the amount of moisture or sweat on the employee and surroundings, and on the particular electrical equipment involved. Class B helmets, consequently, should certainly be used by electrical workers and individuals whose work may put them near electrical components. Class B helmets are also water-resistant and slow-burning.

Class C helmets offer some impact protection but no voltage protection. They're usually made of aluminum. Consequently, some Class C helmets can even conduct electricity. Naturally, they should never be used when electrical contact is possible.

Class D helmets are fire-resistant and will not conduct electricity. They're made chiefly for firefighters.

### Bump caps

A step or two below full-size helmets are "bump caps" (see Fig. 18.3). Without a full-size hard hat to compare them with, bump caps can easily be mistaken for Type II helmets because they come in similar models, styles, and even colors. Smaller and less expensive than conventional hard hats, and usually constructed of lightweight high-density polyethylene, bump caps are scaled-down protective helmets designed to provide a minimal level of head protection. Although manufacturers push them as ideal for work areas not requiring ANSI hard-hat protection, don't even consider them for construction use. The financial savings aren't a fair trade for the greatly reduced levels of protection they offer. They're usually advertised "for use in areas with low or obstructed head clearance that do not require safety helmet protection," or "to provide protection against minor bumps and abrasions. Not designed for situations that require safety helmet protection."

## Helmet Construction

Protective helmets for use in the construction and manufacturing trades were first introduced near the end of World War I; these early prototypes were meticulously pieced together by the layering of resin-impregnated canvas strips. About a decade later, "hard hats" gained widespread acceptance when San Francisco's Golden Gate Bridge was publicized in 1939 as the world's first "start-to-finish" hard-hat construction project. The qualifying "start-to-finish" may have been added because several other large construction projects of the 1930s, including Hoover Dam, also used hard hats. Over the years, newer materials and advanced computer-aided design technologies have replaced the original hand-built canvas hard hat with the well-engineered, lightweight, color-impregnated, injection-molded polyethylene helmet available today (see Fig. 18.4). Indeed, helmets consisting of high-density polyethylene (HDPE) have been available for more than 30 years, and are still the most popular and widely used.

Today's safety helmet is actually a combination of two separate components working together to help absorb the force generated when a falling item strikes the helmet (or when your head and helmet strike some sharp, heavy,

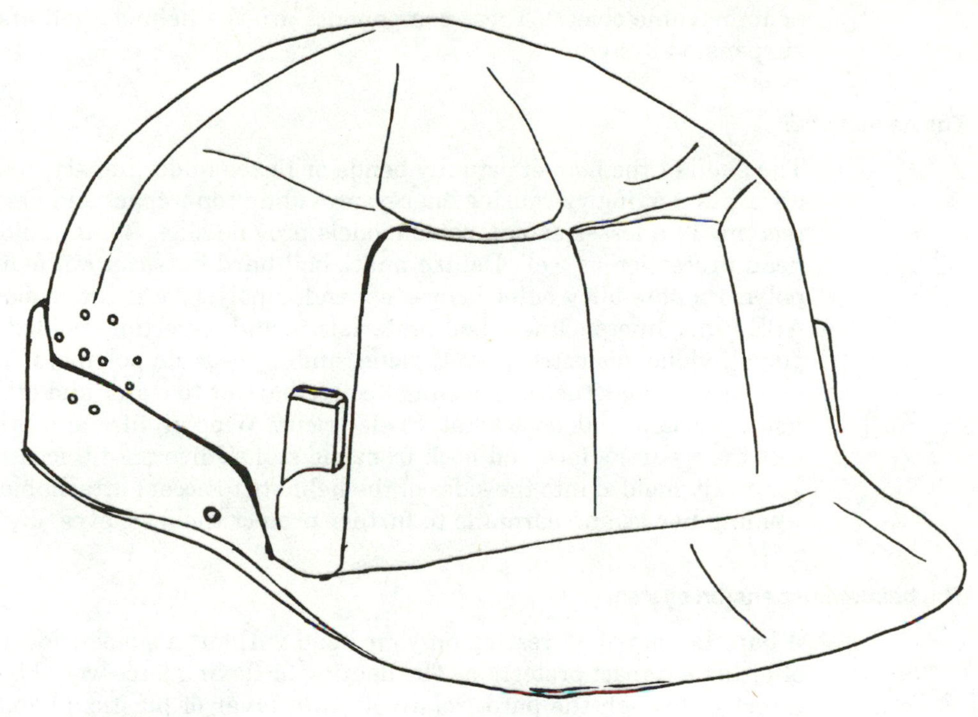

**Figure 18.3** Bump Cap.

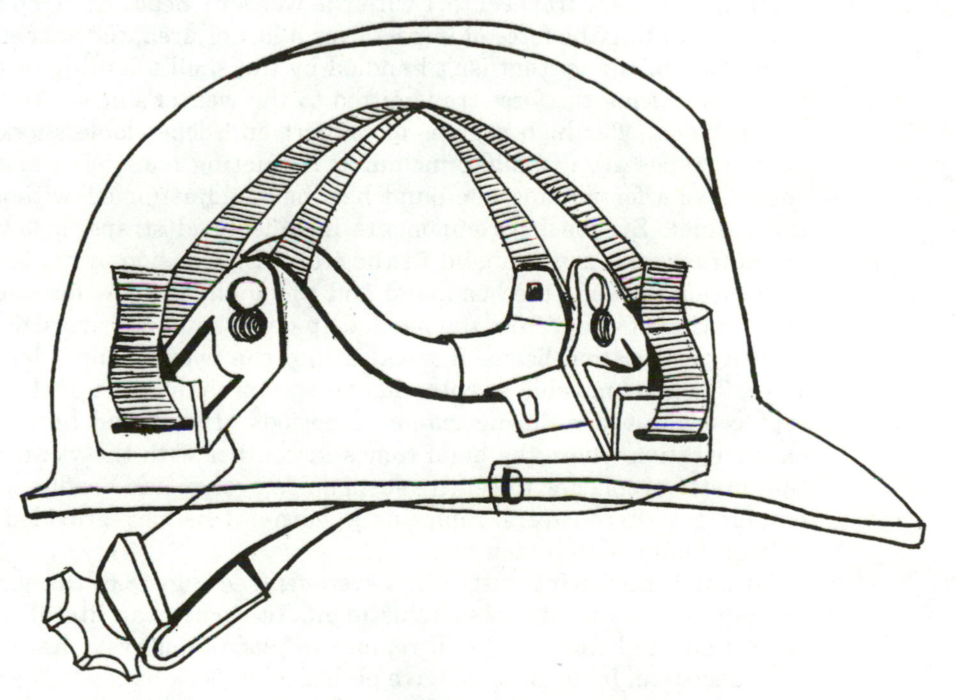

**Figure 18.4** Helmet Construction.

or immovable object). These components are the helmet shell and the helmet suspension system.

### The helmet shell

The shell of the helmet actually bends or flexes under impact, resulting in the absorption of injury-causing energy, providing top-impact and penetration protection. To a lesser extent, some models provide side, frontal, and back-of-the-head protection as well. Deluxe multishell hard hats contain features such as polycarbonate alloy outer layers, expanded polystyrene inner foam cores, and ABS inner liners. Other shell materials include injection-molded high-density polyethylene, dielectric polyethylene, and high-grade polymers. The shell also acts, in certain situations, as an effective barrier to water and other liquids, to hot materials, and, somewhat, to electricity. Wide profiles and brims help protect the wearer's face and neck from the sun. Universal-fit accessory slots are generally molded into the sides of the helmets to accept attachable face shields, welding hoods, and earmuffs to further protect the face, eyes, and ears.

### The helmet suspension system

A bare helmet shell resting on your head without a suspension system would offer little impact protection. The denting or flexing force would be transmitted directly through the hard, relatively thin layer of plastic against your skull. Thus the need for an inner suspension system—a framework that holds the shell up and away from contact with the wearer's head. By stretching slightly and distributing the force of impact over a larger area, the suspension system helps absorb energy that isn't handled by the shell's denting or flexing, thus further reducing the force transmitted to the wearer's head. All told, suspension systems offer high degrees of comfort and dependable shock absorption. Several types are available, including ratcheting four-, six-, and eight-point suspensions for making one-hand headband adjustments without removing the helmet. Suspension components include head straps, a low nape strap in the rear, and a sweatband in the front. Suspension systems also provide ventilation between the headband and the shell, to allow for cooling and the evaporation of sweat from the head. Clip-on sweatbands are also available, to absorb perspiration before it trickles into the eyes or onto glasses (see Fig. 18.5). These removable sweatbands on the forehand part of the band can be replaced as needed during extended periods of heat and humidity, to avoid skin irritation where the band comes in contact with the wearer's head. For that matter, replacing the entire suspension system and headband provides an economical alternative to replacing a worn-out helmet, provided the original helmet shell is still intact.

Although ratcheting suspension systems (see Fig. 18.6) can provide almost unlimited size and tightness adjustments (a wearer can literally bend upside down and the helmet will still remain in place), not every hard hat features such a system. Instead, some have pinlock slide/lock bands that are adjustable in $^1/_8$ hat-size increments (see Fig. 18.7).

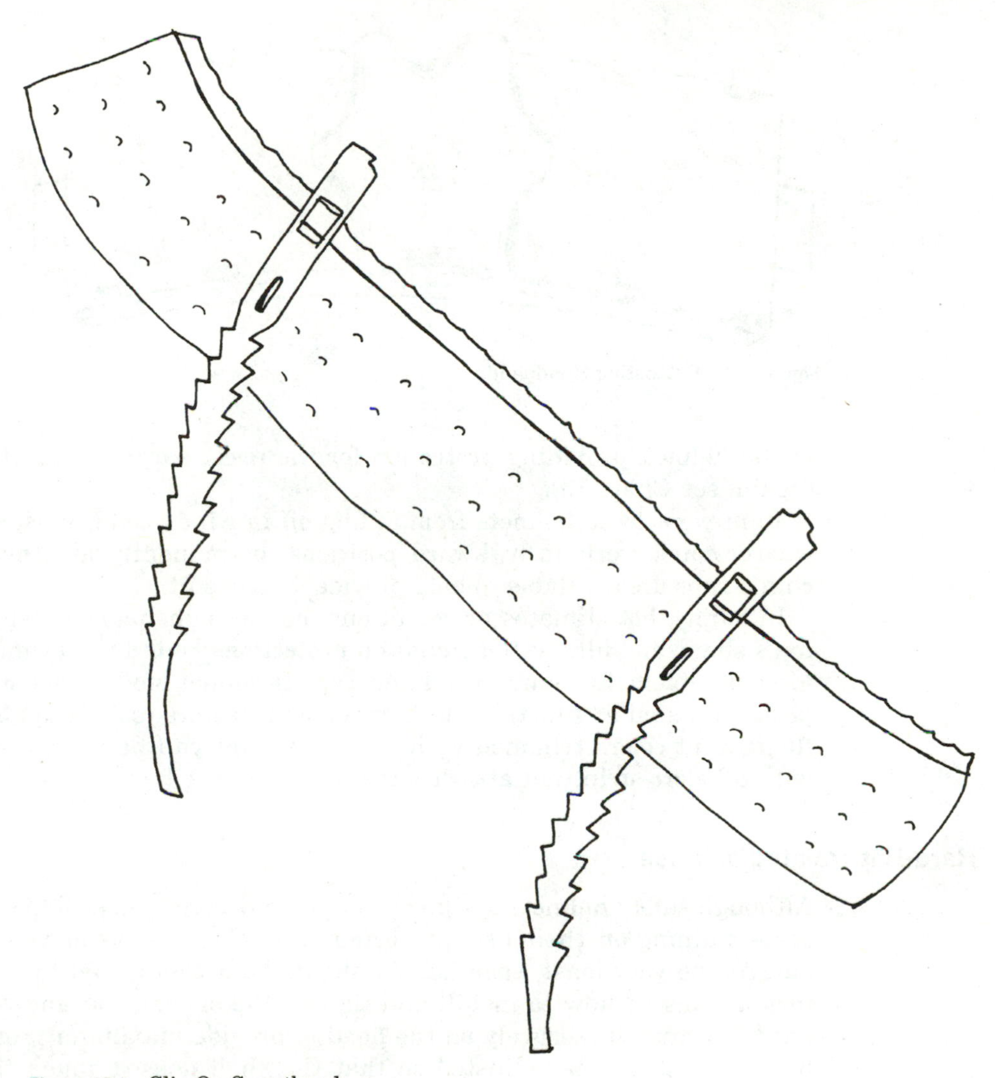

**Figure 18.5** Clip On Sweatband.

## Helmet Accessories

For cold-weather conditions, twill, fleece, and other thermal liners are designed to be worn beneath hard hats to keep the wearers warm (see Chap. 36). The liners slip onto the helmet's suspension system in a way that permits air to circulate through the material, creating a warm-air barrier between the wearer's head and shell without bulky weight. Winter liners protect the forehead, ears, part of the cheeks, and nape of the neck. Some have nose- and mouthpieces that can easily be fastened in place with Velcro tabs. Liners are simple to install and remove, and they are available for practically all suspension systems. A different kind of "exterior" liner for keeping warm with hard hats is a lightweight knitted head warmer that pulls over the top

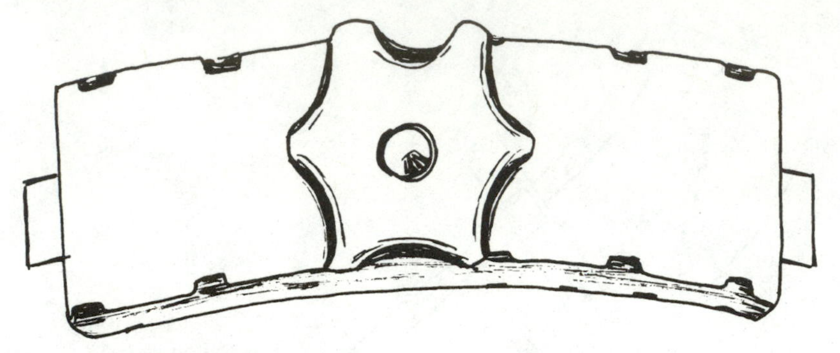

**Figure 18.6** Ratcheting Headband.

of the helmet, providing protection for the neck, ears, and most of the face (again, see Chap. 36).

To prevent loose helmets from falling off in windy conditions, or when the wearer must work in awkward positions, permanently attached two-point chin straps are available to help provide a secure fit.

In sunny, hot climates or conditions, helmet sunshades extend the hard-hat's shade and ultraviolet radiation protection around the brim. Sunshades are available to fit many Type I and Type II helmet models (see Fig. 18.8). To perform a similar function, neck or nape protectors (see Chap. 36), some in fluorescent colors trimmed with reflective tape, can be purchased complete with a Velcro strip that attaches to the back of the hard-hat shell.

## Hard-Hat Training and Use

Although safety helmets are fairly simple in design, you still need to provide some training on their use, inspection, and care whenever you hand them out. At the very least, each helmet should be accompanied by a copy of the instructions on how to install and tighten the suspension and headband. A hard hat must fit securely on the head to provide maximum protection. The headband should be adjusted so that the shell doesn't touch the wearer's head. A helmet should be worn with the bill facing forward to provide additional protection for the face, not backwards (unless worn with a welding hood). ANSI performs its experimental rating tests with the helmet facing forward, so it won't certify hard hats when they're worn backwards. Also, helmets should not be worn over sweatshirt hoods or other hats—the incorrect fit will greatly reduce the helmet's ability to absorb energy.

Proper helmet use makes a good topic for toolbox meetings. Explain to your employees how protective helmets will resist penetration, absorb the shock of a blow, and, with Class B models, provide some protection against high-voltage electrical shock. It's a good idea to periodically refresh employees on hard-hat use, and especially to remind them about the importance of wearing their hard hats whenever they perform specific tasks where hazards exist. Plus, as with all safety training, retraining on helmet usage should occur whenever an employee displays, through his or her behavior, a lack of understanding about when and how a hard hat should be worn or maintained.

## Helmet Inspections

Frequent helmet inspections are a necessity. Insist on them, and periodically do some checking yourself. Have wearers inspect their helmets daily for dents, cracks, or other damages and imperfections. The ultraviolet rays in sunlight will gradually weaken a plastic helmet, and the color will tend to fade. Signs that helmets are at the end of their useful life include discoloration, a chalky appearance, or brittleness. Check for brittleness by flexing the brim. If there's no give or flex, there could be a loss of impact, penetration, or electrical resistance. A replacement helmet should be issued to take the place of any helmet that doesn't pass inspection, or that has sustained a heavy blow (even if no damage is apparent). Destroy or dispose of the discarded helmet so that no one else picks it up without realizing that it's been taken out of service.

One odd problem with hard hats is a tendency of some employees to become attached to their helmet, and never want to replace it. Their hard hat becomes a familiar friend, like a well-broken-in shoe, helping to keep wind, snow, rain, sunlight, falling objects, and too-close-for-comfort fixtures from smacking into their noggins. This ding, they'll point out, is from the morning that Clyde's pliers bounced dead-center from the tip of a roof top; and that hairline crack is from when the helmet fell onto a concrete driveway and *almost* got flattened by a front-end loader, and so on.

## Helmet Care

Helmet care is simple. Wearers should:

- Remove the sweatband periodically, and wash or replace it.
- Clean the shell and suspension system by dipping them in warm, mild soapy water, scrubbing, rinsing, and drying.

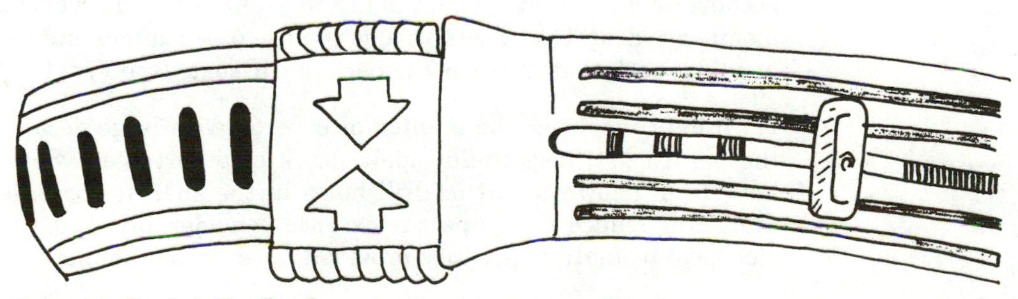

**Figure 18.7** Pin-lock Headband.

**Figure 18.8** Hard Hat Sunshade.

- Store hard hats away from sunlight and high heat. It's a big mistake to keep helmets on the rear window shelf of a car, because sunlight and extreme heat may cause degradation that will lessen the strength of the shells and the amount of protection they afford. Hard hats stored like that can also become dangerous flying objects in the event of an emergency stop or automobile accident. One alternative is the use of a commercially manufactured hard-hat rack that fits over the back of car seats (see Fig. 18.9).

- Hard hats should not be painted or covered with large decals or stickers—the paint or stickers could hide a crack or other defect. Some individuals have been known to cut or drill holes in the shell to increase ventilation. That will reduce the impact-resistance considerably, and will negate the electrical insulation qualities provided by an intact helmet.

## Helmet Limitations

Caution: Although hard hats are great pieces of protective equipment, they should *never* give wearers a false sense of security. Safety helmets have their limitations. Even manufacturers admit that while helmets have been engineered to provide as much protection as possible, realistically, they can only be expected to protect their wearers against relatively minor hazards such as head bumps against fixed items, falling small tools and objects, sparks, and accidental minor brushes with electricity. To think that a helmet will protect a wearer from a metal toolbox that falls from a second-story level is a totally unreasonable expectation. Nor will hard hats provide much protection from side impacts. Because of those limitations, hard hats cannot be substituted for effective safety practices at worksites having any of the hazards mentioned previously in this chapter.

When all is said and done, the greatest problem with hard hats is usually one of behavior—when and how they're being worn, and if they're being worn at all.

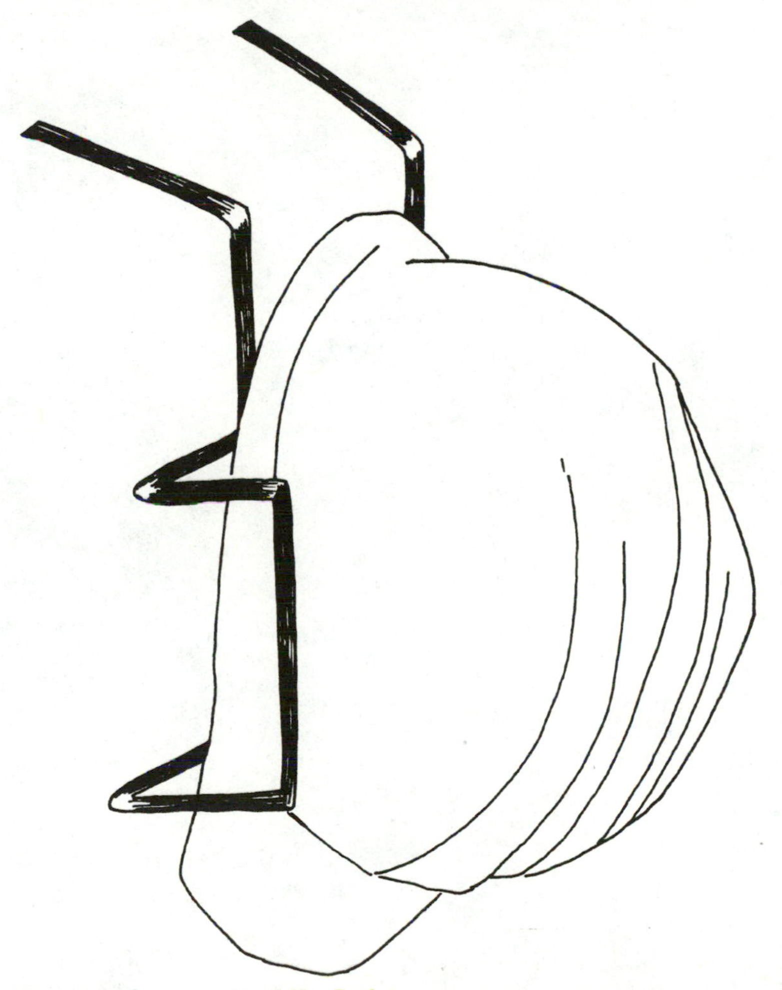

**Figure 18.9** Seat-back Hard Hat Rack.

Chapter

# 19

# Eye, Face, and Respiratory Protection

## Quick Scan

1. Employees, subcontractors, vendors, and visitors should wear (at least) safety glasses with side shields while working or visiting an active construction site.
2. Select and provide appropriate eye and face protection for employees working with or near striking tools, grinding or abrasive tools, burning or welding tools, dusts or airborne particulates, chemicals, and all other equipment, materials, and methods that pose eye hazards.
3. *Insist* that the safety eyewear in Quick Scans Nos. 1 and 2 above be worn when needed.
4. Instruct employees to wear dust masks or respirators wherever there is a possibility of nuisance or harmful dust, fumes, smoke, gases, mists, or similar respiratory hazards.

## Why Eye Protection?

Despite what you may think, there's limited recourse for recovery from serious eye injuries. Many kinds of incidents can cause irreversible damage to a person's sight. Indeed, the eyes are rather fragile—and especially vulnerable in a worksite environment where fragments of metal or concrete or wood fly about.

Most severe worksite optical injuries are caused by foreign bodies entering the eyes: bits or pieces of metal, wood, plastic, stone, masonry, or whatever material is being worked with. You know what it feels like to get a stray eyelash anywhere on the whites of your eyes. Well, foreign bodies in the cornea are infinitely more painful and potentially damaging. When injured, the cornea heals extremely slowly because of its limited blood supply; and once its integrity is "run through," bacteria can enter the inner eye and cause additional infection and damage.

Many eye injuries happen too quickly for victims to react; they take place quicker, in fact, than the "blink of an eye." And foreign bodies can be propelled unexpectedly from practically any direction.

It's ironic that, for such small organs, the eyes seem like big targets at work. Look closely into a mirror at one of your own eyes. There's nothing much in the way of natural protection, is there? Oh, you can close your eyelids—but there's not much substance to an eyelid, either.

According to the Bureau of Labor Statistics, the top cause of eye injuries in American workplaces is objects or particles—foreign bodies—entering the eye.

### How the eyes work

Simply put, a human eye is shaped like a tennis ball, hollow inside with three different layers of tissue surrounding the main fluid-filled cavity. The entire visual system resembles a closed-circuit television or monitoring system. The eyes receive and are stimulated by light energy from what is being "seen." Signals from the images formed on the retina are then sent through optic nerves to the brain. While sight occurs in the eyes alone, vision is the interplay between the eyes and the brain. The information "seen" by the eyes is conveyed to the brain not as pictures but as nonvisual electrical impulses carried by the optic nerves. The brain then reconstructs these impulses and interprets the images. The picture we see as we look out into the world is actually not in our eyes, but in our brains. How that occurs is an incredibly complex process.

The positioning of our eyes is significant as well. Human eyes are located in the front of the head to provide effective binocular vision. Binocular vision—seeing in three dimensions—is the product of the two eyes working together. It's what allows depth perception. That's why it's so important to have *both* eyes in working order. People who lack binocular vision must judge distance by using clues from other sources—space visible between moving cars, for instance, or other facts known from past trials and errors. Peripheral vision—our ability to see to the left and right of normal straight ahead vision—is also vital for normal functioning in the work environment. This is especially critical for backhoe and heavy equipment operators, and for automobile, truck, and other over-the-road drivers.

## Face and Eye Protection

There are many different kinds and models of face and eye protection. Some of the more common ones include safety glasses with side shields, goggles, and face shields.

### Safety glasses

Safety glasses are designed primarily to protect the wearer from impact hazards. The frames and lenses of most safety glasses are more rugged than those of ordinary dress glasses. Numerous safety lens options are available, including single-vision lenses, bifocals, trifocals, and "no-line" progressive power lenses. Not many years ago, wearers of safety glasses frequently complained that the lenses distorted their peripheral vision, causing headaches and dizziness. New wraparound designs (see Fig. 19.1) for lenses have eliminated those earlier problems of peripheral distortion, and consequently this style is growing rapidly in popularity. Wraparound safety glasses are also becoming more common for home use, with many employees preferring to use the same type of lens in both settings.

Numerous functional lens coatings are designed to meet the needs of special work environments. The various coatings help resist abrasion, block ultraviolet radiation, and reduce reflections and glare. Choices of lens material typically include glass, hard resin plastics, and polycarbonate. Wearers used to favor glass, which is fairly scratch resistant. Today the plastic alternatives are growing in popularity; plastics not only have much better impact resistance than glass but are lighter in weight.

The next consideration for safety glasses is the use of side shields. Detachable side shields may be as strong as the front safety glasses lenses, and fit the frame stems well, allowing no large gaps around the eyes. Half side shields provide side coverage, while full side shields afford both side and top protection. Safety glasses are available with a variety of shielding options:

- without side shields
- with detachable half side shields
- with integral or built-in half side shields

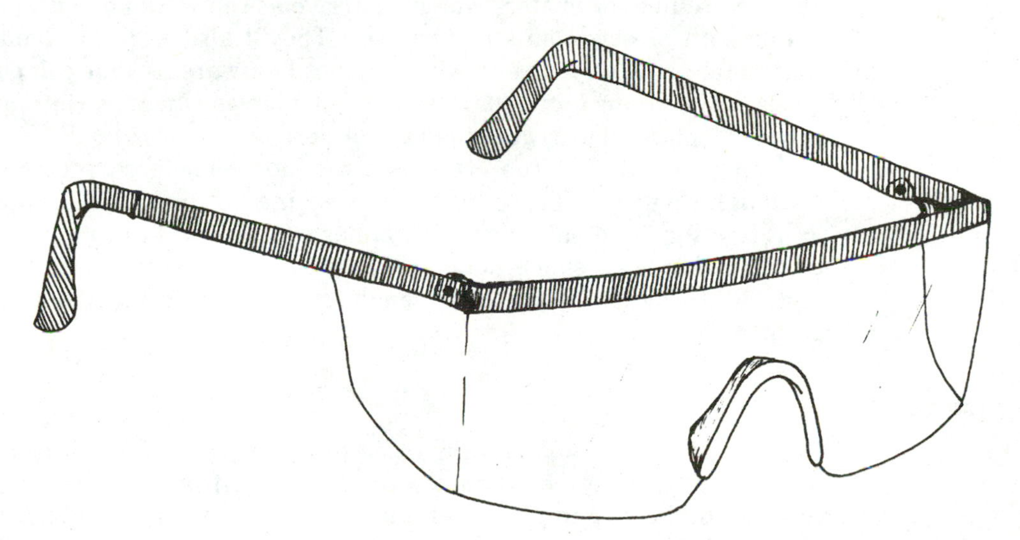

**Figure 19.1** Safety glasses.

- with detachable full side shields
- with integral full side shields
- in the wraparound style having side shield protection designed into the one-piece lens.

Pure and simple, employees on worksites should always wear side shields on their glasses, for the added impact prevention these provide. This extra protection is too valuable to do without. The Bureau of Labor Statistics has concluded that most eye injuries occur because the eye protection being used did not provide sufficient angular coverage. Because the laws of probability recognize that greater space exists from which airborne particles and objects can come at an angle, it's more likely that an incident will not be caused from directly in front of the eyes, but will come from the side, from beneath, or from above.

An important consideration is that of choosing between permanently fixed and removable side shields. There are pros and cons to both types. Individuals who always wear glasses anyway because of the need for corrective lenses are more likely to wear their safety glasses for home and all-around use if they've got removable side shields. That's a plus. On the other hand, removable side shields can easily be forgotten, misplaced, or lost. And because they're removable, there's a temptation for some workers to go without them. The workplace issue here can be liability if a worker who removes the side shields or fails to wear them becomes injured. Thus, from an employer's point of view, permanently affixed full-coverage side shield protection is more desirable.

An alternative to regular safety glasses or prescription safety glasses are inexpensive plain or nonprescription safety glasses that are designed to be worn alone or over regular prescription glasses (see Fig. 19.2). These safety glasses are fine for visitors who may arrive at the worksite with no glasses at all, or with prescription street glasses. They'll also work for vendors or subcontractors new to your site who may not be aware of your company's safety glasses rule, or for individuals on your own crews who may (infrequently) forget or misplace their regular safety glasses.

Keep in mind that contact lenses are not considered protective devices. Although they're slightly better than wearing nothing at all, contacts are too light and flimsy to help much. If eye hazards are present, regular eye protection (glasses with side shields, or goggles) must be worn in addition to or instead of contact lenses. Dust caught underneath contact lenses can cause painful abrasions.

## Goggles

Goggles are also available in many types and styles. They provide a tighter, more comprehensive fit than safety glasses with side shields, because they conform to the face immediately surrounding the eyes, shielding them from a variety of additional hazards, including dust and chemicals. Many goggles—

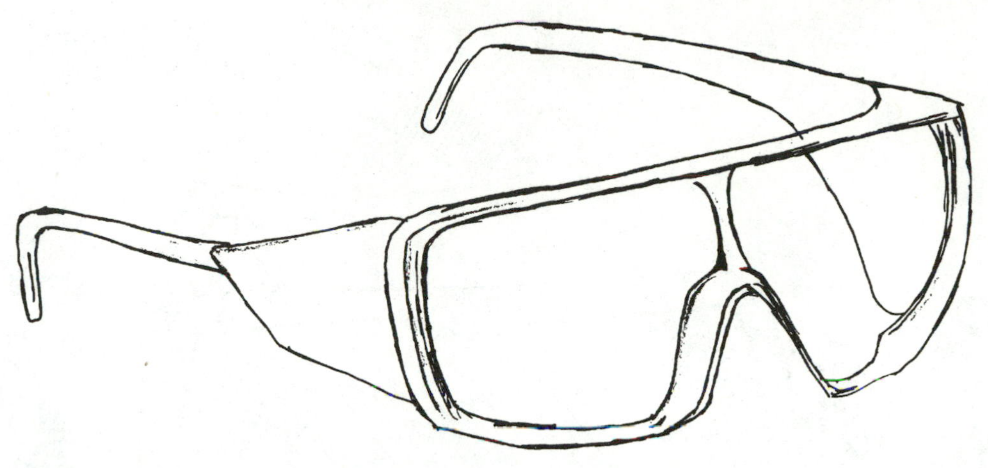

**Figure 19.2** Plain safety glasses.

but not all—offer impact protection. "Cover" goggles contain one-piece replaceable lenses (see Fig. 19.3), while "cup" goggles feature separate, usually round lenses, for each eye (see Fig. 19.4). In general, goggles are available in the following permutations:

- Cover goggles with or without ventilation
- Cup goggles with or without ventilation
- Tinted cover burning or welding goggle/lenses
- Tinted cup burning or welding goggles/lenses

Some goggles are designed to fit over regular prescription glasses and safety glasses when the hazard of dust, particles, or splashing liquid is present.

Here's a sample advertisement for the Encon 160-degree Chemical-Splash/Impact Goggles: "Top-of-the-line comfort and protection. Styled to offer long-lasting protection from impact and splash. Lenses screen out 99 percent of harmful UV radiation up to 380 nm. Fog-free lenses are chemically treated to prevent fogging in extreme temperatures and humidity. Specifications: Curved, polycarbonate 0.060-in lens resists scratches and provides full 160-degree peripheral vision. Fits over large prescription glasses. Choice of standard or fog-free models. Fog-free goggles are available in a variety of body and lens colors to balance the light intensity of your workplace."

And another for Unvex's "Stealth Goggles": "Sleek, revolutionary low-profile design. Unique toric lens system uses both a horizontal and vertical radius. The toric lens optimizes peripheral vision, plus it gives a close and optically correct fit. Snap-in lens replacement system. Specifications: Indirect ventilation system has upper and lower vents that channel a constant flow of air over the lens—reducing fogging and improving vision. Molded from a soft, flexible plastic, the Stealth gives you a customized fit by conforming and sealing to a wide variety of facial contours. Impact-resistant polycarbonate lens has 4C+ coating

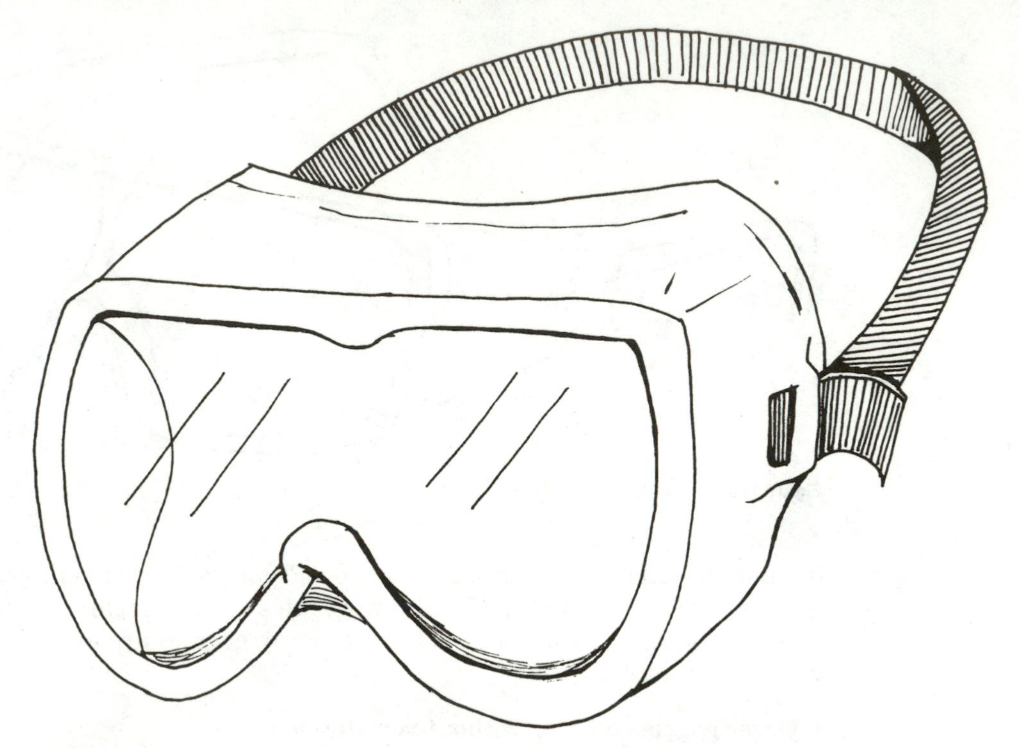

**Figure 19.3** Cover goggles.

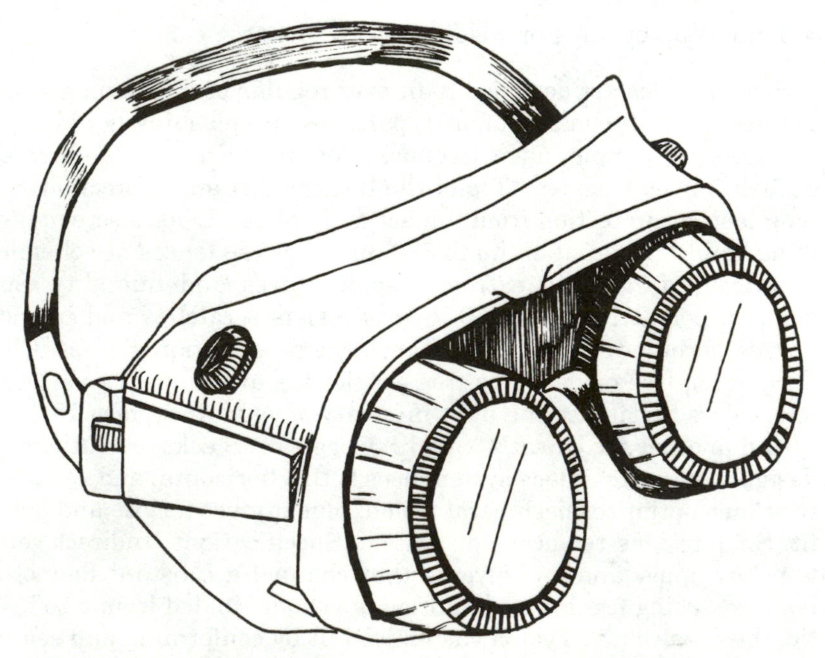

**Figure 19.4** Cup goggles.

to minimize fogging, scratching, static, and provide extra UV protection. Headband is easily adjusted using hinges. This eliminates any gaps when worn with hard hats, helmets or other safety headgear."

## Face shields

Clear polycarbonate, acetate, and other plastic face shields (see Fig. 19.5) are protective devices intended to shield the wearer's face, or portions of the face in addition to the eyes, from impact, dust, particles, and splash hazards. Face shields are considered secondary protectors and must always be used along with other, primary protectors such as safety glasses or goggles. The face shield is usually held on the wearer's head by an adjustable band that allows the shield to be conveniently flipped up and down, or it can be inserted similarly onto the universal attachment slots of a hard hat. A face shield is a must when working with power tools and equipment such as a grinder or concrete hammer; if a grinding wheel or chisel or other working tool should snag or break and be thrown into the air, a full shield would protect the wearer's face

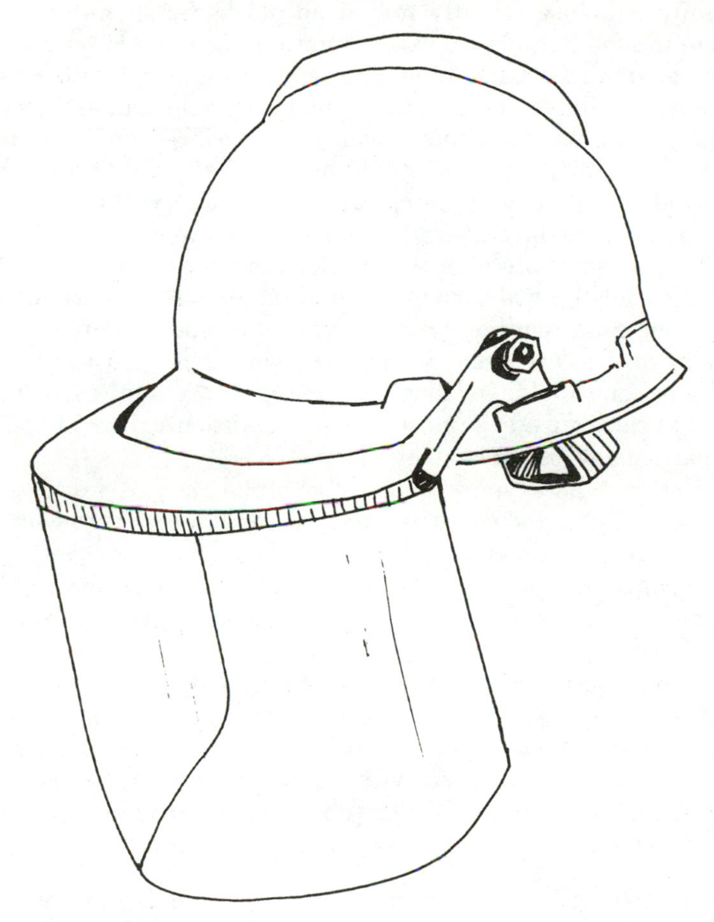

**Figure 19.5** Face shield.

and neck. Although they're adequate for blocking chips, shavings, and particles, they don't offer protection against heavy impact or from objects that may fly up around the shield. Because the face shields are considered less "personal" in nature than glasses or goggles, they're typically kept near the tools or equipment they will be used with, and shared by whomever happens to be using the tools or equipment at the time. Some of the larger shields extend far enough to protect the neck as well. Others include chin guards for added protection.

## Optical Radiation Protection

Optical radiation concerns in the workplace come mainly from welding and torching operations. Electric arc welding, gas welding, cutting, and torching produce ultraviolet (UV) and infrared (IR) radiation that can quickly injure the eye. I know something about radiation hazards from experience: one day a number of years ago my six- and eight-year-old daughters came home crying, their eyes covered with emergency room medical patches and their faces reddened from what appeared to be sunburn. They couldn't see. My wife and the kids had visited a family that had recently rented an old barbershop-turned-apartment. While there the children played with miniature toy animals set up in a bay window illuminated by a barber's "blue" lamp—once used for sterilizing scissors and other hair-cutting implements. Luckily, no permanent damage was done. How could their optical "sunburn" have been prevented? Through the use of tinted lenses. (Of course, the "sunburn" would have still affected their faces.)

Studies indicate that long hours spent in the sun without proper eye protection increase a person's chances of developing eye disease. A rule of thumb is that whenever someone is in the sun long enough to get a suntan or a sunburn, ultraviolet-light-absorbent tinted safety lenses should be worn. Ordinary sunglasses cannot protect eyes from special situations in which intense light sources are encountered. Be aware that exposure to arc welding, tanning lights, snowfields, or looking directly at the sun (especially during a solar eclipse) can cause a painful corneal condition called photokeratitis, or worse, a permanent loss of central vision.

Depending on where your company's employees are working, and what they're doing, shading of the protective lenses is necessary when protection from optical radiation is needed. That can mean protection from sunlight, from welding, burning, and torching. There are tinted burning and welding goggles and glasses with side shields, and hot works helmets with tinted and impact protective lenses.

Certain construction crews and trades may employ the above-mentioned processes in day-to-day tasks. Or the hot-works tasks may be associated more closely with vehicle and heavy equipment repair or other maintenance work. Hazards to be concerned with while performing the above activities include impact of particles/foreign objects, radiant energy or heat, and intense light sources:

- *Impact of particles / foreign objects.* Cover-plate lenses are designed to protect filter-plate lenses, so filter-plate lenses won't become pitted, scratched, and difficult to see through. While affording some impact resistance, cover plates must not be used instead of safety plates or safety glasses for eye protection. They are designed for use *with* safety glasses, shields, or lenses that can be worn inside or outside of the bright-light filtering lens, if necessary.

- *Radiant energy or heat.* Typical welding and burning lenses provide sufficient protection against heat from most hot-work processes.
- *Bright light created by arcing or burning.* Special tinted filters can prevent the eye from damages possible from exposure to concentrated bright points of light. The choice of filter shades may be made on the basis of visual acuity, and can vary from individual to individual, depending upon how sensitive to light his or her eyes are. The lens selector guide accompanying this chapter contains suggested shade numbers for various hot-work tasks. The higher the shade number, the greater (or darker) the protection. Shades darker or more dense than those shown for various operations may be needed for individuals more sensitive to light.

When installing lens filter plates, safety plates, or other lenses, the markings on those lenses/plates must be readable from the outside of the product. In other words, you should be able to read the identifying words or numbers on the goggle or mask lenses by looking from the outside. If ID markings are readable from the inside of a mask or pair of goggles, the lenses have been installed backwards.

### What OSHA says

OSHA has prepared some tables focusing on eye protection. The first one we'll include is Table E-1 (see Fig. 19.6), which presents various eyewear options. The accompanying OSHA table on applications (see Fig. 19.7) references those eyewear options with various operations or activities and eye hazards. The third OSHA table, Table E-2 (see Fig. 19.8) supplies filter lens shade numbers recommended for protection against radiant energy.

### Prescription lenses

Employees who wear corrective lenses should have their prescription incorporated in the glasses if required. Prescription lenses must meet the same requirements for impact protection as nonprescription lenses, which may limit selections. Other forms of eye protection such as goggles should permit corrective lenses to be worn without loss of vision. I don't know about you, but I'd be practically worthless without my corrective lenses. Many employees are in a similar situation.

Some companies take care of everything. They send their employees to an optometrist, select safety frames, and provide the complete package. Other companies require employees to get their own prescriptions, then select frames. Some eyewear manufacturers offer devices into which a prescription lens can be placed behind the protective lens. The employer is responsible for purchasing prescription eye protection when it is warranted. In some cases,

**TABLE E-1 - Eye and Face Protector Selection Guide**

| | |
|---|---|
| 1. | GOGGLES, Flexible Fitting - Regular Ventilation |
| 2. | GOGGLES, Flexible Fitting - Hooded Ventilation |
| 3. | GOGGLES, Cushioned Fitting - Rigid Body |
| 4. | SPECTACLES, Metal Frame, with Sideshields [1] |
| 5. | SPECTACLES, Plastic Frame - with Sideshields [2] |
| 6. | SPECTACLES, Metal-Plastic Frame - with Sideshields [1] |
| 7. | WELDING GOGGLES, Eyecup Type - Tinted Lenses [2] |
| 7A. | CHIPPING GOGGLES, Eyecup Type - Clear Safety Lenses |
| 8. | WELDING GOGGLES, Coversepc Type - Tinted Lenses [2] |
| 8A. | CHIPPING GOGGLES, Coverspec Type - Clear Safety Lenses |
| 9. | WELDING GOGGLES, Coverspec Type - Tinted Plate Lens [2] |
| 10. | FACE SHIELD (Available with Plastic or Mesh Window) |
| 11. | WELDING HELMETS[2] |

[1] Non-side shield spectacles are available for limited hazard use requiring only frontal protection.
[2] See Table E-2, in paragraph (b) of this section, Filter Lens Shade Numbers for Protection Against Radiant Energy.

**Figure 19.6** OSHA Table E-1, Eye protection options.

the employer will pay all of the prescription costs. In many cases, the employer will pay for basic selections. Beyond those, the employee pays the balance for more expensive frames.

### Fit, comfort, and style

No matter what kind of protective eyewear is chosen, a proper fit is essential. Otherwise the protection itself could inhibit the use and could even result in an injury. An improper fit will make it more likely the individual will choose not to wear the eyewear.

Weight, shape, and fit are important. Negative characteristics decrease the likelihood that employees will wear their glasses. The lens material can be plastic, polycarbonate, or glass. Glass is the heaviest and least popular. The shape and fit has changed dramatically, at least during the past decade; modern protective eyewear features a fixed bridge, adjustable frames, and adjustable temples (which run behind the ears and hold the eyewear in place). The degree of safety provided across the industry has been established, so the manufacturers are now concentrating on style. People won't wear protective eyeglasses that are ugly; they want models similar to street wear, with fashionable frames and lenses. Wraparound products, which offer great peripheral protection, are especially popular now.

The days of the "Buddy Holly look-alike glasses"—the ones with small thick lenses and heavy black frames—have given way to new designs that have been finely engineered with fashion in mind. The most welcome development to the employee is the fashion revolution in safety eyewear. The availability of new fashionable styles that are indistinguishable from everyday prescription glasses has transformed employee's attitudes about safety eyewear. A side benefit is that employees start to wear safety eyewear at home, helping to reduce off-the-job injuries, lost time, and medical claims. Popular frame designs, fashion tints, and a variety of lens styles are just a few of the available options.

Here is some advertising copy for one of the more modern models: "Bouton's New 7000 UFO Mercury. User friendly optics (UFO) so advanced, they're out of

**Applications**

| Operation | Hazards | Recommended protectors |
|---|---|---|
| Acetylene-Burning, Acetylene-Cutting, Acetylene-Welding | Sparks, harmful rays, molten metal, flying particles | 7, 8, 9 |
| Chemical Handling | Splash, acid burns, fumes | 2, 10 (For sever exposure add 10 over 2) |
| Chipping | Flying particles | 1, 3, 4, 5, 6, 7A, 8A |
| Electric (arc) welding | Sparks, intense rays, molten metal | 9, 11, (11 in combination with 4, 5, 6, in tinted lenses advisable) |
| Furnace operations | Glare, heat, molten metal | 7, 8, 9 (For severe exposure add 10) |
| Grinding-Light | Flying particles | 1, 3, 4, 5, 6, 10 |
| Grinding-Heavy | Flying particles | 1, 3, 7A, 8A (For severe exposure add 10) |
| Laboratory | Chemical splash, glass breakage | 2 (10 when in combination with 4, 5, 6) |
| Machining | Flying particles | 1, 3, 4, 5, 6, 10 |
| Molten metals | Heat, glare, sparks, splash | 7, 8, (10 in combination with 4, 5, 6, in tinted lenses) |
| Spot welding | Flying particles, sparks | 1, 3, 4, 5, 6, 10 |

**Figure 19.7** OSHA Applications for eye protection chart.

**TABLE E-2. - Filter Lens Shade Numbers for Protection Against Radiant Energy**

| Welding operation | Shade number |
|---|---|
| Shielded metal-arc welding 1/16-, 3/32-, 1/8-, 5/32- inch diameter electrodes | 10 |
| Gas-shielded arc welding (nonferrous) 1/16-, 3/32-, 1/8-, 5/32-inch diameter electrodes | 11 |
| Gas-shielded arc welding (ferrous) 1/16-, 3/32-, 1/8-,5/32-inch diameter electrodes | 12 |
| Shielded metal-arc welding 3/16-, 7/32-, 1/4-inch diameter electrodes | 12 |
| 5/16-, 3/8-inch diameter electrodes | 14 |
| Atomic hydrogen welding | 10-14 |
| Carbon-arc welding | 14 |
| Soldering | 2 |
| Torch brazing | 3 or 4 |
| Light cutting, up to 1 inch | 3 or 4 |
| Medium cutting, 1 inch to 6 inches | 4 or 5 |
| Heavy cutting, over 6 inches | 5 or 6 |
| Gas welding (light), up to 1/8-inch | 4 or 5 |
| Gas welding (medium), 1/8-inch to ½-inch | 5 or 6 |
| Gas welding (heavy), over ½-inch | 6 or 8 |

**Figure 19.8** OSHA Lens Shade Numbers Table E-2.

this world. Never before have you seen or felt anything like the 7000 UFO Mercury. It's sleek, cool, featherlight. The three-way adjustable temple provides the ultimate custom fit while the frameless replaceable lens gives you a feeling of virtual weightlessness. But more than that, the Mercury is also your best line of defense against job-related eye hazards, even under the most extreme conditions. Aggressive new lens designs eliminate potential distractions allowing full peripheral vision. The impact resistant polycarbonate lens protects your eyes from 100 percent of harmful UV radiation and meets ANSI Z87.1 standards." The accompanying illustration calls to mind a skier's sleek sunglasses.

Another advertisement touts the benefits of a new model of frameless safety glasses: "Now you can see clear to the edge. Introducing frameless safety glasses with no distractions in sight. Until now, virtually all protective eyewear has been designed with the same built-in problem. Frames. Irritating lines of distraction that keep you from seeing clearly in all directions. The Phoenix Edge put an end to all that. These completely frameless safety glasses have eliminated those aggravating lines of distraction in a new generation of eyewear that's as handsome as it is practical. Safety glasses so clear and light you feel like you're not wearing glasses at all. A simple, clean style that looks as good as it feels. And for the first time, frameless, unobstructed sight in all directions, with no distractions. That's the new view from the Phoenix Edge."

Here's yet another ad: "Safety looks great in Vail. Peak Performance Eyewear. You'll love the performance. Total optical protection. *Prism perfect* lens correction technology. Superior peripheral optics. Clear, comfortable vision. Sleek, ultralightweight design. Integrated side impact protection.

Action inspired and engineered for durability, Vail meets the Z87.1-1989 ANSI performance standard without side shields. Offered in a variety of frame colors and lens/tint options, Vail is the best choice for indoor or outdoor jobs. Our TSR Gray lens tint meets color traffic signal recognition requirements of ANSI Z80.3-1986. They'll love the look. Vail ensures wearer acceptance, excellent protection and a comfortable fit for every face...."

### Eyewear training

Employees must be trained on whatever eyewear is provided. Supply training on the proper application of equipment—including when to use it, where to use it, how to don/doff and wear the eye protection. Also instruct employees on the limitations of the equipment and in its proper care and maintenance. To meet OSHA requirements, a written record that indicates proper instructions have been provided on the use and care of the PPE must be maintained and signed by the employer and the workers, and kept on file.

### Eyewear care and maintenance

One often-overlooked aspect of a safety eyewear program is the importance of training employees in the proper care of their glasses. For example, an employee who works with abrasive dust particles may get into a habit of cleaning the lenses of his safety glasses by wiping them off with his shirt. That can radically shorten the lenses' useful life. Lens cleaning tissues and fluid should be provided at the worksite.

### Getting compliance

It's not enough to simply provide safety eyewear to employees. It's not enough to train the employees on how and when and where to use safety eyewear. Nor is it enough to tell employees to wear safety glasses. It's management's responsibility to provide eyewear, to provide training on the eyewear, and to see that employees wear their protection while on the job, and aren't just wearing them when someone is looking. Employees have got to understand *why* wearing protective eyewear is so important.

There are lots of reasons why people avoid or refuse to wear safety glasses or goggles:

"They're not comfortable." That was true, years ago—but not now. There are enormous numbers of comfortable safety glasses and goggles on the market.

"They're too distracting," This excuse is often used by individuals who don't routinely wear corrective lenses. Newer, wraparound models of safety glasses offer extremely clear side protection that also allows for unrestricted peripheral vision.

"They give me a headache." Unlikely, but this excuse commonly comes from the memories of apparent distortion from older models.

"They're ugly and make me look stupid." This does not have to be the case anymore. Oh, basic, nerdy safety glasses are still available, but major strides have been made in the development of fashionable safety glasses. Believe it or not, some individuals have been wearing their safety lenses at home, to sports events, even to social gatherings. Employees are more likely to wear safety glasses if the glasses look good. Provide safety eyewear that features style and comfort, and individuals will go from *having* to wear to *wanting* to wear eye protection. Lightweight materials and adjustability in the glasses and goggles have gone a long way to create more user-friendly products. Nylon and polycarbonate take color well, and allow tinted lenses which enable employees working in sunlight to have vision protection equal to sunglasses while still enabling them to see colored materials that are critical to correctly completing their tasks.

Again, key factors that help to ensure worker compliance and establish a successful program are a good, comfortable fit, stylish designs and a wide selection. These personal motivators help workers exercise some individuality and control and are far more successful in obtaining compliance than external regulations and procedures.

To support this individuality, allow employees to pick their own eyewear frame or model out of about four to six styles that you've preapproved. Let them participate in the selection process, and they'll be more likely to wear them on their own.

Ideally, your company eyewear program will fund both the prescription examination and the glasses. Make sure that employees understand that regular streetwear lenses don't afford the protection of their safety glasses. Although OSHA does not require employers to pay for prescription lenses, management has the responsibility to ensure that wearing regular glasses under safety glasses does not disturb the proper fit of either pair, and that prescription lenses fit into the chosen style of protective eyewear frames correctly.

## First Aid for Eye Injuries

Eyes are supersensitive to pain. A loose eyelash, a speck of dirt, or almost any foreign body in the eye can result in extreme irritation. Because eyes are so sensitive, it's difficult to gauge the extent of damage when injuries occur. Here are guidelines for rendering first aid for common eye injuries, as suggested by ophthalmologists:

*Impaled objects.* Do not remove any impaled objects. Whenever caring for a patient with eye injuries, cover both of the injured's eyes, even though only one of the eyes may be injured.

*Specks in the eye.* Avoid rubbing any speck or particle that's in an eye. Instead, lift the upper lid over the lower lid in hopes of allowing the lashes to brush the speck off the inside of the upper lid. Have the subject blink a few times to let the eye move the particle out.

*Blows to the eye.* An ice-cold compress should be applied as soon as possible, and kept on for about 15 minutes to reduce pain and swelling. A black eye or blurred vision can mean internal eye damage. Have the subject visit an ophthalmologist immediately.

*Chemical burns.* The injured eye should be flooded with potable water or with cleansing solution as quickly as possible. When using an eyewash, the eye should be held wide open, and the water or cleansing solution should directly flush the eye for at least 15 minutes. Medical assistance should follow.

*Cuts to the eye and lid.* The affected eye should be bandaged lightly, then medical help sought immediately. Never try to wash a cut eye or remove an impaled or stuck object. Avoid the application of pressure to an injured eye or eyelid and be careful not to rub the eye.

### Portable emergency eyewashes

The most effective first step in treating chemical contamination of the eye or skin is flushing or washing the affected area. Medical experts stress that immediate access to an emergency eyewash is critical. The potential for full recovery from chemical contamination of the eye is excellent—if the victim reaches an eyewash within 10 to 15 seconds.

Again, the duration and volume of flushing is also important, with a minimum time recommended for flushing of 15 minutes. Portable eyewash units are available that can be used at any worksite (see Figs. 19.9 and 19.10). They can be set up on a picnic table, kitchen counter, back end of a pickup truck, or similar location that's close by and fairly simple to access. The portable units contain exhaustible saline solution supplies. Portable units operate by gravity and do not need plumbing or pressurizing. Molded high-density polyethylene plastic reservoirs typically hold between 6 and 16 gallons of eyewash solution—which is available in concentrated form for ready mixing with water. The flushing mixtures are buffered, pH balanced, isotonic, saline solutions with preservatives that discourage bacterial or fungal growth.

Employees must know how to operate the units in areas they frequent, whether the units are kept in the back of a pickup truck or on a portable table set up at the worksite. Training is critical here, because an individual in an emergency situation will hardly be in a relaxed state of mind, and may, in fact, not be able to see clearly at the time. Also remember that emergency eyewash stations are not substitutes for personal protective gear such as safety eyewear, face shields, and protective clothing.

## Respiratory Protection

Respiratory protection is serious business. If your company deals with hazardous materials or concentrations of materials that are harmful to life and health, you need to have those hazards professionally evaluated by competent individuals, and you also need a respiratory protection plan suitable for your worksite(s).

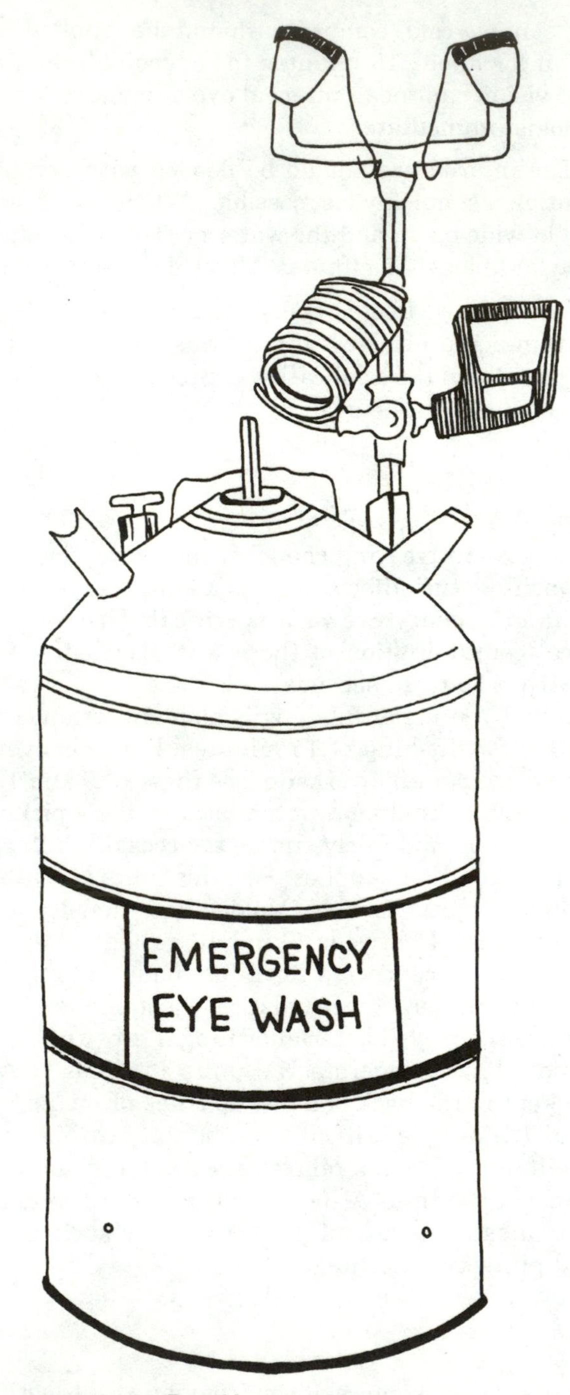


**Figure 19.10** Portable eyewash unit (#2).

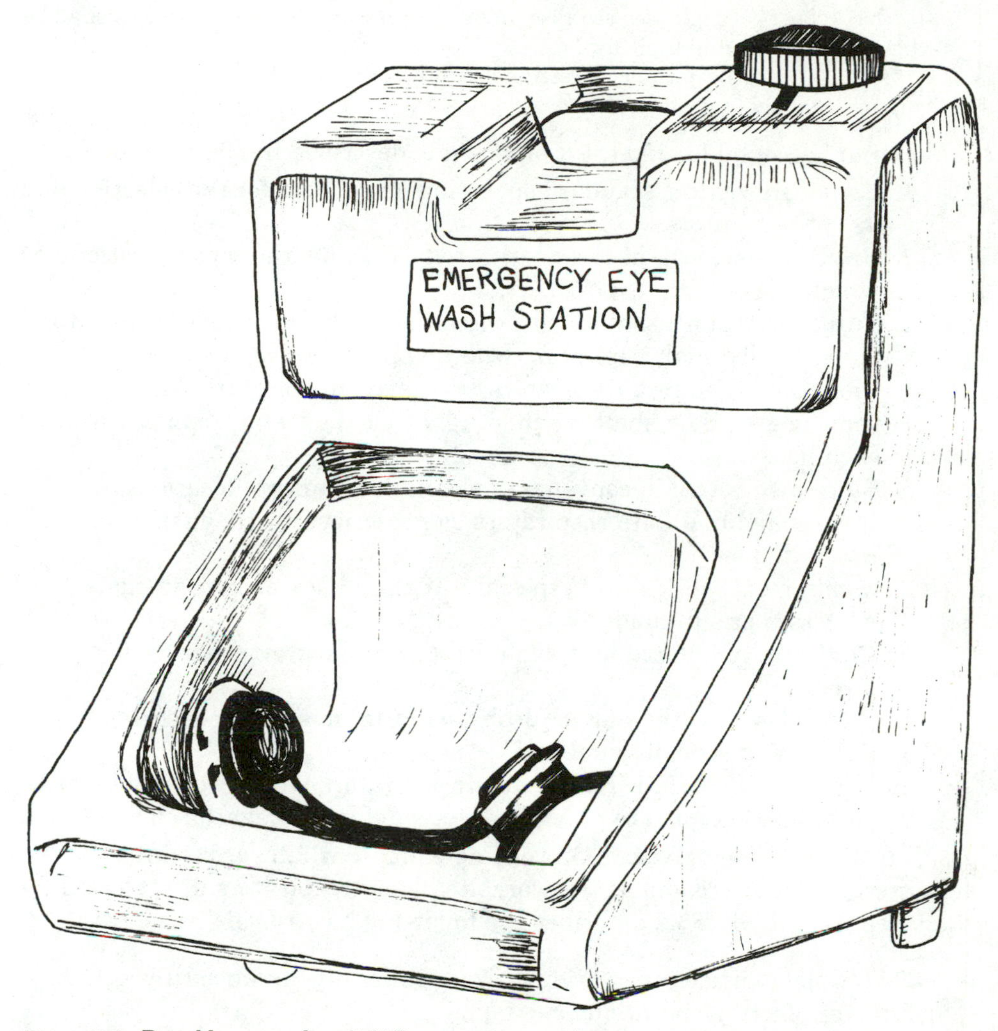


**Figure 19.9** Portable eyewash unit (#1).

## General respiratory considerations

Numerous types of dust masks and respirators are available. The chief respiratory hazard you need to protect your employees from is dust and related particulates. With all of the materials that are proving to be carcinogenic, it's best to require protection whenever in an atmosphere or near the generation of dusts, mists, fumes, gases, and other related respiratory hazards.

## What OSHA says

1926.103(a)(1) In emergencies, or when feasible engineering or administrative controls are not effective in controlling toxic substances, appropriate respiratory protective equipment shall be provided by the employer and shall be used.

1926.103(b)(2) Respiratory protective devices shall be appropriate for the hazardous material involved and the extent and nature of the work requirements and conditions.

1926.103(c)(1) Employees required to use respiratory protective devices shall be thoroughly trained in their use.

1926.103(c)(2) Respiratory protective equipment shall be inspected regularly and maintained in good condition.

A minimal acceptable respirator program is described in 1910.134(b):

1. Establish written, standard operating procedures for the selection and use of respirators.
2. Select respirators based on the hazards and the concentrations to which workers are exposed.
3. Train respirator users in both the uses and limitations of respirators.
4. Assign individual respirators where applicable.
5. Clean and disinfect respirators regularly. Respirators used by more than one worker shall be thoroughly cleaned and disinfected after each use.
6. Store respirators in convenient, clean, and sanitary locations.
7. Inspect and maintain respirators as prescribed by the manufacturer's instructions.
8. Monitor the work area, especially if there is a change in materials, processes or site conditions.
9. Continually enforce and evaluate the respirator program for effectiveness.
10. Work closely with your local physician to make sure each respirator wearer is physically qualified.
11. Use only NIOSH/MSHA certified respirators where applicable, or be otherwise accepted to provide protection for the hazards encountered.

Naturally, you must give careful consideration to whatever respiratory hazards are present or created at the worksite before respirators are selected for employee use. ANSI Z88.2 describes the following hazards as:

- *Dusts.* Solid particles in varying sizes, generally produced by grinding, drilling, or blasting operations.
- *Fumes.* Solid metal particles of extremely small size (generally less than one micrometer in diameter), produced by welding and smelting operations.
- *Mists.* Formed from atomized liquid with varying particle size, produced by spraying, plating, cleaning, and mixing operations.
- *Gases.* Like air, gases can spread freely throughout an area, sometimes displacing oxygen.
- *Vapors.* Gaseous states of liquid formed at room temperature by evaporation. Solvents are one example.

## Types of respirators

**Air-purifying respirators.** Air-purifying respirators are used only in environments that contain enough oxygen to sustain life, or at least 19.5

percent oxygen. They use special filters and cartridges that remove specific particulates, vapors, and gases from the air. In order to be effective, the level of contaminants must be within the concentration limitations of the respirator wherever and whenever it is used. The useful life of these respirators, filters, and cartridges depend on the concentration of contaminants in the air, the breathing volume of the wearer, the capacity of the filtering medium, and even the relative humidity and area ventilation. These respirators include disposable dust and particulate masks (see Fig. 19.11), maintenance-free respirators with permanent, factory attached cartridges (the entire respirator discarded when cartridges are exhausted), low-maintenance respirators with replaceable cartridges and prefilters, and reusable respirators available in both half- (see Fig. 19.12) and full-face styles with replaceable cartridges, canisters, filters, and individual parts (see Fig. 19.13).

**Filters.** Filters remove dust and other particulates, mists, and metal fumes by trapping them within the filter material. Filters should be changed before or when the wearer's breathing becomes difficult.

**Cartridges.** Cartridges usually contain carbon, which absorbs certain toxic vapors and gases. When the wearer detects any taste or smell that indicates the charcoal's absorption capacity has been reached, the cartridge is exhausted and incapable of supplying any more protection, and must be changed. Depending on the contaminants, cartridges can be used alone or in combination with a filter and filter cover. Cartridges are also available with a HEPA (High Efficiency Particulate Air) filter permanently attached.

Equipment using cartridges includes disposable respirators or dust masks designed for use against nontoxic particles and nuisance dusts only, in concentrations not exceeding OSHA permissible exposure limits. These respirators and dust masks must not be used to protect against substances such as silica, asbestos, cotton dust, or any other toxic dusts, fumes, mists, vapors, and gases. There's a huge variety of types and models available, and they're best reviewed with a distributor's sales representative who specializes in personal protective equipment.

**Supplied-air respirators.** These respirators are used in toxic environments, or in environments that are immediately dangerous to life and health (IDLH) as long as they're equipped with an air cylinder for emergency escape. They provide a supply of air, for extended periods of time, through a high-pressure hose that is connected to an external source of air such as a compressor, compressed air cylinder, or portable breathing pump.

**Self-contained breathing apparatus (SCBA).** Similar to the SCUBA concept, but designed for use on land instead of under water, SCBAs provide the highest level of respiratory protection available. They protect workers in oxygen-

**Figure 19.11** Disposable dust mask.

deficient atmospheres, in poorly ventilated or confined spaces, and in atmospheres that are immediately dangerous to life and health.

### Proper respirator face seals

Respirators should not be used when a good face seal cannot be achieved. What can prevent a good seal? Typical cause could be a beard or long sideburns, major scarring of the face, or the simultaneous use of incompatible prescription or safety glasses or goggles.

### Hazards respirators can protect against

Just a few of the hazards that properly matched respirators can protect against include:

- Asbestos
- Paint sprays and mists
- Welding fumes
- Wood dust
- Organic vapors
- Acid gases
- Formaldehyde
- Solvents
- Lead

OSHA has prepared a table to help with respirator selection. Table E-4 (see Fig. 19.14) lists respirator options in response to certain atmospheric hazards.

**Figure 19.12** Half-mask respirator.

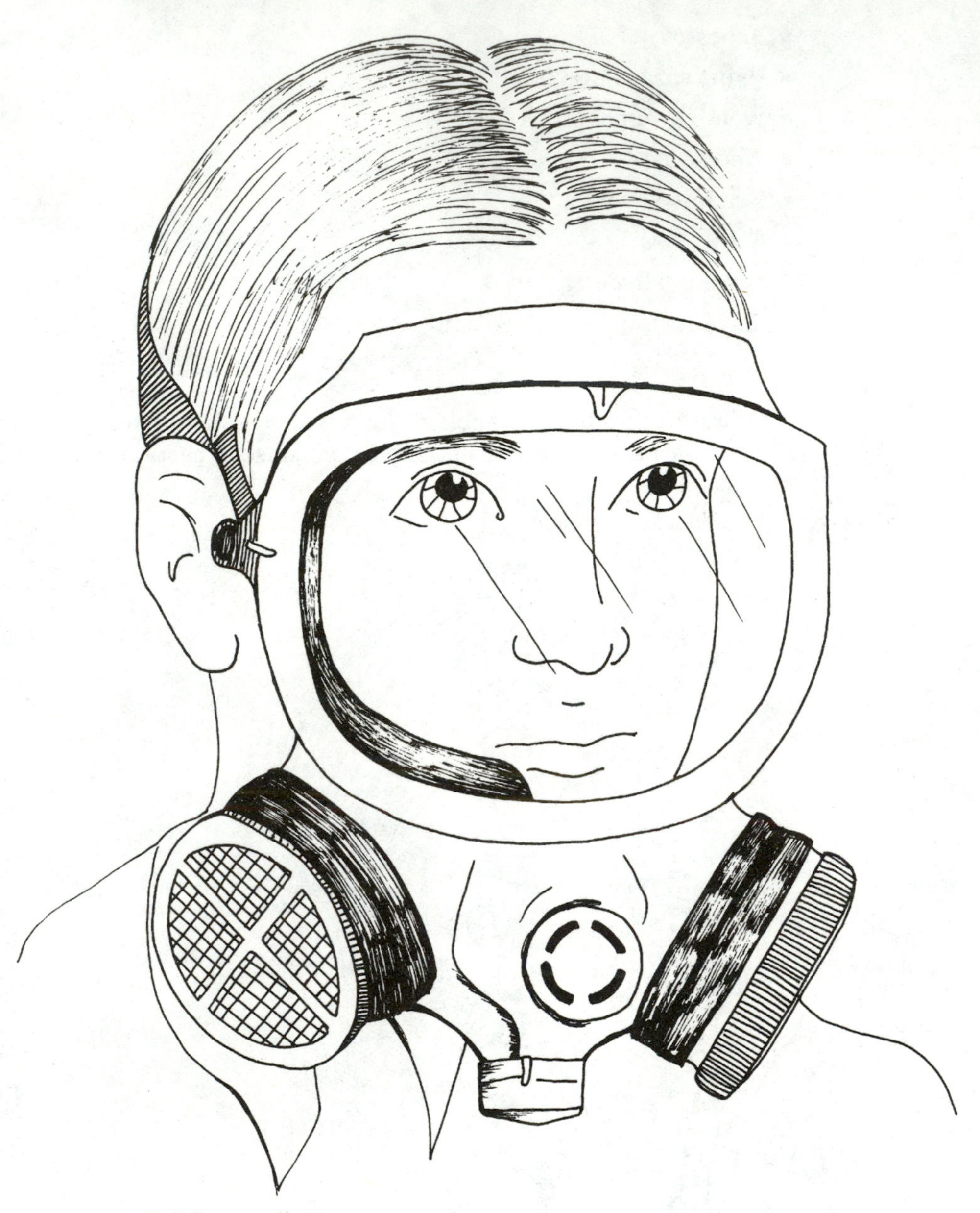

**Figure 19.13** Full-face respirator.

**TABLE E-4. - Selection of Respirators**

| Hazard | Respirator (See Note) |
|---|---|
| Oxygen deficiency | Self-contained breathing apparatus. Hose mask with blower. Combination air-line respirator with auxiliary self-contained air supply or an air-storage receiver with alarm. |
| Gas and vapor contaminants immediately dangerous to life and health | Self-contained breathing apparatus. Hose mask with blower. Air-purifying full facepiece respirator (for escape only). Combination air-line respirator with auxiliary self-contained air supply or an air-storage receiver with alarm. |
| Not immediately dangerous to life and health | Air-line respirator. Hose mask without blower. Air-purifying, half-mask or mouthpiece respirator with chemical cartridge. |
| Particulate contaminants immediately dangerous to life and health | Self contained breathing apparatus. Hose mask with blower. Air purifying, full facepiece respirator with appropriate filter. Self-rescue mouthpiece respirator (for escape only). Combination air-line respirator with auxiliary self-contained air supply or an air-storage receiver with alarm. |
| Not immediately dangerous to life and health | Air-purifying, half-mask or mouthpiece respirator with filter pad or cartridge. Air-line respirator. Air-line abrasive-blasting respirator. Hose-mask without blower. |
| Combination gas, vapor, and particulate contaminants immediately dangerous to life and health | Self-contained breathing apparatus. Hose mask with blower. Air-purifying, full facepiece respirator with chemical canister and appropriate filter (gas mask with filter). Self-rescue mouthpiece respirator (for escape only). Combination air-line respirator with auxiliary self-contained air-supply or an air-storage receiver with alarm. |
| Not immediately dangerous to life and health | Air-line respirator. Hose mask without blower. Air-purifying, half-mask or mouthpiece respirator with chemical cartridge and appropriate filter. |

NOTE: For the purpose of this part, *immediately dangerous to life and health* is defined as a condition that either poses an immediate threat of severe exposure to contaminants such as radioactive materials, which are likely to have adverse delayed effects on health.

**Figure 19.14** OSHA Table E-4 on respirator options.

# Chapter 20

# Hearing Protection

## Quick Scan

1. Train employees on the dangers of excessive noise exposure on the worksite.
2. Conduct a noise survey on worksite tasks having the loudest noise levels. Test sound levels inside mobile equipment cabs and near all loud activities, including operations with and near compressors, engines, and other power tools and equipment.
3. Provide comfortable noise-reduction personal protective devices such as earmuffs, earplugs, and canal caps; train employees on the protective equipment's capabilities, limitations, use, and care.
4. Wherever and whenever appropriate, enforce the use of hearing protection throughout the worksite.
5. Supervisors and other members of management must abide by hearing protection rules while working at or visiting the site.

## Hearing—An Easily Lost Resource

Years ago, construction workers thought nothing of operating all kinds of loud power equipment without wearing hearing protection. Now, generations of retired construction workers need to turn up their televisions and radios to deafening volumes in order to hear them. Their family members need to scream across dinner tables at them, and daily these hearing impaired retirees are missing out on the subtler sounds of life.

For the most part, their conditions developed slowly, over the years, without any perceptible warning. Because typical hearing loss is so incremental, the victims aren't liable to recognize what's happening until it's too late. The same thing is happening to today's rock music band members and concertgoers. Hearing loss from rock concerts and from playing music loudly under headphones is eventually going to take a heavy toll on today's youth. To a certain extent, it's also still happening to construction workers who are bucking hearing protection rules by failing to protect themselves.

### The ears

The human ear can be divided into three main parts: the outer ear, the middle ear, and the inner ear.

The outer ear collects sound waves and channels them through the eardrum to the middle ear. When the sound waves pass through the eardrum, they cause it to vibrate.

The middle ear consists mainly of three connected bones, the hammer, anvil, and stirrup. Vibrations passed along by the eardrum are transmitted through those three bones into the inner ear.

The inner ear, which is also called the *cochlea,* resembles a snail in shape, is filled with fluid, and is also equipped with millions of tiny hairs that act as sensors. As the sound waves pass by, they strike the hairs and cause the hairs to move. The hairs are attached to nerve endings that are, in turn, connected to the brain by an auditory nerve. The wave movements are thus turned into electrical impulses that are interpreted by the brain as sound.

### Sound

Sound is any pressure variation in air, water, or other substances that the human ear can detect. Sound travels in three-dimensional waves. It is measured in decibels (dBA) for loudness, and in hertz (Hz) for frequency. Not all sounds are detectable by the human ear. *Ultrasound* is an extremely high-frequency sound (like a dog whistle), and *intrasound* is sound at an extremely low frequency, commonly referred to as vibration. Noise levels for some common sounds are:

|  |  |
|---|---|
| Threshold of youthful hearing | 0 decibel |
| Threshold of good hearing | 10 decibels |
| Whisper | 20 decibels |
| Conversational speech | 60 decibels |
| Noisy office | 80 decibels |
| Alarm clock | 80 decibels |
| Factory | 90 decibels |
| Passing truck | 100 decibels |
| Auto horn | 120 decibels |
| Night club with live rock band | 120 decibels |
| Riveter, chipper | 130 decibels |
| Threshold of pain | 140 decibels |
| Aircraft carrier deck | 140 decibels |
| Jet engine | 160 decibels |
| Saturn rocket | 195 decibels |

There are several different types of noise:

Continuous noise, which is steady and never stops.

Intermittent noise, which occurs off and on.

Impact noise, which is a sharp burst of sound.

## Noise causes hearing loss

The damage caused by excessive noise, the most common cause of hearing loss, is often painless and gradual. The amount of hearing loss depends on how loud the noise is, and on how long the individual is exposed to it.

Excessive exposure to high levels of noise damages the tiny hairs in the inner ear—permanently bending them down so they can no longer spring back into their original position. Once the hairs are laid down, they can't be reactivated or reenergized. The noise can be damaging in several forms, including loud and prolonged sound, or a single very loud noise such as an explosion.

Experts also say that excessive noise may harm overall health by contributing to mental and physical stress. There's a cumulative effect that noisy surroundings may have on a person's mental state. Some individuals find it difficult to think when working in loud environments. The concentration needed to block out the noise while thinking takes a generous amount of their mental effort, so not much is left to deal with the actual tasks at hand. When they're not able to fully concentrate on their work, anxiety sets in. The anxiety breeds stress, and the background noise seems even louder to them. The situation escalates until physical symptoms manifest themselves, and the person simply cannot function properly.

Excessive noise can also be a major contributing factor to accidents. Loud environments tend to mask certain sound cues that when normally heard, help reduce hazard exposures at the workplace. An employee working near jackhammers or other power tool activities is less likely to hear another employee call out a warning, or to notice a request for help. Other employees will hurry their own work so they can leave a high-noise area as soon as possible.

## What OSHA says

OSHA has traditionally been a huge proponent of hearing protection, and it still is, with good reason:

> CFR 1926.52: Occupational Noise Exposure
>
> Protection against the effects of noise exposure shall be provided when the sound levels exceed those shown in Table D-2 of this chapter when measured on the A-scale of a standard sound level meter at slow response (see Fig. 20.1).
>
> When employees are subjected to sound levels exceeding those table levels, feasible administrative or engineering controls shall be used. If such controls fail to reduce sound levels within the levels of the table, personal protective equipment as required in subpart E, shall be provided and used to reduce sound levels with-

## TABLE D-2
## Permissible Noise Exposures

| Duration per Day, Hours | Sound Level dBA Slow Response |
|---|---|
| 8 | 90 |
| 6 | 92 |
| 4 | 95 |
| 3 | 97 |
| 2 | 100 |
| 1 1/2 | 102 |
| 1 | 105 |
| 1/2 | 110 |
| 1/4 or less | 115 |

**Figure 20.1** OSHA Sound level Table D-2.

in the levels of the table. If the variations in noise level involve maxima at intervals of 1 second or less, it is to be considered continuous.

In all cases where the sound levels exceed the values of the table, a continuing, effective hearing conservation program shall be administered.

When the daily noise exposure is composed of two or more periods of noise exposure at different levels, their combined effect should be considered rather than the individual effect of each. Exposure to different levels for various periods of time shall be computed according to the formula set forth in paragraph (d)(2)(ii) of this chapter.

1926.101 Hearing Protection

Wherever it is not feasible to reduce the noise levels or duration of exposures to those specified in Table D-2, ear protective devices shall be provided and used.

Ear protective devices, also called hearing protection devices (HPDs) are therefore mandated in high-noise environments, and must be fitted or determined individually by competent persons. Plain cotton, though once commonplace, is no longer acceptable for use as hearing protection.

Wearing hearing protective devices definitely results in an altered world of sound. Few people realize the dramatic effect cutting out just 10 to 20 decibels of sound can have on their normal hearing function.

Humans depend on their hearing as they consciously and unconsciously monitor the incessant world of sound and acoustic events that surround them. Reduce the ease with which a person can detect and hear this multitude of sound sources, and the effect can be profound.

### Recording hearing loss

Hearing loss can be recorded as an injury or an illness, depending on the cause. If a hearing loss results from an instantaneous event such as an explosion, it's considered an injury. The hearing loss is recordable, however, only if the affected employee receives medical treatment, loses consciousness, becomes restricted in work or motion, or is transferred to another job as a result of the injury.

If hearing loss results from anything other than an instantaneous event, it is considered an illness regardless of whether these qualifiers for injury recordability occur. All illnesses are recordable if there is a causal relationship between the job and the hearing loss.

To determine if a hearing loss suffered by an employee becomes a recordable incident, the injury or illness must meet the standard criteria, or there must be a work-related shift in hearing ability (an average reduction of 25 decibels or more at the 2,000, 3,000, and 4,000 Hz levels in either or both ears). That means to determine recordability in the latter manner, a baseline hearing ability evaluation must be done on an employee, to serve as a reference point.

## Hearing Protection

For all practical purposes, there are three main types of hearing protection: earmuffs, earplugs, and canal caps.

### Earmuffs

Most earmuffs (see Fig. 20.2) consist of a pair of plastic domes that cover the ears, with a connecting spring band that adjusts to various head sizes and ear positions and provides the tension to seal the domes against the head. Bands may be worn over the top of the head, to the rear of the head, or under the chin. Attenuation or performance characteristics vary in accordance with the position of the headband.

Cushions are attached to the domes where they contact the head. These may be foam-, air-, or liquid-filled and should conform to the irregularities of the surfaces they contact. The dome opening and inside dimension should be large enough to encompass the outer ear so that the earmuffs will properly fit.

Earmuffs possess several distinct advantages over other sound-reducing measures:

- Generally provide the greatest amount of attenuation.
- Are easier to fit.
- Are easier for some employees to use.
- Are easily monitored.
- Can be used with infected or collapsed ear canals.
- Can be reused.

Earmuffs also possess several distinct disadvantages:

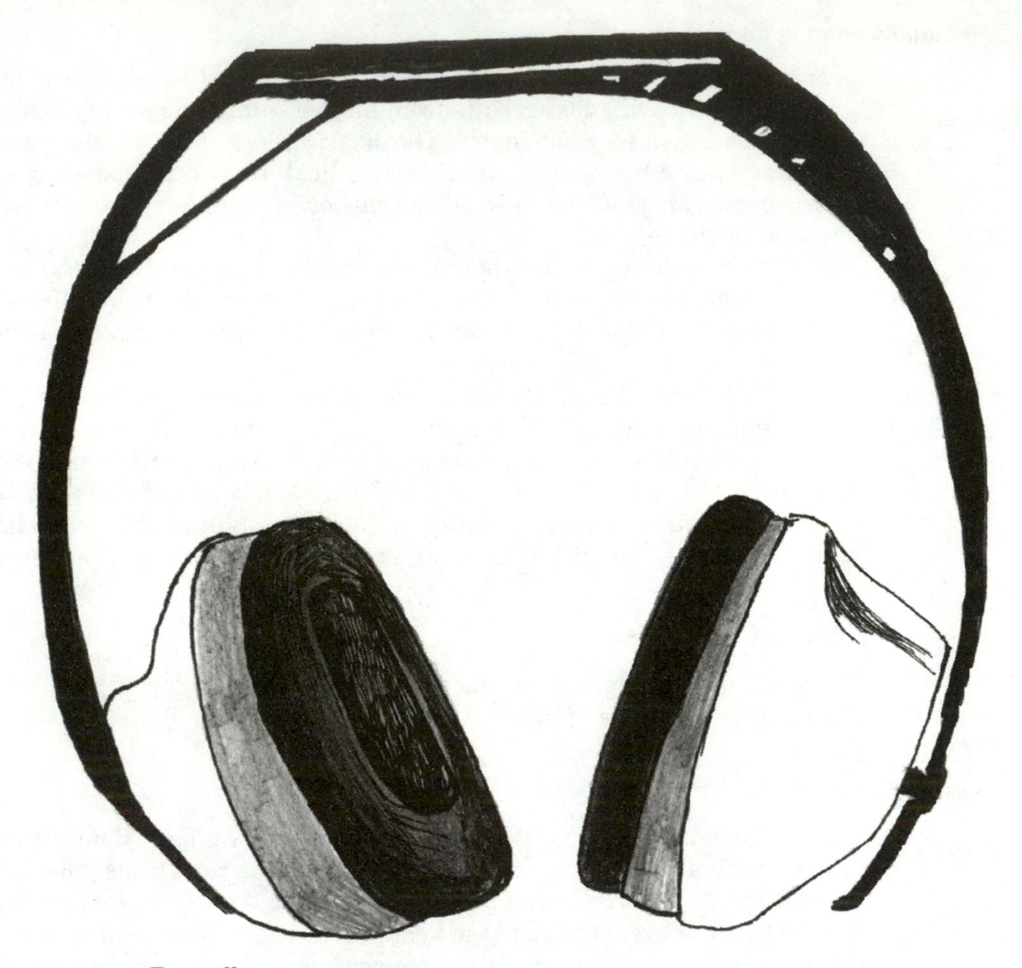

**Figure 20.2** Earmuffs.

- Are initially more expensive.
- Require a tight seal against the head and may be uncomfortable. (In actual use, earmuffs can yield as little as 60 percent of their rated protection because of wear and tear on the clamping force and ear seals.)
- Are usually bulky, heavy, and hot.
- May not be effective when worn with glasses or a hard hat.
- Can cause dermatitis where the cushion contacts the skin.
- May lose attenuation if the cushions become hard, cracked, or lose the fluid or air filling, or if the headband loses its tension.

Most modern earmuffs feature soft ear cushions, low headband force, a soft head pad, and even pressure distribution. Flexible suspensions alleviate pressure hot spots. Many earmuffs are designed for long-term wear in worksites where low frequency noise is a concern.

### Premolded earplugs

Insert- or plug-type protectors (see Fig. 20.3) fit directly into the ear canal. They come in many configurations and are made of rubber, plastic or silicones. Proper fit depends on the ability to make contact along the entire circumference of the ear canal. This generally requires an outward pressure be exerted on the insert to form a proper seal. Earplugs or inserts are supplied in two general configurations:

1. Premolded, sized, in bullet or cylindrical shapes. Some are custom-fitted to the exact shape of the typical human ear canal opening. Some come in standard sizes, while others are individually moldable. Many styles are disposable and should be discarded after one or several uses, depending on how much dirt they're exposed to.
2. Premolded, universal fit. Manufactured with two or more flanges of progressively larger diameters on the same stem to accommodate practically any size ear canal.

Like earmuffs, earplugs possess several distinct advantages:

- Are less expensive than muffs or canal cap protectors.
- Are less cumbersome to wear, carry, and store.

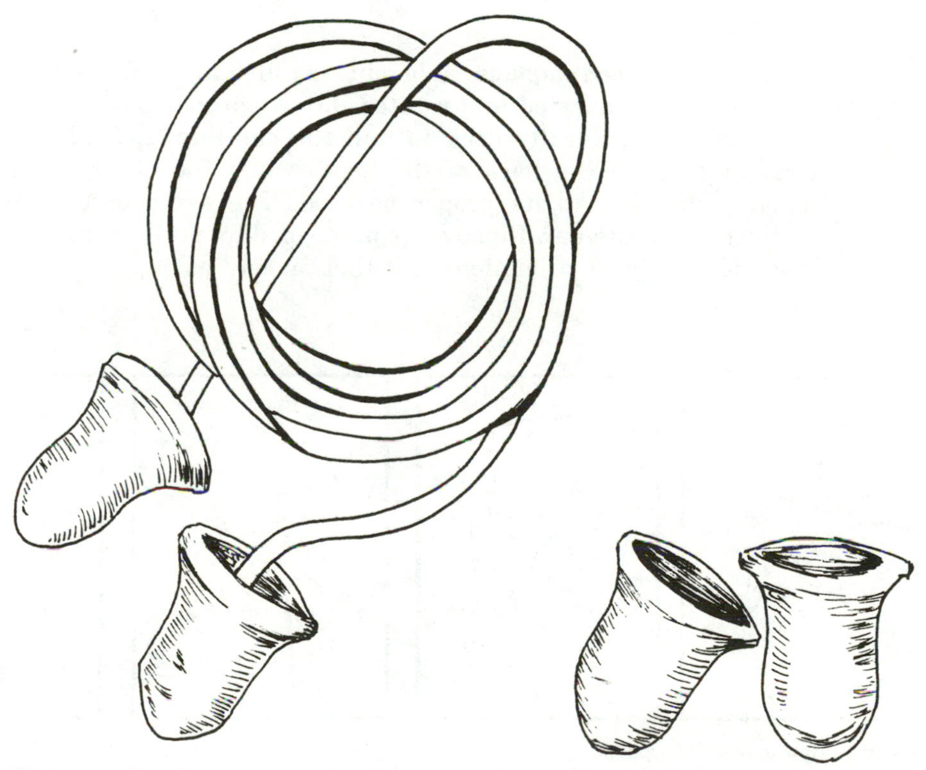

**Figure 20.3** Earplugs.

- Can be worn with hard hats and glasses.
- Are more comfortable in hot, humid workplaces.
- Are available in a range of sizes and shapes.

Earplugs also posses several distinct disadvantages:

- Require a tight seal of the ear that may be uncomfortable.
- Are easily lost and are hard to monitor.
- Become hard or shrink if not replaced periodically.
- Require sizing for each ear.

One particular problem is that in actual use earplugs, on average, yield as low as only about 25 percent of their rated values. Improper insertion is the chief cause of the lower ratings.

- Need to be reseated periodically.
- Require some degree of manual dexterity for insertion, and some individuals experience difficulty in inserting them correctly. Moreover, careless use can transfer dirt into the ear canals.
- Should not be used in infected ear canals.

**Inserting premolded earplugs.** The plug is rolled in the fingers and placed into the ear canal, inserted and rotated into its proper position (see Fig. 20.4). One hand is generally used to pull the ear flap up and out—usually by reaching across the back of the head) while the other hand inserts and rotates the plug to its proper position. The same procedure is followed with the opposite ear. Removal is performed by pulling the insert or plug to one side to avoid a suction, and then gently easing the inserts from the canals.

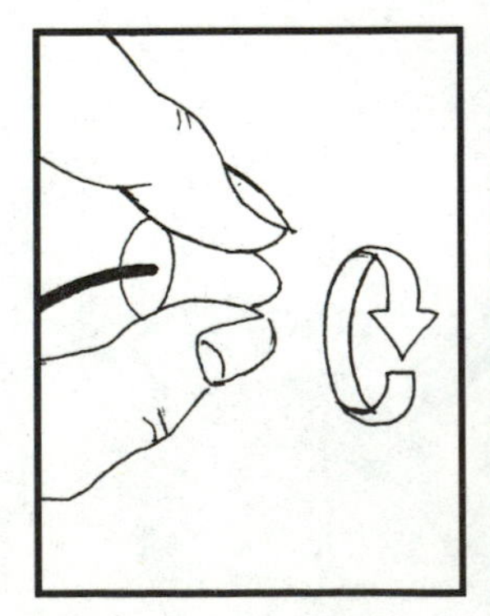
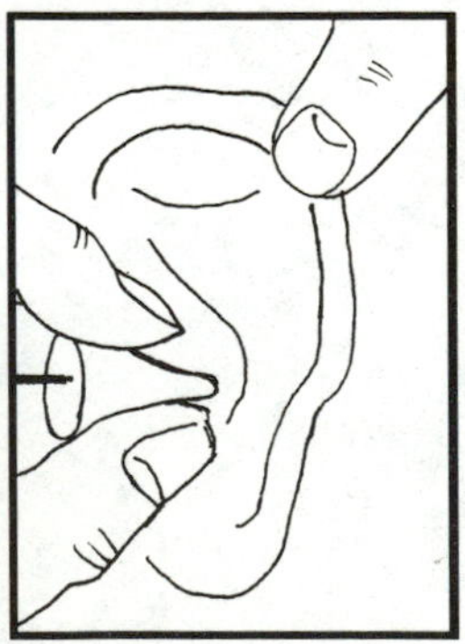
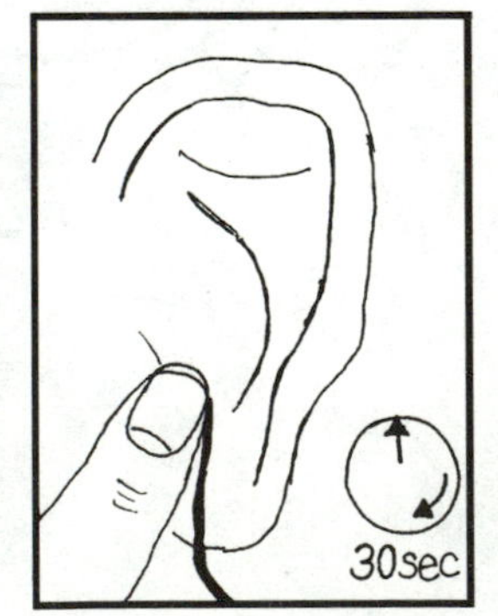

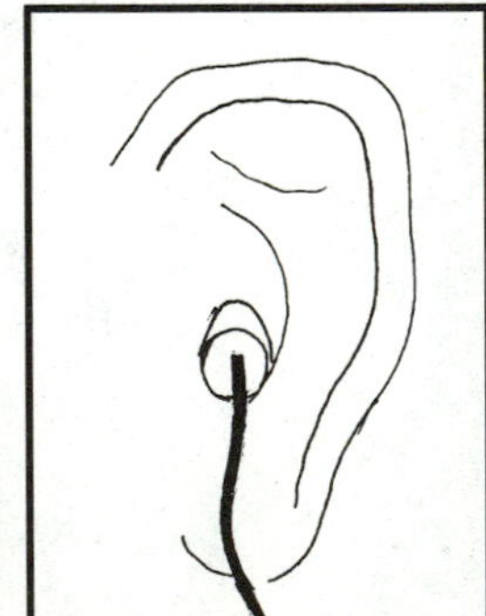

**Figure 20.4** How to install earplugs.

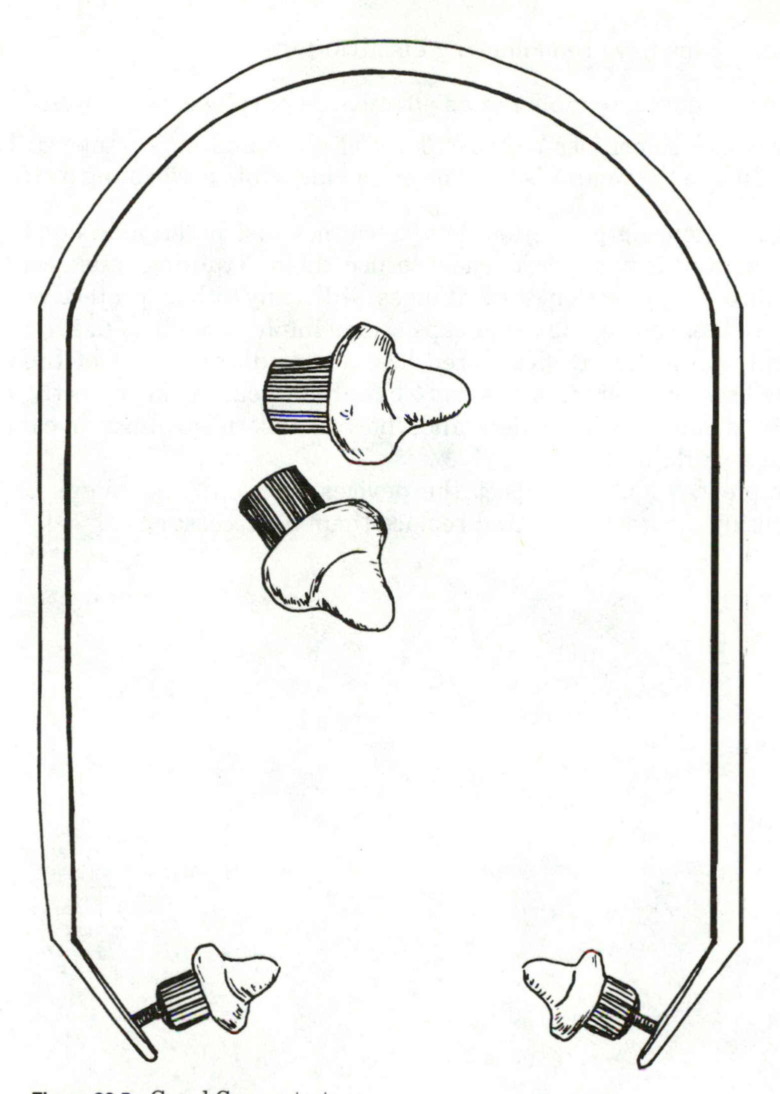

**Figure 20.5** Canal Cap protectors.

### Canal cap protectors

Canal caps (see Fig. 20.5) consist of soft pads fastened to a springy headband. They attempt to seal the ear canal from the outside, without actually entering it.

Canal caps have the following advantages:

- They're inexpensive.
- They're lightweight and comfortable.
- They're cool, and won't cause perspiring in hot weather conditions.

Canal caps have the following disadvantages:

- They're not as reliable nor as effective as earplugs or earmuffs.
- They can sometimes be twisted out of place and may temporarily lose their seal while the wearer is bending or moving while performing a strenuous task.

Again, hearing protection devices cannot just be handed out to employees without any instruction on how to use them. Training must accompany the issuance of protection, like it does with any other protective equipment. Although earplugs and canal caps seem simple enough to use, remember that hearing protection devices often lose a considerable part of their protective value because their wearers don't install or wear them correctly. Basic training by someone who understands how to insert earplugs or canal caps can make a world of difference.

Employees should inspect the devices frequently for signs of hardening, shrinking, or fracturing, and replace them as necessary.

Chapter

# 21

# Hand Protection

## Quick Scan

1. Most hand injuries result from boredom, lack of attention, and worksite distractions.
2. Identify hand hazards at the worksite, then train employees on hazard recognition and prevention.
3. Appropriate, well-fitted gloves should be worn by employees whenever possible while facing point-of-contact hazards.
4. Employees should know how to select, wear, inspect, and maintain their gloves.
5. Because so many types, styles, and models of gloves are available, it's best to consult with a safety distributor who has extensive knowledge in glove applications and selections, and order all company gloves through that individual.

## Hands—Our Most Important Tools

For a moment, hold your hands out about 10 or 12 inches from your face, palms down. Study your wrists first, then the backs of your hands, along the knuckles to your fingertips and nails.

Anthropologists claim that our fingernails are nothing but vestiges of claws used thousands of years ago by ancestors too primitive to be called men and women. Back then, claws—and the rough, strong hands they were attached to—supplied practically all the skills those speechless creatures had. If anything disabled one of their hands or arms, chances are they would have soon perished in the hostile world they faced.

Even today, when it comes to the most-important-appendage-of-the-year award, guess what wins, hands down, time after time? Indeed, if it weren't for our hands, how could a hammer be grasped? A nail be placed? A two-by-four carried? Hands, with their finely manipulative fingers and opposing thumbs, are what enable much of the work to get done. Weighing only about a pound each, our hands are designed to perform almost limitless specialized tasks. Hands can be considered the most basic tools for practically all trades. Can you think of an activity in which hands play no role? Without them, how could you tie your shoes, buckle your belt, button your shirt, or insert a key in your auto's ignition? Just how important are your hands and fingers? How much are they worth to you?

## Hand Anatomy

Most people take the ease with which they work with their hands for granted. In truth, hands rank toward the top level of complexity of all our body parts. Somewhat less complex than the brain, they're still full of nerves and tendons and bones, and require a good deal of brainpower devoted to their manipulation. Unfortunately, hands are easy to injure—they're no match for steel, masonry, wood, and similar materials and tools. Moreover, hands are quite difficult to restore to their original condition once they've been seriously damaged.

### Bones

Twenty-seven individual hand and wrist bones (see Fig. 21.1) are considered a working system in hand anatomy. That's 27 bones to the left hand, and 27 to the right. Each group of 27 hand and wrist bones is further classified into *phalanges, metacarpals,* and *carpals.* The phalanges are finger bones: three in each finger and two in each thumb. The metacarpals together make up the body of each hand, with five on the left, and five on the right. The carpals are wrist bones, with eight to each hand. The carpals are, in turn, joined to the forearms by means of soft-tissue ligaments. Damages to hand or finger bones are generally accompanied by severe pain and the necessity to immobilize (remember restricted motion from Chap. 4?) whatever joint or part is affected—usually for a considerable time while healing takes place.

### Soft tissues

For our purposes, soft tissues of the hands include ligaments, tendons, nerves, and skin (see Fig. 21.2). Ligaments are strong tissues that connect bone to bone, and help make stronger and steadier joints. Tendons are rubbery tissues that connect muscles to bones and enable our joints to flex and move. Nerves provide "feeling" or a sense of touch, and help coordinate hand and finger movement as directed by the brain. Skin, of course, helps keep harmful materials from entering the body's systems and protects against heat and cold. The skin on the back of the hand is thin and stretchable, while the layers of skin

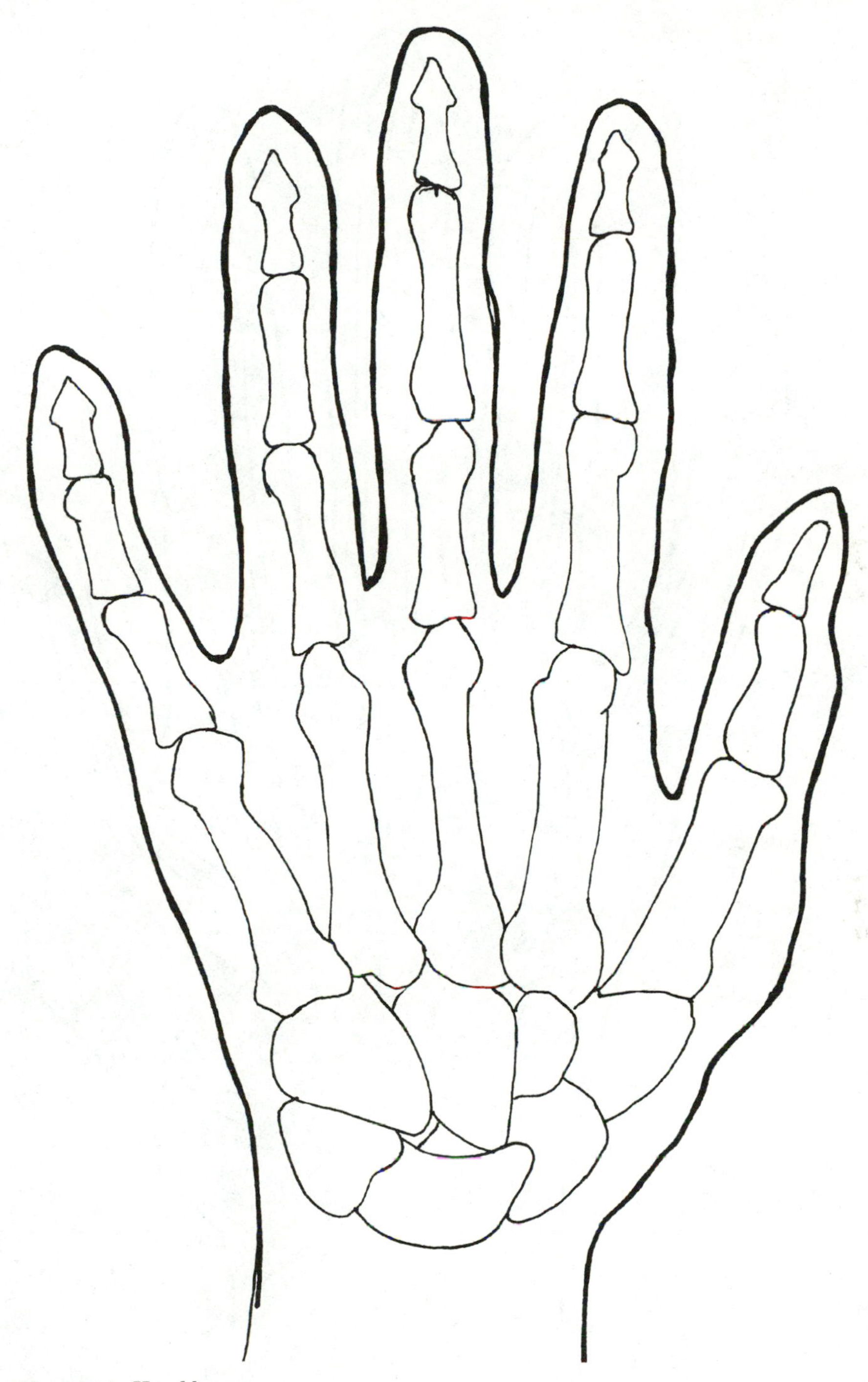

**Figure 21.1** Hand bones.

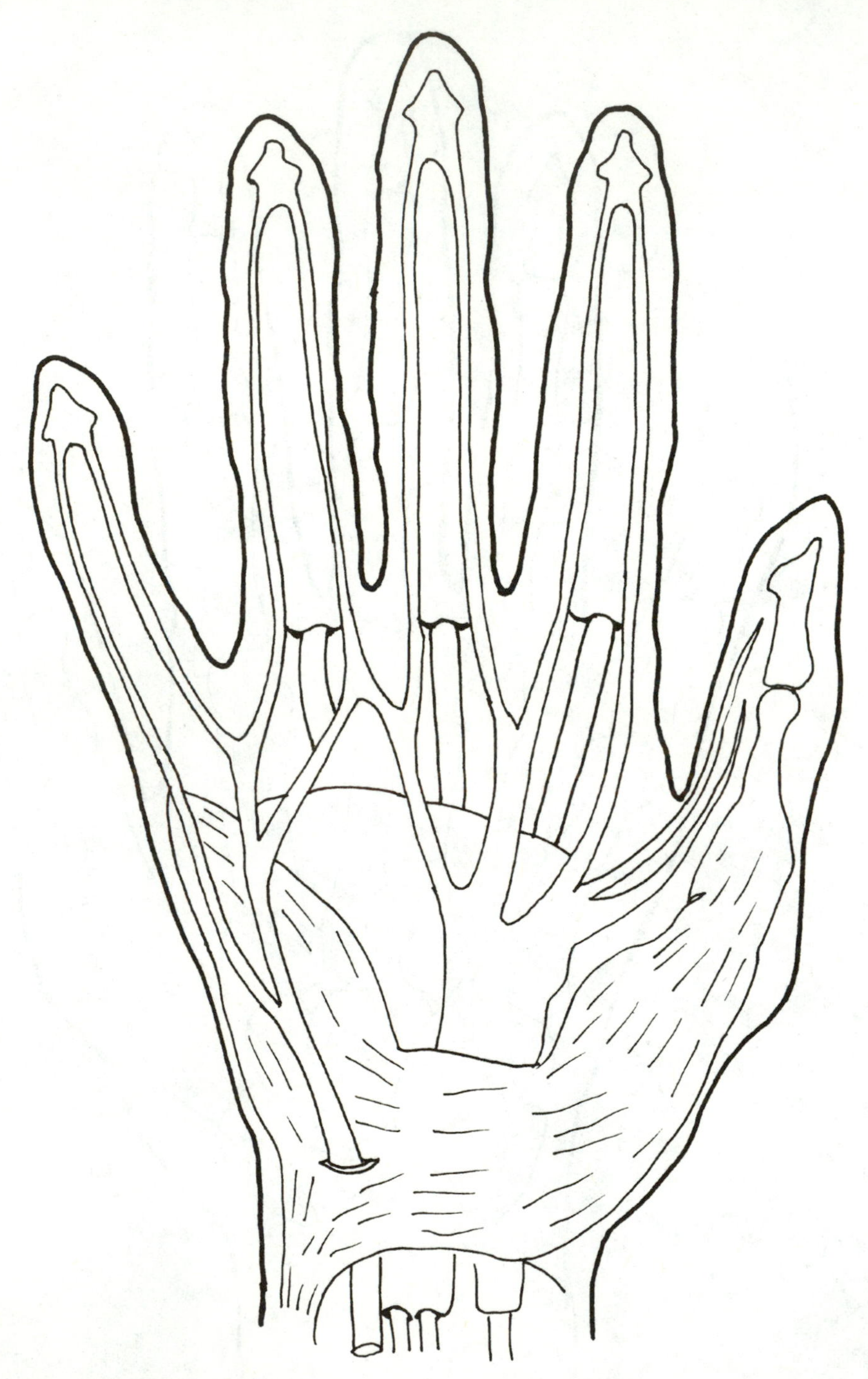

**Figure 21.2** Hand soft tissues.

on the palm or underside of the hands and fingers are considerably thicker and tougher because that's where the work gets done, with the skin providing a means to help cushion and insulate the meaty parts of the fingers and hands.

## Hand Injuries

Because our hands are so vital to almost every kind of task, and because they're used so much—out there at the business end of things day after day—they're vulnerable to injuries. The National Safety Council and Bureau of Labor Statistics estimated that about 17 percent of all body parts injured in work accidents in the United States in 1995 occurred to workers' hands or fingers. The only body part injured more often is the back (24 percent). Major kinds of injuries most prevalent to the hands and wrists include:

### Abrasions

A more common term for an abrasion is a "scrape." Abrasions are often caused by friction from rotating or vibrating tools such as saws, drills, grinders, sanders, or equipment belts and pulleys that chafe and wear away the skin. Maintenance tasks on mobile equipment are notorious for causing abrasions.

Although many scrapes may seem like minor problems to be laughed off by rough and tough construction workers, a loss of surface skin acts like an open window for harmful bacteria and other substances to enter the body. In short, any abrasion requires first aid to make sure it won't develop into a more serious condition.

Prevention considerations include keeping tool and equipment guards and shields in place and in good condition, wearing the right kind of protective clothing and gloves when necessary, and realizing the danger of working around moving parts and equipment.

### Cuts and lacerations

Cuts and lacerations continue to be a serious problem at numerous construction sites. Butt knives are big offenders (see Fig. 21.3). Workers are injured just opening up supplies, cutting banding on skidded materials, or performing such nonwork tasks as opening a can of soup. Other kinds of sharp knives and tools cause equally painful cuts as well. Dull cutting instruments require the hands to apply more force than should be typically applied—with a corresponding loss of control that can easily result in serious lacerations. Sheet metal takes its toll, and so do glass and brittle plastic, when cracked and broken. Add all sorts of tools and equipment to the list—and don't forget masonry, with unexpected sharp edges exposed in blocks and brick.

As a rule, minor cuts heal easily and require only first aid, but deep lacerations may damage nerves and tendons that are needed for normal hand use. Accidental amputations are a nightmare that, even with today's sophisticated microsurgery techniques, are extremely difficult or impossible to repair.

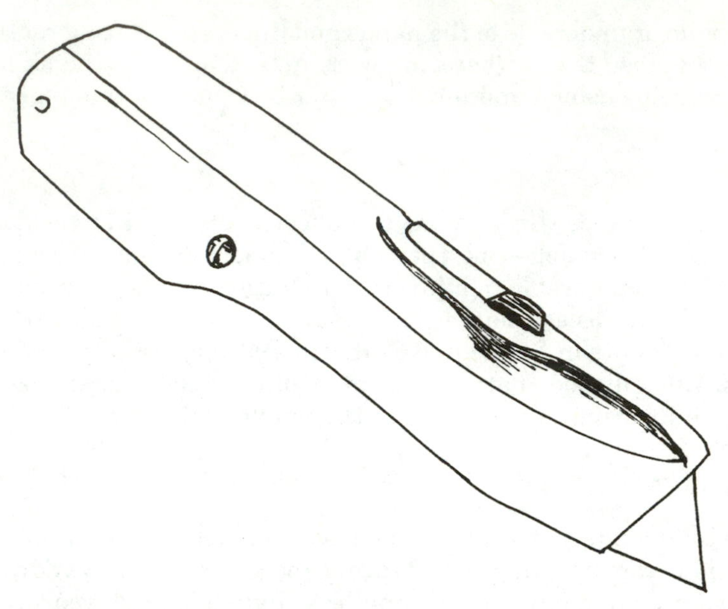

**Figure 21.3** Butt knife.

Prevention considerations for cuts and lacerations include wearing appropriate gloves for the work, keeping hands and fingers away from the paths of blades and other sharp tools, maintaining sharp edges on cutting tools, keeping guards in place, and using hazardous-energy controls before starting any maintenance or adjusting of powered tools or equipment. Extra care is needed around saws.

Watch out for employees using the wrong tool for the job, or using the correct tool in the wrong way. No ill-fitting wrenches. No holding the work in one hand while pushing a screwdriver with the other hand. Again, be especially careful of knives—especially butt knives. Knives account for more hand injuries than most other causes. Instruct employees to cut *away* from the body, instead of toward it, and never to work on the same piece of material with another knife-wielder. Avoid allowing knives to be used as screwdrivers. Store knives separately from other tools—not loose in drawers with other items. Sheaths should be fixed so the knife is pointing backward over the right hip to prevent severing a leg artery if the employee falls while wearing it.

Rotating machinery, tools, and equipment such as drill bits, saw blades, sanders and cutting wheels, and milling cutters and routers can all take their share of flesh, tendon, and bone in an instant. Make it a habit to instruct employees to remove jewelry and roll up long shirtsleeves when working around moving machinery.

Table saws and jigsaws can be hand killers—*always* have employees use a pusher stick, even when making the shortest, simplest cut. They've got to get into the habit of using the stick. With power drills, have the work clamped or affixed so it can't spin around into unsuspecting hands or elsewhere. Instruct employees to use pliers or tool holders to set items closely against grinding wheels or saws.

If possible, use a box-end wrench instead of an open-end wrench to avoid slipping. Employees should pull on the wrench for more control whenever possible, instead of pushing on it.

Cuts and lacerations, in short, can result from any number of unsafe behaviors.

### Puncture wounds

Punctures can range from a shallow poke just beneath the skin, to a major impalement that goes far into the deep tissues of the hand. While puncture wounds to the hand will rarely cause life-threatening blood-loss conditions, they are likely candidates for infection unless cleaned and treated quickly. Again, unless your employees receive training on first aid, and are taught the potential consequences of infection, many may think too lightly of reporting and taking proper care of simple punctures. Typical worksite punctures result from mishaps with nails, splinters of wood, metal burrs or glass, and hand tools such as screwdrivers, punches, drill bits, chisels, knives, and awls.

Prevention considerations for punctures include using hand and power tools with care (see Chap. 30), plus keeping a sharp eye out for nail ends, discarded boards with nail ends, sliver and burr ends, plus sharp edges and protuberances on various construction materials and parts.

### Sprains

With sprains, muscles and ligaments that help hold joints in place get stretched, torn, or bruised. Sprains often result from a person attempting to suddenly lift a load that's too heavy, from trying to apply too much force on a wrench or other tool, or from attempting to break a fall.

Prevention considerations include getting mechanical assistance—from a tool, a piece of equipment, or a fellow employee—when lifting, pulling, or pushing heavy objects, by not attempting to move items while off-balance, by using caution on stairs and uneven surfaces near foundations and on basement floors under construction (especially when carrying a load) and by using correctly set ladders instead of makeshift climbing platforms. If a particular hand tool is undersized for a job, it's best to replace it with a larger version where more leverage is available, or with a power tool.

Discourage employees from wearing jewelry—it's not a good idea to have bracelets, heavy watches, and rings on hands that are working with power tools and other equipment. Loose clothing is also unsafe when working near or with rotating equipment. Jewelry and loose clothing can snag against rotating equipment and twist or wrench the hands.

### Fractures and crushing or pinch injuries

These injuries can occur when hard, heavy items fall on hands or fingers, when a fall occurs at any level, when a vehicle door or building door slams shut, or when hands or fingers are caught in moving machinery such as vehicle motors, conveyors, or other equipment. The latter may be called "pinch-point injuries," which are among the most traumatic, painful, and serious kinds of hand injuries. Pinch points exist whenever a moving object and a non-moving object, or two constantly moving objects, either meet or almost meet. Smashed fingers, amputations, lacerations, bruises, and other lesser injuries can all result from pinch-point incidents. Hand injuries resulting from mobile equipment operations are frequently of the pinch-point type. Fractures and other broken bones often result from pinch points. These, and other causes of fractures, result in some of the most painful injuries at construction worksites. In addition to broken bones, other damages can occur to tendons, ligaments, and nerves. Deep tissue of the hands can be severely bruised as well.

Prevention considerations include body positioning so that work is done away from where objects could fall, and where hands and fingers cannot become caught in moving equipment and pinch points. Slipping, tripping, and falling must also be defended against.

### Contact injuries

Contact injuries include first-, second-, and third-degree burns, cold-temperature burns, electrical burns, chemical burns, and dermatitis and similar rashes such as those caused by touching poison ivy and oak.

Contact injuries can usually be prevented by wearing proper gloves and by not taking chances while working with hot or cold materials or equipment, with electrically charged equipment, or with harsh chemicals. Train employees to use the electrician's one-hand rule whenever possible: Work with one hand only, while keeping the other hand safely hooked to a pocket or at your side. That way the worker's two hands will not complete an electrical circuit that could run through the vital organs in his or her chest. If the job requires two hands, then wear certified linesman's rubber gloves, sleeves, and leather protectors.

Some individuals are much more chemical- and contact-affected than others, and may react more adversely than others do to high and low temperatures, electricity, and materials such as cleaning solutions, fuels and other flammables, metals, and similar substances. Cold-temperature environments, fire and hot-works procedures, and electrical safety are discussed in other chapters in this book. One of the purposes of hazard communication (see Chap. 27) is making employees aware of the risks associated with chemical products they work with.

### Repetitive-motion injuries or illnesses

Repetitive-motion problems, including carpal tunnel syndrome, tendonitis, and "white finger" may occur to employees who frequently perform tasks requiring repetitive fast or forceful hand motions, such as hammering nails or using screwdrivers for long periods of time. Tasks in which employees must frequently grasp and squeeze objects can result in tendon or nerve difficulties.

Pressure on nerves, tendons, and blood vessels in the hands can also have negative effect on hand health. Tool handles that are much too small or too large or that have sharp edges or finger grooves can cause cumulative damage—especially handles on screwdrivers, pliers, scrapers, scissors, and certain brushes. Vibration from tools like grinders, jackhammers, drills, and chain saws can also affect the hands, fingers, and wrists—especially when those tools are used repeatedly, for relatively long periods of time.

Prevention includes finding ways to get the work done using different or varied tools or grips or motions. Also useful are administrative controls such as rotating employees to other tasks intermittently or spreading out the work by employing shorter work periods over longer time periods. Antivibration gloves (see Figs. 21.4 and 21.5) found in the ergonomics section in catalogs and in safety distributor's showrooms, are making their way into construction worksites and being appreciated by employees. Knowing the proper grips to use can sometimes prevent repetitive-motion problems. To a certain extent, hand- and wrist-stretching exercises may also be helpful when performed just before starting the work. The main idea is to prevent individuals from performing the same motions over and over again—overusing specific muscle groups in the hands, or from applying too much force to overcome a poor grip. Resort to alternative procedures, have employees change positions frequently, rotate tasks, insist upon frequent breaks to relieve ergonomic stresses, select ergonomically designed tools, and keep all tools in proper working order. Refer to Chap. 15, on ergonomics, for additional information.

## Why Hands Get Injured

Yes, hands are out there taking constant risks, at or near the business end of many a task. But why exactly *do* they get injured, mostly? Could it be a worksite full of sharp and abrasive materials, thermal hazards, chemical exposures, and unsafe electrical currents? Sure, those are some of the more universal hand hazards encountered at a worksite. But those items are not the chief cause of hand injury. The number-one cause of hand injury is (surprise!) human error—caused by not paying attention, by boredom, by complacency, by inattention. Certainly, while doing repetitious jobs, the mind can wander. An employee becomes preoccupied with other things—social plans, financial concerns, family discord, you name it. Almost anything can take the mind off whatever work business is at hand. And when things become so automatic the employee hardly has to think about what he or she is doing—that's when a hand accident is most likely to occur.

What are some danger signals to watch for? Take notice if you see a lot of nicks, cuts, and abrasions on your employees' hands. Those are little warning signals, flags indicating that there are plenty of near misses and first-aid incidents involving the hands. Of course, minor injuries can occasionally turn into nasty infections that before long become recordable incidents and worse.

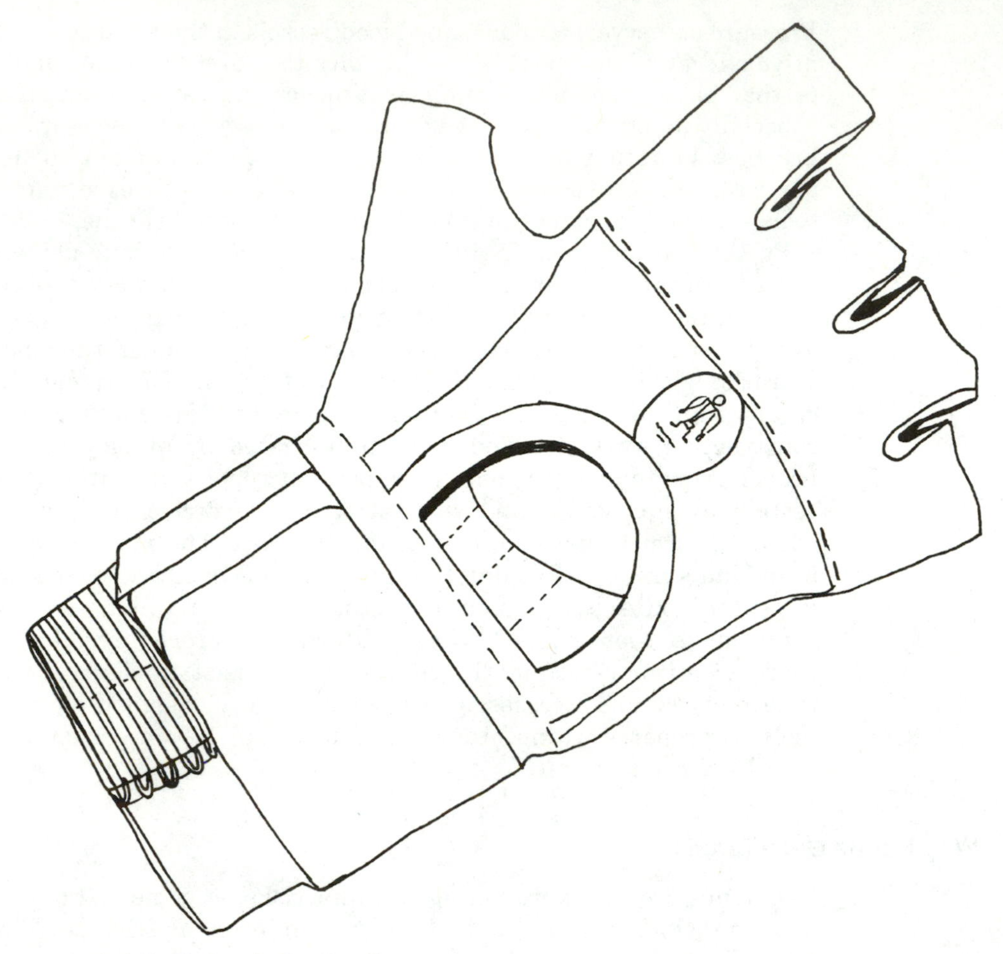

**Figure 21.4** Padded anti-vibration glove.

## Gloves

If there's one universal piece of personal protective equipment for the hands, it has to be gloves. You needn't be in construction to appreciate how many types of gloves are used for work and play. How many types of gloves can you name? Boxing gloves, baseball gloves, driving gloves, mittens, cycling gloves, gardening gloves. There are specialized gloves for almost every purpose imaginable. OSHA recognizes the protection that gloves afford, and requires the selection and use of gloves to protect against:

- Absorption of harmful substances by the skin
- Severe cuts
- Lacerations
- Severe abrasions

- Punctures
- Chemical burns
- Thermal burns
- Harmful temperature extremes

There are hundreds, if not thousands of different gloves designed and manufactured to protect against almost any risk encountered. Indeed, they'll prevent the skin from absorbing chemicals. They'll shield the skin from heat or cold. They'll protect from severe cuts and lacerations and abrasions. They'll help deaden vibrations and shock. They'll help a worker avoid chemical burns. They can guard against the transmission of disease, and have really come into their own with blood-borne pathogens such as HIV and Hepatitis B.

On the downside, gloves also reduce the wearer's dexterity, and increase the amount of strength needed to swing a hammer or grasp a tool.

## Glove types

You'll find gloves made of canvas, leather, rubber, cloth, and numerous modern man-made materials, with textured palms and fingers for better grabbing power. No particular pair of gloves can protect against every possible hand hazard. Most gloves fall into four categories or types:

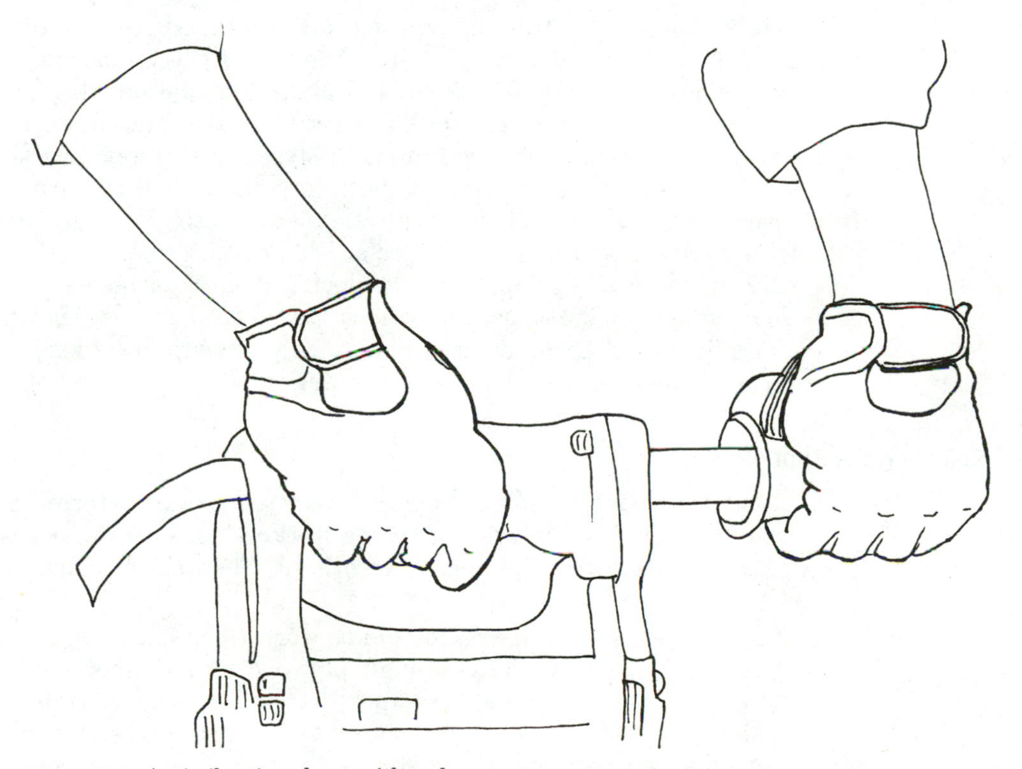

**Figure 21.5** Anti-vibration gloves with tool.

1. *Chemical-resistant gloves* are designed to protect against a wide range of chemicals as well as from nuisance hand injuries.
2. *General-purpose gloves* are designed to protect against cuts, snags, punctures, and abrasions, but not chemicals and liquids. They're usually made of leather or cotton, and can also protect against temperature extremes.

   Cotton gloves, while not as long-lasting as leather, are usually more comfortable and can be specially coated for an improved grip; can be woven thin enough to allow for the handling of small components; or can be used with oily parts to absorb oil and provide a sure grip.

   Leather gloves provide protection against rough surfaces, heat, and sparks. Leather-padded palms protect inner gloves from cuts and punctures.

   Cut-resistant gloves often contain Kevlar, Aramid, or Spectra yarns. Kevlar gloves provide superior comfort and thermal protection. Steel-wire Spectra gloves are designed for use with sharp knives and blades. They're flexible and cut-resistant, but don't protect against punctures.
3. *Product-protection or cleanroom gloves* are designed to provide a barrier between hands and product to help protect the worker from the product and/or the product from the worker.
4. *Special-purpose gloves* are designed for applications where gloves are needed for hot and cold temperatures.

Within these categories, gloves are differentiated further by the type of material they're made of (e.g., natural latex, nitrile buna rubber, neoprene, polyvinyl chloride or PVC, polyvinyl alcohol) and whether they are supported or unsupported (lined or unlined), rolled-cuffed, straight-cuffed, safety-cuffed, knitted wrist, or slip-ons or gauntlet in style. The type of grip the glove has also is a differentiating factor. If the wearer is at risk of burns or cuts, the choice would likely be leather or canvas. Metal-mesh gloves are effective when working with sharp or rough objects. Electricians select rubber gloves and insulating sleeves to protect against electrical shock or burns.

Your employees must also *know when not to wear gloves*—as when working near equipment that could easily and quickly draw their hands into moving or rotating gears or similar tools or machinery.

### Selecting the right gloves

1. *Consider the physical hazards to be encountered.* Determine what risks are most likely to be present in the workers' tasks: cuts, abrasions, punctures, temperature, chemicals, or others. Then select glove types accordingly. For example:
   - Abrasion-resistant gloves should be worn when handling rough materials such as masonry blocks, rough plywood and lumber, and steel cable. When wrists and arms are exposed to the same materials having abrasion or puncture risks, abrasion-resistant arm protection should also be used such as gauntlets or sleeves.

- Electricians should wear approved gloves when working on or near exposed energized, electrical parts. Gloves must be inspected and tested at intervals specified by the manufacturer. Required voltage ratings of the gloves should be determined depending on the voltages of potential hazards. Written electrical safe work practices designate which tasks require gloves providing electrical protection. Protective overgloves should be worn to preserve the integrity of the rubber gloves.
- Chemical-protective gloves should be worn when handling hazardous chemicals and corrosives, depending on the substance handled and the manufacturer's recommendations.
- Welder's gloves should be worn while welding or burning to protect against heat, sparks, rough objects, and flash burns.
- Latex or nitrile gloves should be available for individuals trained for emergency response and first aid where there is a potential for exposure to blood-borne pathogens.

2. *Decide on the most important features the gloves should possess.* Is grip the most important characteristic? Or dexterity? Comfort? Insulation? Slip-resistant surface? Impervious coating?

3. *Select the glove that offers the best resistance to the physical wear and tear of the application.* Remember that no single type or model of glove can protect an employee from everything. You may need to provide special gloves for specific work tasks.

4. *Select the cuff style and length based on the expected hazards of the application.* Choices include band top, knit wrist, safety cuff, bell gauntlet, slip-on, and open cuff. For jobs such as welding or working with metal, a glove with a gauntlet cuff will help protect the wrist and forearm too. For working in the cold, an insulated glove with a knit wrist protects hands and wrists from exposure. For operating equipment, slip-on or open-cuff gloves work well, as they can be put on or taken off quickly.

5. *For chemical resistance, consider permeation and degradation of the material (nitrile, neoprene, rubber) from which the glove is made.* Using a degradation and chemical resistance chart as a guide, select the glove type with the highest rating for the particular chemical you will encounter. In addition, test the glove at the worksite by dipping the glove into the actual chemicals encountered for the particular application. Does it swell, harden, change color, soften, weaken, or become brittle? If so, look for a different glove for the job.

6. *Other general guidelines:*
   - Disposable gloves offer both hand and product protection for one-time, nonstrenuous tasks.
   - Ventilated gloves offer coolness and comfort.
   - Insulated gloves are used for cold and hot work.
   - Thin-gauged, smooth-finished gloves give tactile sensitivity and extra dexterity.

- Rough-finished or embossed gloves offer a strong, nonslip grip.
- Lightweight coated gloves offer resistance to liquid.
- Heavyweight coated gloves offer chemical and abrasion resistance.
- Thicker-gauged supported gloves give snag, puncture, and abrasion resistance.

7. *Select the correct size.* This is important, especially if the wearers are planning to do skilled or detailed work. Manufacturers insist that a glove that fits its owner well will provide better wear and will last longer than a glove that's too small or too large. When incorrectly sized, most gloves tend to be too large, and they're awkward to wear and will make the wearer more clumsy. The clumsiness leads to snags and misdirected grasping that can, in turn, lead to costly mistakes and accidents. On the other hand, gloves that are too small will be uncomfortable and can lead to hand fatigue.

   Many gloves, including most leather, cotton, and synthetics, use alphabet letters for sizes: XS, S, M, L, XL. To determine a glove size, measure the circumference around the palm of the hand at the knuckles (including the thumb).

   Glove Hand Sizes:

   | | |
   |---|---|
   | 6–7 inches | XS (extra small) |
   | 7–8 inches | S (small) |
   | 8–9 inches | M (medium) |
   | 9–10 inches | L (large) |
   | 10–11 inches | XL (extra large) |

   Some gloves, however, use number sizes, but the manufacturer will usually have some reference equivalent in alphabetical sizes as well. If possible, allow your workers to try on several sizes to get the most comfortable fit.

8. *Use the gloves' color to identify contamination or even to link with certain tasks.* Nitrile, neoprene, and rubber gloves are often available in black and in bright colors such as yellow, green, and orange.

9. *It's advisable to purchase all of your gloves from one or two manufacturer's reps.* This will not only give you more buying power, it will make things easier because you'll probably get to try various gloves for free, and the rep will likely be able to suggest different types or styles that closely suit your needs. A knowledgeable salesperson knows what's new on the market, has connections with the manufacturer, and can supply catalogs and educational training materials as well.

   The manufacturers will also help you select the best gloves for your needs. They've got product hotlines; they'll do a glove survey; and they'll supply samples.

Again, the range of types and styles of gloves is mind-boggling. There are gloves made from aluminized fabric, insulated cloth, stainless-steel mesh, and voltage-resistant materials. There are latex and vinyl gloves so thin that even the smallest objects can be handled comfortably. There are gloves with special coatings and textures and long-use, multiuse, and disposable gloves. There are new glove materials that are practically indestructible. Name a hazard, a chemical, a temperature extreme, or a challenge, from cut-resistance to vibration protection to preventing the spread of infection, and you'll find a glove to match. The manufacturers just love to come up with new and better gloves to face the narrowest of situations—it's a way for the glove industry to grow and it's also a way to get improved protection for specific tasks.

Here's a sample list of gloves available:

- Cotton flannel (canvas).
- Insulated canvas.
- Vinyl-impregnated cotton, coated for liquid resistance.
- Canvas plastic dot, for increased gripping.
- Jersey knit.
- Natural rubber.
- Latex.
- Butyl.
- Viton.
- PVC.
- Nitrile.
- Vinyl.
- Neoprene.
- Insulated nitrile.
- String knit.
- Leather palm.
- Insulated leather.
- Driver's leather.
- Steel-reinforced leather.
- Kevlar seamless knit, for cut resistance.
- PVC-coated Kevlar.
- Heat-resistant.
- Cold-insulated.
- Welding.
- Lineman's, for electrical work.
- Leather hand guards, for handling cable and glass where palm and back of hand require protection from cuts and abrasions.
- Antivibration, with new styles coming out as this is being read. Some are made of split leather with gel-filled or visco-elastic polyurethane polymer palm pads for absorbing vibration. See Chap. 15, on ergonomics.

## Glove training

It may sound ridiculous, but hand protection is considered personal protective equipment, and as such, employees should be trained in the proper selection, use, and care of gloves. They need general safety rules for wearing gloves, including:

- How to inspect gloves for flaws.
- How to tell when gloves have reached the end of their protective lives.
- How to safely remove chemically contaminated gloves.
- How to properly decontaminate or clean gloves.
- Understanding what hazards individual types of gloves will not protect employees from.
- How to wash leather gloves in mild soap and water, rubbing gently to clean and allow air to dry.
- Why and how to protect electrical gloves and Kevlar gloves from excessive sunlight.

After the training, employees should be able to answer the following questions:

- Are they using the right gloves for the specific hazards encountered at the worksite?
- Do they know the reasons for wearing gloves at the worksite and what types of exposures they are trying to prevent?
- Are their gloves in good condition, free from damage or chemical saturation? *Glove degradation* generally refers to how quickly chemicals dissolve, break through, or break down glove material. Degraded gloves may get stiffer, more brittle, softer, and weaker, eventually cracking and losing their protective qualities.
- Do their gloves fit properly?
- Do the employees use one type of glove for any and all situations? (They probably shouldn't.)
- Are the employees knowledgeable about how to obtain new gloves or different types of gloves, as the hazards change?

The employer is clearly responsible for assessing hand-related hazards, for providing protective equipment, and for communicating to and training employees how to wear, clean, and inspect the gloves. You need to verify that each affected employee has received and understood the required training through a written certification. As for that training certification, your documentation must specify the employee's name, the training dates, and the subject of certification.

## Protective Skin Creams

Another means of protection for the hands and fingers is skin cream. Various kinds of skin creams can function as "invisible" gloves to protect against excessive water contact, and against substances that could otherwise harm the skin. Grease-guard protective creams are available to apply to arms, hands, and especially around fingernails to protect the skin from grease and oil. Solvent-resistant creams are available that form a solvent-resistant barrier that will prevent dirt, grime, paint, oils, inks, and epoxy resins from adhering to the skin. Water-resistant creams protect skin from water-soluble irritants such as acids, alkalis, plating solutions, and cutting solutions.

Greaseless moisturizing creams will soften, condition, and maintain the skin's natural oils, keeping it from cracking and bleeding in harsh climates and working conditions.

Overall, the way to prevent hand and wrist injuries is to recognize the potential hazards and practice awareness recognition and avoidance. Gloves should be used whenever possible. Tool and equipment guards must be kept in place. "Zero energy"—shutting off power and preventing the unexpected release of hazardous energy—should be practiced while making adjustments or repairs on equipment, and hands and fingers must be kept away from pinch points.

Again, while primitive man and woman had only their crude hands to work with, we've got lots of hand-helpers available: special tools, devices, procedures, and personal protective equipment. Best of all, we have our minds to use. Don't let your hands—or the hands of your fellow employees—get clipped.

Chapter

# 22

# Body and Foot Protection

## Quick Scan

1. Employees should wear close-fitting or medium-fitting jeans or durable trousers.
2. Allow short work pants only for tasks where no threats from lacerations, abrasions, splinters, sunburn, poison ivy, or similar hazards exist.
3. Recommend long-sleeved work shirts; accept shirts having nothing less than four-inch sleeves.
4. Company supervisors should set a good example by what they wear at the worksite.
5. Safety work shoes or boots are a must.
6. Safety footwear having worn soles and uppers must be discarded and replaced.
7. Improperly laced shoes and boots have caused more falls and injuries than construction employees care to admit. Insist that employees maintain their laces at all times.
8. Prohibit the wearing of light-duty home-use waterproof boots that are worn over socks. Require waterproof boots possessing similar characteristics to safety shoes or boots, or lightweight pullover boots that are worn over regular safety footwear.

## Work Clothing—Where Safety and Neatness Do Count

Some individuals tend to dismiss clothes as not really being part of an employee's personal protective equipment. That's a mistake. First of all, bare skin is not very durable. As you'll see in the chapter about working in hot weather, tanning exposure from the sun—while considered a handsome asset by some—is definitely not healthy over the long term. Ask any dermatologist. But that aside, there are other reasons why clothes *can* make a difference on the worksite.

After safety and hygiene, there's appearance. Believe it or not, neat, appropriately dressed employees lend credibility to the operation. Imagine the impression you'd get of Bob Vila's TV show, or of the PBS feature on restoring old homes, if the participants walked around with soiled T-shirts full of holes, or if they wore dress trousers with penny loafers. Sure, employees may be able to get away with wearing shorts and sleeveless shirts for certain tasks, in the warmest weather, but that should be the exception. Typically, employees should wear a minimum of socks, safety shoes, medium-fit jeans, and a short-sleeve shirt. A long-sleeved shirt is better. The clothes should be neat and—at least at the beginning of the day—fairly clean. There are just too many opportunities for scratches, lacerations, punctures, abrasions, and infection at the worksite. Bare skin may at times be a little more pleasant for employees to work in, but it increases the risk of injury and does little for your company's image.

There will be individuals who disagree, but I have yet to hear an argument that would change my mind on the importance of appearance. Most people, including potential customers of your company, equate professionalism with appearance. You could have the best, most skillful workers, but if they're not given the opportunity to show what they can do, their skills are lost. Real estate agents, government officials, vendors, subcontractors, utility workers, insurance agents, neighbors, passers-by, and numerous visitors will interact at various times with your company's employees. First impressions on how things and people look can be lasting. If your employees are dressed like everyone believes construction workers are *supposed* to be dressed, those workers will go a long way toward helping your company's credibility. And that credibility may ultimately be reflected in additional business, which, you can explain to them, ultimately helps make their positions more secure.

The fact that it takes many small things to make a successful business is well pointed out here, when workers will scoff at the importance of wearing long pants, shirts, and semi-neat clothes on the job. Your business, as does that of other construction and contract companies, needs all the positive factors it can muster while competing in the open marketplace. After all, your company is probably going up against some fly-by-night operators who will be in (and out) of the business for relatively brief periods, undercutting you and the rest of their competition while not working safely, not insisting on quality, and not knowing from day to day what their costs are or even if they're making a profit. Your employees may very well point to some of these workers and say, "They're working in sneakers and gym shorts and no shirt. Why can't I?"

Answer with the information from this chapter, and if that doesn't convince them of the need for a neat, expected, and safe appearance, then you've got to start wondering about their future with your company.

### Clothing considerations

- There are certain types of modern materials that are made expressly for working in warm temperatures. Inquire at your local safety products distributor for samples or literature on the latest examples.
- High-visibility vests with a minimum of 200 square inches of blaze orange or green material should be worn by employees when operating and working around heavy equipment or when exposed to traffic hazards (see Fig. 22.1).

## Foot Protection

Feet are incredibly complex appendages. They certainly come in handy for walking, climbing, balancing, driving, and just plain standing. They're also awfully close to ground level—the place that heavy objects just happen to be landing.

According to studies undertaken by the Bureau of Labor Statistics, most workers who suffer foot injuries are not wearing protective footwear at the time. Not surprisingly, the lion's share of injured employees work for employers who do not require that their workers wear safety shoes or boots on the job.

**Figure 22.1** High-visibility vest.

Well, if an employee's work exposes the employee to injury from falling, rolling, or sharp objects, or slippery surfaces, he or she should use appropriate foot guards, safety shoes, or boots and leggings, pure and simple.

What to wear? Footwear is getting more specialized all the time. Look at athletic shoes. There seems to be a different shoe for every sport: for walking, for crosstraining, for tennis, for basketball, for baseball, for golf, for cycling, for wrestling, for hiking, for rock climbing, for kayaking, for skiing. You probably aren't surprised that there also are lots of shoes and boots specifically designed and manufactured for construction use. In addition to offering protection from falling and crushing items, construction worksite footwear should also help prevent the wearers from slipping. The soles should resist puncture, and carry enough tread for positive, secure traction.

### Safety footwear construction

Leather has traditionally been the shoe and boot uppers material of choice, because it breathes like skin and can mold to form-fit the wearer's foot (see Fig. 22.2). Insoles should be cushioned to help alleviate the daylong stress of standing, walking, and climbing. Additional comfort features include arch supports to help distribute weight over a broader area, and rigid shanks to give additional support. Soles, including heels, should be strong and nonslip, never leather.

There are individuals who will say that substantial leather work shoes or boots are good enough. And for most of the time, they very well might be. But there's little lost in comfort in some of the better safety-toed shoes and boots, and the added protection is a worthwhile investment—especially if the safety toes are ever really needed. They can be put to the test when working around heavy equipment, when performing maintenance on vehicles or equipment, while using heavy power tools, and while handling heavy materials such as concrete blocks, bricks, and lumber.

What about the height of the shoes or boots? Low-cut safety shoes still leave the ankle at risk. It's better to go with higher styles that give the ankle support. The most popular styles are six-inch and eight-inch work boots. The

**Figure 22.2** Leather work boot.

eight-inch models offer considerably more support and protection for only a slight tradeoff in weight and comfort.

Involve safety vendors to search for the right shoes for your employees. Contact shoe manufacturers' area sales managers or representatives. Perhaps you can establish an arrangement where you'll pay a certain amount per year per employee toward safety shoes, and maybe you can arrange a group discount pricing. Involve employees—especially the ones who are the hardest to fit and the hardest to please—in the shoe selection process. This gives them a sense of ownership, keeps them involved, and shows them that the company supports the foot protection program.

### Safety footwear fit and condition

A comfortable fit is critical with work shoes and boots. Never buy cheap footwear, and never buy footwear that's too tight, on the assumption that it will stretch with time. Tight shoes or boots hinder blood circulation, and during cold weather they prevent an insulating layer of warm air from protecting the feet. Conversely, shoes that are too large will cause blisters and sore heels, and may cause employees to invert their ankles or trip while climbing or walking. Also, because feet tend to spread out as we age, don't assume you'll be wearing the exact same size year after year. Have your feet measured every time you buy shoes. And both feet should be measured because often one will be considerably larger than the other. Get the shoes that fit the larger foot. Because feet tend to swell by the end of the workday, it's best to purchase your work shoes or boots late in the afternoon—when your feet will have swollen to the largest size they'll be.

The age and condition of shoes are important safety considerations—especially the condition of the soles. Fortunately, the surfaces a worksite safety-toe shoe usually encounters are not as abrasive as those found in manufacturing facilities, where continuous walking on concrete and asphalt surfaces is common. At the worksite, soles typically last quite a while. They should have nonskid treads that will not slip on bare wood flooring or wet concrete.

Sometimes it's painfully obvious to everyone but the wearer that his or her shoes have worn way past their prime, and should be discarded and replaced. But nothing fits quite like old shoes. They're comfortable and supple and familiar.

### Foot and leg protection considerations

- Leggings protect the legs and feet from welding sparks; safety snaps allow leggings to be rapidly removed.
- Saw chaps or similar protective leggings should be worn by employees operating a chain saw.
- Padded aluminum alloy, fiberglass, or galvanized steel metatarsal foot guards can be worn over normal work shoes, but they're bulky and awkward to wear and may cause tripping hazards if employees wearing them aren't

careful. Plastic or composite knee pads make kneeling tasks a lot safer and more comfortable for the kneecaps (see Fig. 22.3).

- Heat-resistant safety shoes are available to protect roofers, pavers, and other workers exposed to hot temperatures. Safety shoes should be sturdy and have impact-resistant toes. Many come with metal insoles and other guards to better protect against puncture wounds.
- Again, six- and eight-inch boots are preferred over the lower models because they'll provide support and protection past the ankles.
- Although safety-toed sneakers exist (see Fig. 22.4), it's difficult to tell them apart from regular sneakers at a distance, and other employees may then believe that regular sneakers can be worn on the job. Make sure the difference between safety sneakers and regular sneakers is known throughout the worksite if someone chooses the former.
- As employees work longer hours and as they get older, shoe comfort becomes an increasingly important factor in reducing fatigue.
- Protective footwear purchased after July 5, 1994 shall comply with ANSI Z41-1991, "American National Standard for Personal Protection—Protective Footwear." Protective footwear purchased before July 5, 1994 shall comply with ANSI standard "USA Standard for Men's Safety-Toe Footwear," Z41.1-1967.

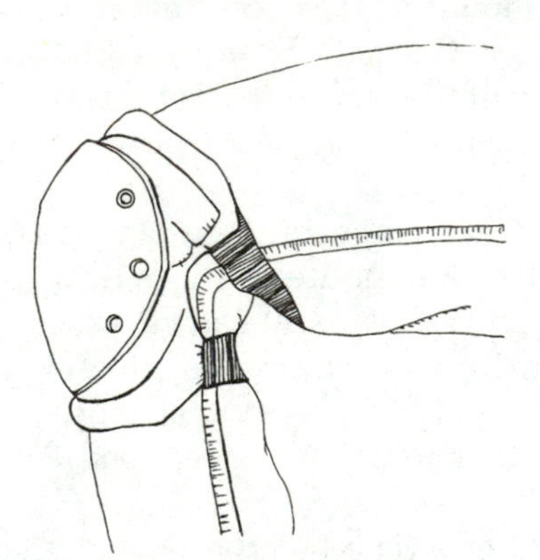

**Figure 22.3** Knee protector.

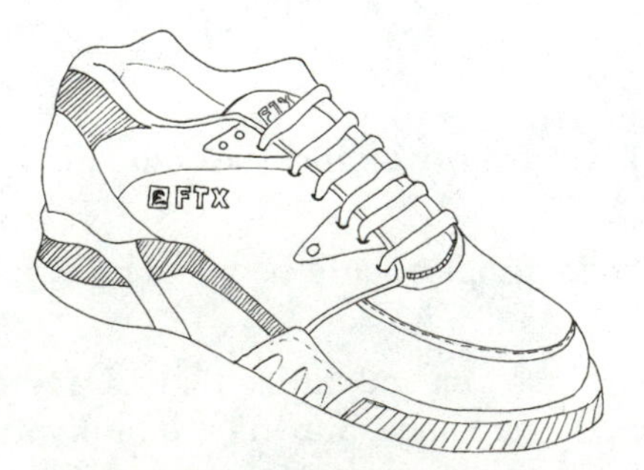


**Figure 22.4** Safety-toed sneaker.

- What about the footwear of visitors, nonworkers, and others who may access the worksite? Generally, the hazards they'll encounter should not be as risky as those faced by your company's employees, as long as the visitors and other nonemployees stay out of active heavy work areas. But open-toed, medium- or high-heeled shoes, or sandals must not be allowed on the worksite.
- Be careful that employees, when faced with wet and muddy conditions, don't elect to wear inexpensive waterproof boots designed for home use. The soles of these boots offer little or no puncture resistance, they have no inside support or lining, and they possess dangerously poor traction patterns. If the boots are to be worn *over* the regular safety footwear, that's a different story. Stand-alone boots, to be worn over socks alone, should have safety characteristics similar to those of regular safety shoes or boots.
- Watch the laces on safety footwear. Some laces attach near the top on metal eyelets that eventually bend so the laces will come loose and flop around. Woven laces are notorious for doing this. This sounds almost too embarrassing to mention, but people are out there tripping over their own laces. If a boot's laces come loose, it's usually a sign that either the laces are too long or the metal eyelets need attention. The eyelets can usually be bent back into their original configuration with a pair of pliers. Another problem with woven laces is that the little plastic nub on the end will come off and the lace will fray into a stringy mess that's almost impossible to stick through an eyelet. Then employees will tuck the frayed, unthreadable laces down into the sides of their boots and walk around with the loose footwear slipping and flopping all day. Frayed laces can, of course, be replaced. But they can also be repaired by cutting back the frayed ends, twisting the ends together with glue, and allowing the glue to dry overnight.

Unfortunately, many individuals will not recognize the need to correct these problems on their own, and will have to be politely reminded before they take action.

Chapter

# 23

# Fall Protection

## Quick Scan

1. Set a goal of 100 percent fall protection for your employees, using personal protective equipment and safe work practices and procedures.
2. Install guardrails around open floors, walls, and platforms, and wherever else falls are possible. The toprails must be sturdy enough to withstand a 200-pound load.
3. Train workers in safe work practices before allowing them to work on foundation walls, roofs, trusses, and exterior wall and floor erections and installations.
4. Whenever possible, employ fall-protection systems like slide guards and roof anchors, and alternative work practices when a guardrail system cannot be used.
5. Train all employees to recognize fall hazards, and support the training of competent persons in your crews who will be able to set up and maintain effective fall-protection systems.
6. Through local trade associations and regulatory agencies, understand and abide by or exceed current minimum fall-protection regulations.

## Fall Protection—A Major Safety Concern

Although this book is not one to dwell on statistics, falls consistently cause a proportionally high number of serious injuries and deaths in the construction industry, year after year. What can be done to protect your company's employees? Certainly, you need to take a practical look at both the hazards and what's available to protect your employees. Consider every employee as you would a son or daughter or brother. Then what lengths would you go through to make sure they're protected? Taking a "people first" approach will generally direct you to a safe resolution, and will likely result in more than the minimum protection required by law.

Fall-protection regulations caused controversy within the construction industry for a number of years because rules developed primarily for heavy construction had been applied to residential construction as well. In 1994, for example, OSHA issued a rule that lowered the threshold at which fall protection was needed to six feet. Residential construction companies trying to comply with this rule found that they experienced *more* fall and trip hazards because of what had to be done to satisfy the rule, so OSHA later agreed to reopen the rule so residential worksites could be studied further.

All of this give-and-take over minimum heights and fall-protection system requirements, however, can sometimes miss the true mark—which is: *you want your employees to avoid falls.* Period. A fall from $5^1/_2$ feet can be just as dangerous as one from 6 feet, or 10 feet. Again, don't shift all of your attention to meeting *minimum requirements.* Ask yourself, is there a potential for falling at the worksite? If there is, how can you protect your employees? Think of what's best for your employees. Think of the consequences a serious injury will bring—both to the injured and to the company. That, in itself, is a strong argument for 100 percent fall protection. Some construction companies in business now are practicing 100 percent fall protection, and so can your company.

### Where fall hazards exist

There are plenty of places in the typical construction worksite from which employees can fall. Some of these hazards include:

- Unprotected leading edges of floors, roofs, and building components under construction.
- Edges of trenches and excavations.
- Roofs of all pitches.
- Finished and unfinished skylights.
- Stairways, ladders, scaffolds, and other work platforms.
- Wall openings.

## Fall-Protection Equipment

There are lots of fall-protection systems and pieces of safety equipment on the market, with more being developed all the time. One good source of information comes from the roofing industry, and another comes from companies that specialize in confined-space entry equipment and rescue operations. The topic is also frequently covered in publications prepared by safety organizations such as The National Safety Council, and in the journals of numerous construction trades. Your local OSHA office can provide input on types of systems, and so can sales representatives from industrial supply companies—who can often put you in direct contact with several major manufacturers of fall-protection equipment. Here are some kinds of equipment currently available:

### Full body harnesses and belts

There are numerous reasons to use full body harnesses (see Fig. 23.1) instead of belts for personal fall protection. It's not just because OSHA has made belts illegal as of January 1, 1998. It's for safety's sake. When a fall occurs, belts deliver a vicious jerk and severe impact to a person's midsection—which has had disastrous effects to many workers. A full body harness, however, will suspend a person in a sitting position where the buttocks will absorb most of the load. The harness will distribute the falling shock forces fairly evenly throughout the body through the use of a sliding D-ring and subpelvic straps. Harnesses offer shock-absorbing designs for the body, and they are adjustable for good fits, with buckles and hardware designed not to cause injury upon impact. Imagine how it would be to be suspended by a single waistline belt—the uncomfortable and dangerous positions an employee could end up in.

### Shock-absorbing lanyards

Shock-absorbing lanyards are fall-protection devices whose jobs begin once a fall occurs (see Fig. 23.2). They're attached at both ends—one end to a full body harness that someone is wearing, and the other end to a sturdy anchor point. If a person falls and an impact load is placed on the lanyard, the "woof" material is progressively torn apart, helping to absorb the initial shock of the fall, as well as quickly and smoothly decelerating the person.

Manufacturers, however, recommend about 20 feet of clearance to safely use shock-absorbing lanyards. This is needed to provide enough room for the

**Figure 23.1** Full body harness.

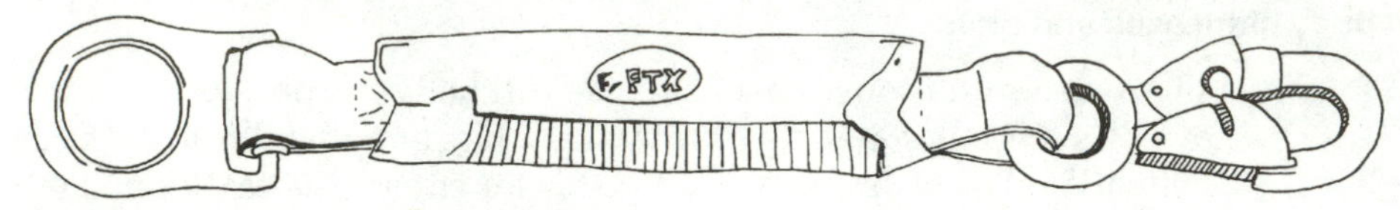

**Figure 23.2** Shock-absorbing lanyard.

lanyard's braking action, and so that the falling individual will not come into contact with anything on the line of fall.

### Self-retracting lifelines

Self-retracting lifelines are fall-arrest devices using self-retracting cables. They combine locking mechanisms with energy-dissipating components that allow almost immediate—yet smooth—fall arrest. They're fully automatic, and require no adjustments by the users. Self-retracting lifelines are typically anchored to a point above the back attachment D-ring of a full body harness. They're designed for use on jobs on vertical, horizontal, or inclined planes. Users need to consider the strength of the anchor point and also the pendular swing effects before using a self-retracting lifeline. Self-retracting lifelines operate on the principle of centrifugal force. Their cables give the worker the freedom and comfort to do whatever job is at hand, while providing protection that will stop a fall in two feet or less.

Because self-retracting lifelines involve components that are not able to be fully inspected by the users, the units should be inspected by the manufacturer at least once per year under normal operating conditions. They're easy to maintain, but certain precautions must be taken to ensure safety of operation:

- Lifelines must be inspected before each use. The user should pull out the entire length of cable to check for damage, and should pull sharply on the cable several times in order to verify that the braking mechanism is working properly.
- The system must only be used by workers who have been trained by a competent person in its operation and maintenance.
- To limit the possibility of a swing or pendulum effect, the system must be anchored as close as possible to the user's head.
- Self-retracting lifelines must not be used with a body belt of any type. They must be attached to the back D-ring of a full body harness for maximum safety.
- When the system is not in use, the cable should be fully rewound into the housing. It should always be rewound slowly. Never let it rewind freely even for a short distance.
- If an impact load is generated against the unit, it should be taken out of service immediately and sent back to the manufacturer for inspection and recertification.

### Positioning lanyards

Positioning lanyards are generally constructed of stretch-resistant three- to six-foot lengths that help a person with positioning only—they'll prevent someone from getting too close to an edge or other hazard. They are available with a variety of anchorage hooks.

### Snaphooks

Snaphooks (see Fig. 23.3) can be used as anchoring connectors on positioning devices and on shock-absorbing lanyards. They often contain double-locking mechanisms to ensure against rollout or other forms of accidental opening.

As with all fall-protection devices, snaphooks should be inspected before each use:

- Determine if the hook gate can be opened without depressing the lock mechanism. If it can, discard the hook immediately.
- Check the integrity of the spring mechanism. If a reasonable amount of force is not needed to open the gate, or if the gate does not snap back into place, the spring may be worn or damaged. If so, discard the hook immediately.
- Check for visible signs of damage or wear. If the hook is cracked or otherwise damaged, or has been subject to a severe impact load, discard the hook immediately. Bear in mind that if the hook has been subject to a load, the device to which it is attached may need to be discarded as well, depending on the device.

## Recommendations for Using Fall-Protection Equipment

Such equipment is designed to protect people's lives. It must be selected, used, maintained, and stored by competent and properly trained people who will carefully:

- Determine and evaluate fall risks on the jobsite.
- Define an appropriate fall-arrest system for the job to be performed, and provide an appropriate plan for rescue.
- Preferably select all system components from the same manufacturer.
- Ensure that the equipment selected is in compliance with the laws and standards currently in force.
- Check the conditions of use of the system by reading and complying with the instructions supplied with the components.

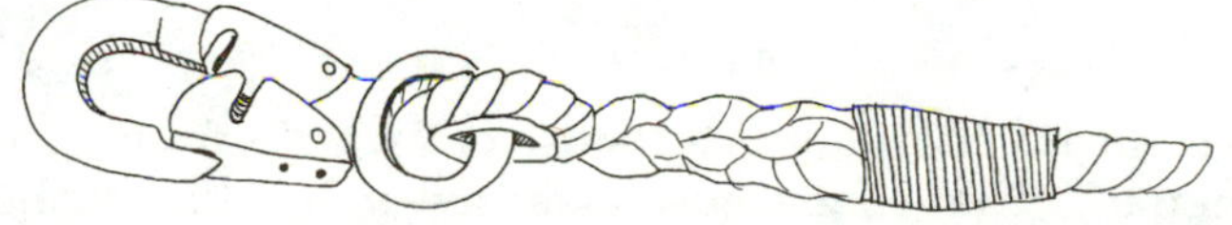

**Figure 23.3** Snaphook.

- Provide the user with initial and continuous training necessary for handling, using, maintaining, and storing the equipment entrusted to him or her.
- Select reliable anchorage points located as close as possible to the user, and if possible above the user's head, having a minimum static strength at failure of at least 5,000 pounds.
- Carefully check each of the system components before each use.
- Avoid the user's working alone.
- Store the equipment under conditions that will not alter the system's components.
- Prevent any modification of a system component without prior agreement from the manufacturer.
- Report any defect, anomaly, wear, or fall that might affect a system component.
- Immediately remove from service any component that has been subjected to an impact load.

More specifically, for residential fall protection:

#### Fall-proofing floor and wall openings

- Install guardrails around open floors and walls where the fall distance to the next level is six feet or greater. The top rails must be able to withstand a 200-pound load.
- Construct guardrails with a top rail at 42 inches of height, with a midrail about half that high (21 inches).
- Install toeboards when other employees are working below the work area.
- Cover floor openings larger than two inches square with material that can safely support the working load.

#### Alternatives to constructing guardrails

- Use other fall-protection systems like slide guards or roof anchors, or alternative safe-work practices when a guardrail system cannot be used.
- Wear proper shoes or footwear to lessen slipping hazards.
- Train workers on safe work practices before performing work on foundation walls, roofs, trusses, and where performing exterior wall erections and floor installations. An example of a safe work practice is to have employees wrap their feet and legs around the webs of trusses to lessen the possibility of a fall.

### Working on Roofs

Employees can get lulled into a false sense of security from spending considerable time on roofs. That edge is just so far away. But the truth is that many employees simply forget where they are, and fail to remember that being off-balance or out of position anywhere on the surface of a sloping roof can mean trouble. Or they elect to work on a roof during weather that makes such work extremely dangerous, in icy or windy conditions, for instance. Some of the most important guidelines for working on roofs include:

- Check for slipping hazards before getting onto roof surfaces. Frost can make certain roofs impossible to walk on safely. Temperatures that hover around the freezing point can turn trickles of rainwater or snowmelt into clear ice before employees become aware of it. Slimy algae-like moss or similar growths can also create dangerous slipping hazards. So can buildups of snow, dirt, leaves, and other debris.
- Cover and secure all skylights and openings, or install guardrails to keep workers from falling through the openings.

  Every year it happens. Someone goes up on a roof to perform a particular job, gets caught up in some detail, forgets about the skylight he or she just walked past, then accidentally steps backwards through it, falling all the way to a lower level. It happens more often than you might think.
- When the roof pitch is over 4:12, and up to 6:12, install slide guards along the roof eave after the first three rows of roofing material are installed.

  Manufactured slide-guard brackets are roof brackets that support two-by-six boards at a 90-degree angle. Together the brackets and boards are designed to stop something or someone from sliding down the roof surface. When the roof pitch exceeds a 6:12 pitch, additional slide guards should be installed every eight feet up from the first slide-guard installation.
- The excellent *Jobsite Safety Handbook,* prepared in a joint effort by the National Association of Home Builders and OSHA, says to use a safety harness system with a solid anchor point on roofs with pitch greater than 8:12, or if the ground-to-eave height exceeds 25 feet. Naturally, that doesn't mean that you can't use the same system with a ground-to-eave height of 22 feet. Or 16 feet. Again, remember that some rules, regulations, and suggested practices you'll read and hear about are *minimum* precautions. You've got to study the particular situation to arrive at the best solution.

  Anyway, to not use fall-protection equipment where steep roofs or high ground-to-eave distances are involved is just begging for disaster to strike.
- When storms, lightning, hail, high winds, or other severe weather conditions create unsafe conditions, roofs must be *off limits* to employees.

  No amount of scheduling pressure should override nasty weather when it comes to roofing operations, or performing work that can be accessed only by roof.

## OSHA's Subpart M, Appendix A

This appendix helps explain how roof widths can be determined, and supplies nonmandatory guidelines for complying with 1926.501(b)(10), and has been borrowed from the same standard. The illustrations present OSHA examples A through F, in Figs. 23.4 (OSHA Example A), 23.5 (OSHA Example B), 23.6 (OSHA Example C), 23.7 (OSHA Example D), 23.8 (OSHA Example E), and 23.9 (OSHA Example F).

**Example A - Rectangular Shaped Roofs**

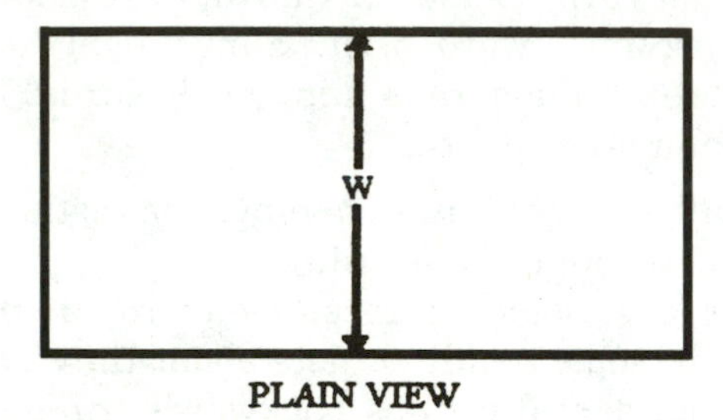


**Figure 23.4** OSHA Roof widths, Example A.

**Example B - Sloped Rectangular Shaped Roofs**

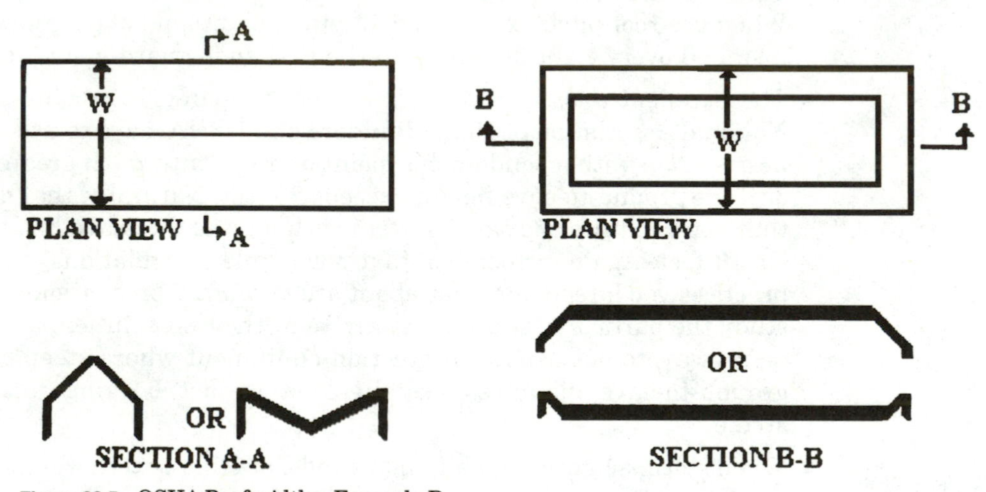


**Figure 23.5** OSHA Roof widths, Example B.

## Example C - - Irregularly Shaped Roofs With Rectangular Shaped Sections

Such roofs are to be divided into sub-areas by using dividing lines of minimum length to minimize the size and number of the areas which are potentially less than or equal to 50 feet (15.25 meters) in width, in order to limit the size of roof areas where the safety monitoring system alone can be used [1926.502(b)(10)]. Dotted lines are used in the examples to show the location of dividing lines. W denotes incorrect measurements of width.

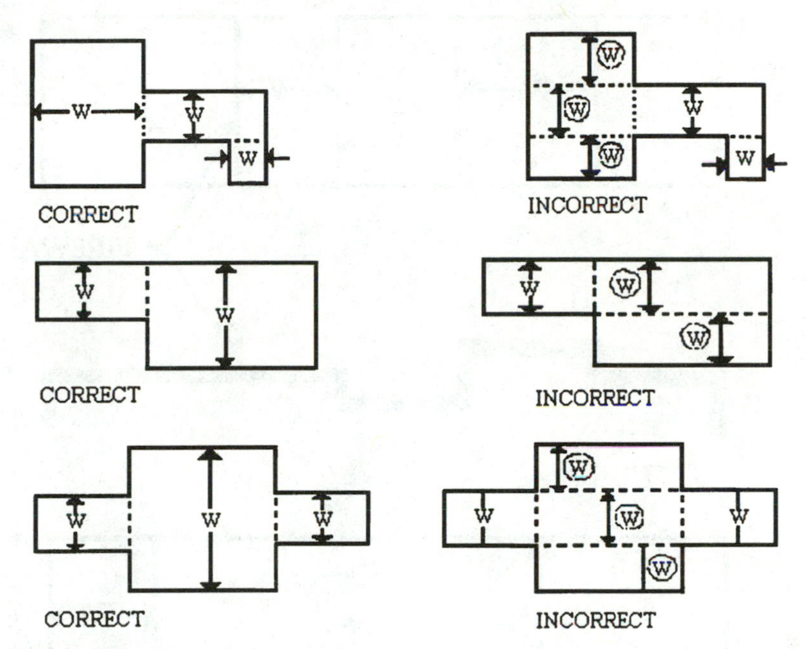


**Figure 23.6** OSHA Roof widths, Example C.

## Example D
## Separate, Non-Contiguous Roof Areas

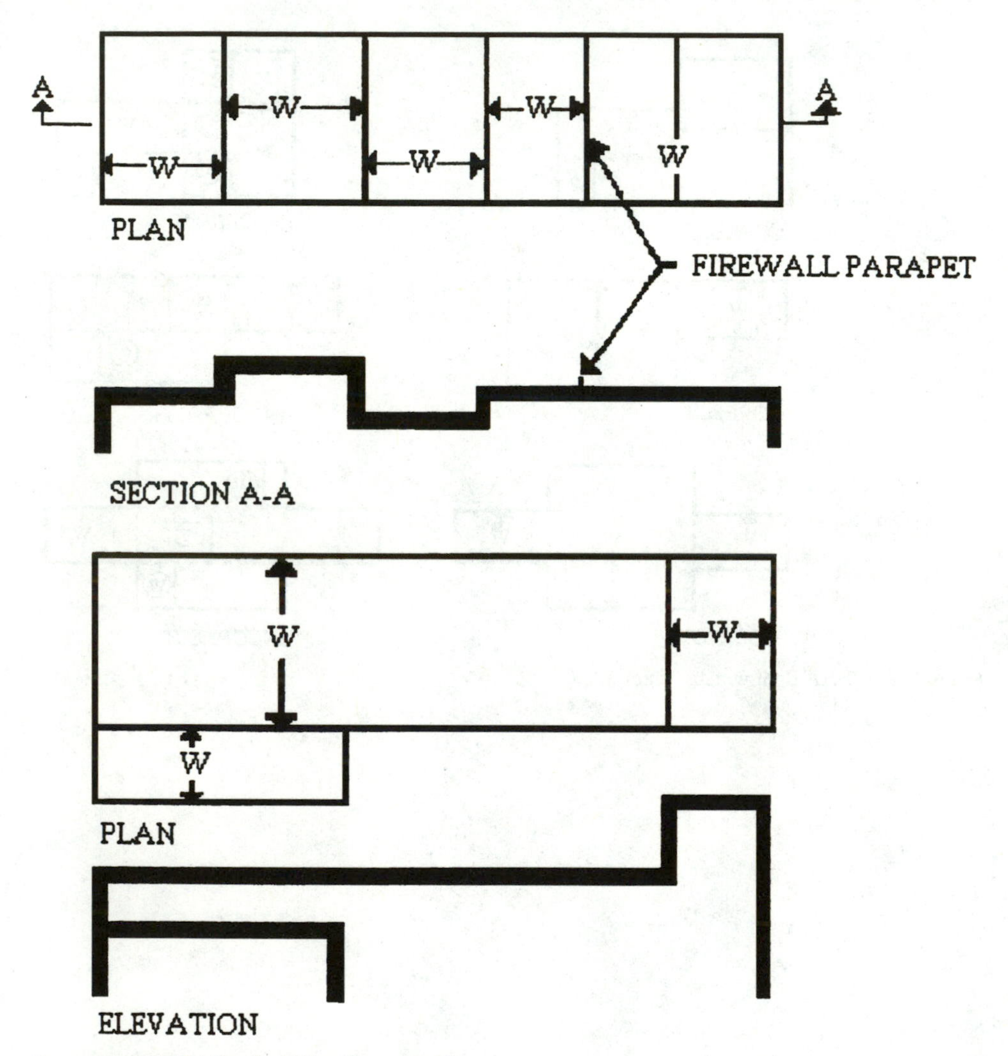


**Figure 23.7** OSHA Roof widths, Example D.

### Example E - Roofs With Penthouses, Open Courtyards, Additional Floors, etc.

Such roofs are to be divided into sub-areas by using dividing lines of minimum length to minimize the size and number of the areas which are potentially less than or equal to 50 feet (15.25 meters) in width, in order to limit the size of roof areas where the safety monitoring system alone can be used [1926.502(b)(10)]. Dotted lines are used in the examples to show the location of dividing lines. W denotes incorrect measurements of width.

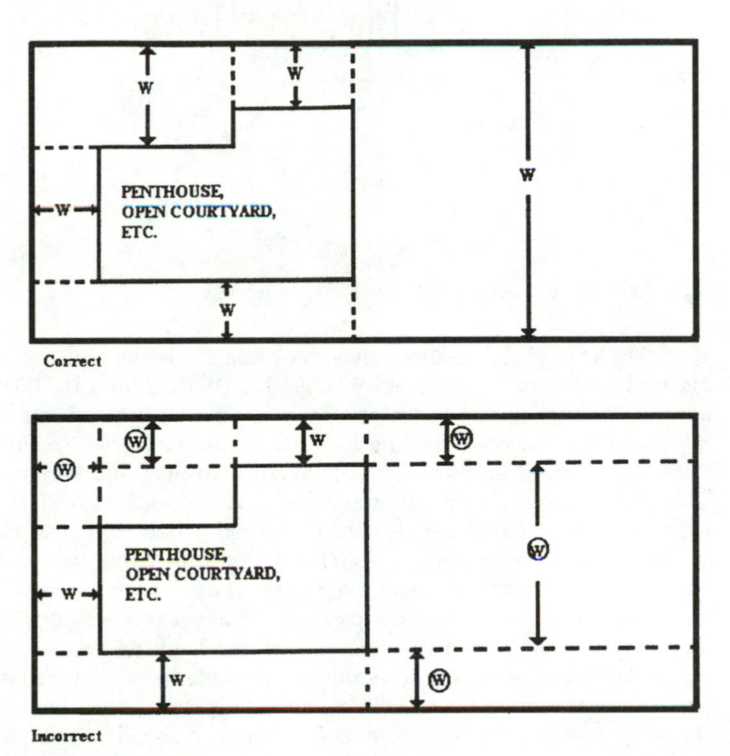


**Figure 23.8** OSHA Roof widths, Example E.

## Example F - Irregular, Non-Rectangular Shaped Roofs

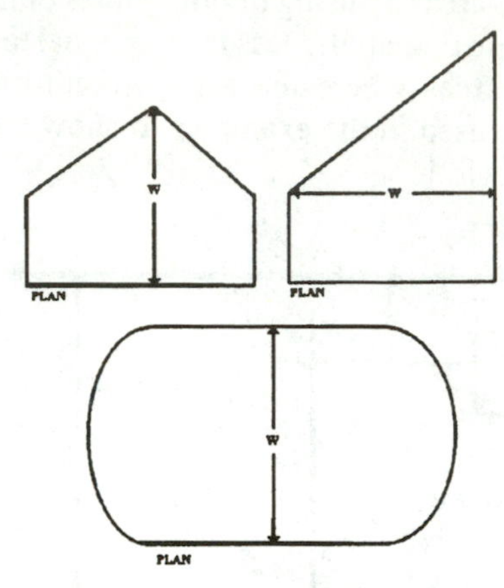


**Figure 23.9** OSHA Roof widths, Example F.

1926.501(b)(10)(1) This Appendix serves as a guideline to assist employers complying with the requirements of 1926.501(b)(10). Section 1910.501(b)(10) allows the use of a safety monitoring system alone as a means of providing fall protection during the performance of roofing operations on low-sloped (means a roof having a slope less than or equal to 4 in 12, vertical to horizontal) roofs 50 feet or less in width. Each example in the appendix shows a roof plan or plans and indicates where each roof or roof area is to be measured to determine its width. Section views or elevation views are shown where appropriate. Some examples show "correct" and "incorrect" subdivisions of irregularly shaped roofs divided into smaller, regularly shaped areas. In all examples, the dimension selected to be the width of an area is the lesser of the two primary dimensions of the area, as viewed from above. Example A (see Fig. 23.4) shows that on a simple rectangular roof, width is the lesser of the two primary overall dimensions. This is also the case with roofs which are sloped toward or away from the roof center, as shown in Example B (see Fig. 23.5).

(2) Many roofs are not simple rectangles. Such roofs may be broken down into subareas as shown in Example C (see Fig. 23.6). The process of dividing a roof area can produce many different configurations. Example C gives the general rule of using dividing lines of minimum length to minimize the size and number of the areas which are potentially less than 50 feet wide. The intent is to minimize the number of roof areas where safety monitoring systems alone are sufficient protection.

(3) Roofs which are comprised of several separate, noncontiguous roof areas, as in Example D (see Fig. 23.7), may be considered as a series of individual roofs. Some roofs have penthouses, additional floors, courtyard openings, or similar architectural features; Example E (see Fig. 23.8) shows how the rule for dividing roofs into subareas is applied to such configurations. Irregular, non-rectangular roofs must be considered on an individual basis, as shown in Example F (see Fig. 23.9).

While this chapter's intent is to provide some fall-protection guidelines, it cannot take the place of competent persons, formal training, and a close reading of the standards involved. Fall protection is one of the most important necessities of the construction industry; it should be given full attention and support by your company's management.

Part

# 4

# Recognizing and Dealing with Jobsite Conditions and Hazards

*Section 4 gets into the hazards. Slips, trips, and falls may sound like sissy stuff to many employees, but the end results are sobering. Housekeeping is another topic that's given reluctant attention by many companies, often coming into play only at the tail end of a typical job. Hazardous energy control—if this is a phrase your company's employees are unfamiliar with—it will be a critical new addition to your safety program. Part 4 also reviews electrical work practices, hazard communication, dealing with chemicals at the worksite, confined spaces, and fire protection. These are important topics that demand hazard recognition and prevention through meaningful training and effective work procedures. Other topics include tools and equipment, ladders, stairs, and scaffolding; material handling and mobile equipment operations; plus excavation and trenching activities. This part concludes with material on working in hot and cold environments, critter exposure, and site security.*

Chapter

# 24

# Slips, Trips, and Falls

## Quick Scan

1. Train employees to recognize slipping, tripping, and fall hazards.
2. Insist that those hazards either be removed or barricaded when found.
3. Frequently audit for worksite behaviors and conditions that could lead to falls.
4. Allow reasonable time periods for the completion of individual work tasks.
5. Train employees on proper housekeeping and material-handling procedures.

Slips, trips, and falls, as simple as they seem, can be deadly. Overall, only traffic accidents cause more accidental deaths. Most falls are slips or trips that occur at ground- or floor-level, not falls from high places. Unfortunately, they often result in expensive, even disabling injuries—especially broken bones and torn or sprained muscles and ligaments. And most of them could have been easily prevented in advance.

Slips often occur when hazardous walking conditions are not recognized until too late. Typical culprits include smooth or uneven surfaces; wet, oily, greasy, or icy spots; loose flooring; and partially completed stairs, steps, and flooring. Trips are generally associated with poor housekeeping and careless storage of materials. Examples include crowded staircases and aisleways, cluttered floors, poorly strung extension cords, and improper temporary storing of tools and materials.

But sometimes it's not that people don't see materials placed in aisleways or on steps; they do. They see the hazard, but instead of eliminating the condition, they'll step around or over or through it. They'll allow the hazardous condition to remain while they count on consciously remembering that it's there so it can be avoided. That, of course, is not the safe way of handling things. It's too easy to forget about a tripping hazard. Or simply to not pay attention. No, the way to greatly reduce the likelihood of slips, trips, and falls is through following proper housekeeping and material-handling procedures, actively auditing for slip and trip hazards, and arranging for refresher slip, trip, and fall training.

## Slip, Trip, and Fall Training

The following information can be stressed at tool-box safety meetings. Employees should be reminded to:

- *Wear good safety shoes.* Nonskid soles are a must, and shoes should not be worn if they have become threadbare and smooth on the bottom. Leather heels are a no-no; rubber or man-made soles that grab should be chosen instead. The uppers should provide sturdy ankle support, and the laces must be kept tied. Some types of lacing are notorious for easily coming loose; these include the kinds that are drawn around open-sided eyelets instead of being laced through holes.
- *Pay particular attention to the installation of staircases.* Be especially careful with staircases that are put up for temporary use; there's a tendency to cut corners here, and that must be avoided. Treads that are cracked or that spread beyond their stringer supports must be repaired or replaced. Watch that stringers don't spread and create a hazardous condition that could lead to failure under a load.
- *Recognize dangerously low levels of work lighting.* Inadequate lighting during construction of basement, garage, and attic areas has caused many a fall for construction employees. Brightness is especially important in narrow staircases. Employees should not have to grope in the dark. A flashlight or extension light should be provided to improve footing in unlighted areas.
- *Avoid makeshift substitute ladders.* Standing on tool boxes, toilet seats, material packaging, and the like is a quick way to make unplanned floor contact. Review the ladder section in Chap. 31 with all employees.
- *Avoid jumping from elevated surfaces.* Employees cannot be permitted to leap from the bed of a pickup truck, or from the front steps of a front-end loader. The same goes for jumping over trenches and excavations. Likewise, they need to be careful when exiting a floor under construction: even though the vertical leap may be relatively small, landing surfaces are usually not flat. If an angle is involved, the landing may take an unexpected turn, and the impact could easily stretch and stress tendons, muscles, ligaments, and bones that aren't used to those kinds of impact pressures.
- *Instruct employees to follow your company's standard operating procedures for all tasks and processes.* Employees must learn the safety rules for their jobs, whether they're working on roofs or on the ground. They need to know how to use all of the equipment necessary for their tasks, as well.
- *Let employees pace themselves.* Allow enough time for individual tasks so that workers won't have to take shortcuts. Be realistic with your expectations. Don't expect or encourage goal-setting that is unrealistic.
- *Employees need to be reminded simply to be careful.* That includes moving at safe speeds, watching out for other people, changing direction slowly. Indeed, to watch where they're going, not so much where they've been.

- *Falling does not abide by equal-opportunity laws.* Age often plays a role in the outcome of a fall. The older the person is, the more likely the results will be severe. Age is discriminatory when it comes to broken bones, strains, and other ailments.

## How to Fall

If an employee does fall, he or she can reduce the probability of injury by falling the right way. Stunt actors know how to fall and can, in fact, take a tumble with little danger of injury. Of course, they didn't just automatically know how to fall; they had to learn how by studying techniques and practicing, gradually easing into the more difficult falls after mastering easier ones. Guidelines for safe falling include:

1. *Relax.* Try not to stiffen and tense muscles during a fall. A brittle appendage is more likely to be damaged than one that's supple and giving, able to take up the slack easily.
2. *Absorb.* The arms and legs must give or bend like a spring, to absorb the impact of the fall. It might sound crazy, but an hour or so with a martial arts instructor can instill the basics of how to fall. Martial arts students practice falling all the time, and they have the benefit of exercise mats to practice on. Again, falling is an art that must be learned and practiced in order to become a natural and instantaneous reaction. There is some rolling, and some continuity of motion while the energy is expended.
3. *Roll.* Try to move with the direction of the energy of the fall—to minimize injury while harmlessly dissipating the force.

## More Tips on Reducing Falling Hazards

- *Expect the unexpected.* Falls happen because people don't expect them. The more that employees anticipate and guard against falls, the fewer they'll have. That doesn't mean that workers have to go through their daily work tasks *constantly* thinking about falling, but they do have to learn to recognize slipping and tripping hazards and how to avoid or correct them.
- *Be aware of employee abilities and limitations.* Supervisors need to be aware of individual employee capabilities. They shouldn't expect unreasonable amounts of work to be accomplished just to meet construction deadlines.
- *Do things the safe way, even if that way is more difficult and requires a little more time.* This means taking a few extra minutes to properly stack materials; to barricade a slippery spot at the top of a stairwell; and to place a walk-off floor mat at the front entrance, where mud and water have been tracked in from a sloppy foundation excavation. This means recognizing the importance of slip, trip, and fall prevention.

# Chapter 25

# Housekeeping

## Quick Scan

1. Develop a recycling plan for the scrap materials that result from your employees' work.
2. Make sure that your employees know how important it is to practice good housekeeping at their worksites.
3. Provide the containers, equipment, and support needed for an effective housekeeping program.
4. Encourage employees to pay attention to proper housekeeping throughout their workdays. Don't accept sloppy work areas, especially at the end of the workday.

I remember years ago showing a prospective home-buyer a house that was in a state of mid-construction. He walked near the foundation and basement block walls, which were not yet backfilled, and motioned for me to come over to look at something. He pointed to the bottom of the excavation, where I saw numerous pieces of debris—short pieces of two-by-fours, two-by-sixes, cardboard, chunks of block, styrofoam, fast-food take-out bags and wrappers, and other items, including a few crumpled beer cans. The man shook his head and remarked, "This builder's not for me." On the drive to the next house he talked about his concerns—not only that the wood, cardboard, and other organic waste that had been tossed in next to the foundation could eventually decompose and cause the backfilled earth to settle, or could attract wood-destroying insects, but that if the supervisor or builder didn't have any control over where the waste was going, and the workers didn't care, then how could the rest of the work be trusted? After all, a lot of construction work occurs and remains out of sight below ground, within walls, and hidden throughout the structure. In that home-buyer's mind, poor housekeeping habits led to an unfavorable first impression of the builder's entire operation.

To the potential customer, that of course had nothing to do with safety. But it will with us. Housekeeping standards are a proven barometer—indicative of how safe a company's employees tend to be. Show me a sloppy operation, and I'll wager the company's safety statistics are also substandard.

## The Importance of Good Housekeeping

Being neat, tidy, and organized might not seem very important, but they are. Those terms carry a lot of weight in the successful operation of any business, yours included. Poor housekeeping can result in many serious problems:

- *Low morale.* Substantial psychological stress can occur to people (including your employees) who occupy cluttered, disheveled work areas. Hey, it's depressing. Consider your own residence. Don't you feel better when everything is neat and orderly? Consider any kind of operation. How would you feel about a disorderly hospital? Would you trust its physicians and nurses? Or a restaurant, if the kitchen was sloppy, with garbage all over, and the employees were messy? Would you trust the food? Or a bank, if the lobby needed maintenance, the windows were dirty, light bulbs were out, and papers lay all over? Would you trust a bank like that with your life savings? Probably not. Nor do people want to deal with messy contractors.
- *Excessive generation of waste.* Wasted time and materials make it more costly for your company to do business. The more costs you incur, the more you have to charge for your products and work, and the less favorable you'll compare to other contractors. Why handle debris two or more times?
- *Fires.* Piles of waste invite accidental or arson fires.
- *Accidents.* A definite correlation exists between the frequencies of some types of accidental injuries and housekeeping: the poorer the housekeeping, the greater the likelihood of injury. Slips, trips, and falls increase. Strains and sprains closely follow. Lacerations increase from double-handling of scrap materials. Boards with nails and other fasteners get stepped on. Dust and sawdust build up and get blown into eyes when power tools are used or when winds whip through open areas.

## Insist on Good Housekeeping Standards

What, then, can be done? Simple. Instruct your supervisors and employees to maintain proper housekeeping standards by following four simple steps:

1. Clean all tools, equipment, and work areas at the end of each workday, and keep things reasonably clean and orderly while working.

   By keeping at housekeeping chores a little at a time, they won't become overwhelming. This goes for the inside of vehicle cabs as well. A pop bottle rolling around on the cab floor could lodge beneath a brake pedal and prevent an emergency stop.

2. Put all waste in the proper bins or places for efficient, safe removal. Do you really want your employees to chuck everything into one big pile? Wouldn't containers earmarked for specific types of debris be more effective? Is there a separate container for lunch-type trash and garbage? There should be.
3. Keep floors, stairways, and aisles free and unobstructed. Report unsafe conditions to the supervisor. Employees should be asked to help keep the housekeeping operations going if and when things start getting out of control.
4. Properly stack, store, and put away materials, tools, and equipment.

## What OSHA Says

OSHA definitely considers housekeeping to be a major indicator of a company's overall level of management. A sloppily run housekeeping operation will likely mean an inferior safety program as well. Inspectors encountering a disorganized, messy worksite will probably not begin the inspection with a warm, fuzzy feeling.

OSHA's standard on housekeeping follows:

> 1926.25(a) During the course of construction, alteration, or repairs, form and scrap lumber with protruding nails, and all other debris, shall be kept cleared from work areas, passageways, and stairs, in and around buildings or other structures.
>
> (b) Combustible scrap and debris shall be removed at regular intervals during the course of construction. Safe means shall be provided to facilitate such removal.
>
> (c) Containers shall be provided for the collection and separation of waste, trash, oily and used rags, and other refuse. Containers used for garbage and other oily, flammable, or hazardous wastes, such as caustics, acids, harmful dusts, etc. shall be equipped with covers. Garbage and other waste shall be disposed of at frequent and regular intervals.

## Recycling

Whenever possible, set up a system of recycling. Instead of simply burning scrap wood, or hauling mixed debris to a landfill, recycle. Contact the recycling representative in your county for help. By doing so, you can possibly (in addition to benefitting the environment) encourage other contractors to recycle their wastes, too. And that could even gain some public works publicity or promotional exposure—which wouldn't hurt your company's business, either. Small amounts of materials, such as nails, fittings, and pipe ends can be saved and collected at the shop to accumulate enough to recycle.

The key here is to develop a system with input from your employees, who will know why something can or cannot be done, and once a plan is arrived at, to make sure that everyone is trained to follow it. Provide the necessary containers and make arrangements for the materials to be accepted for recycling. Then audit the system frequently, and keep it consistent from worksite to worksite. An additional benefit of recycling is that potential customers, and other members of the community, will recognize that recycling is not a huge money-making endeavor for your company, but it's rather a voluntary way of helping the local environment.

If you have individual employees doing work for others, make sure they clean up after themselves. Leave no scrap unhandled. Make sure that everyone understands what arrangements need to be made, such as where the scrap is to be left, and where the packaging materials or unused building materials should be taken. The same goes for subcontractors. Insist that as part of their services, all remaining products or scraps and debris generated by their operations be properly disposed of or taken off the worksite. Train your company's employees to watch out for debris that subcontractors appear to be abandoning at the site.

Housekeeping must not be left to chance. Instead, use it to your advantage. Maintaining a neat, organized worksite will benefit your company in many ways, including fewer injuries, less waste-generation and costs, improved morale, higher efficiency, and a better public image.

## Worksite Illumination

An added note to housekeeping is proper illumination. Make certain that there's adequate lighting in all work areas—especially during early morning or early evening hours when employees may need to gain access to portions of garages, attics, or basements that may not yet have light fixtures installed.

OSHA, in 1926.26 and .56, states the following on illumination:

> Construction areas, aisles, stairs, ramps, runways, corridors, offices, shops, and storage areas where work is in progress shall be lighted with either natural or artificial illumination. The minimum illumination requirements for work areas are contained in Table D-3 (see Fig. 25.1). The illumination measurement units—called foot-candles—are early units of illuminance, or the illumination resulting when 1 lumen of luminous flux is uniformly distributed over an area of 1 square foot. One foot-candle, in turn, equals 10.76 lux. Further information on lighting, including recommended values of illumination, can be found in the American National Standard A11.1-1965, R1970, Practice for Industrial Lighting.

**TABLE D-3**
**Minimum Illumination Intensities in Foot-Candles**

| Foot-Candles | Area of Operation |
|---|---|
| 5 | General construction area lighting. |
| 3 | General construction areas, concrete placement, excavation and waste areas, access ways, active storage areas, loading platforms, refueling, and field maintenance areas. |
| 5 | Indoors: warehouses, corridors, hallways, and exitways. |
| 5 | Tunnels, shafts, and general underground work areas: (Exception: minimum of 10 foot-candles is required at tunnel and shaft heading during drilling, mucking, and scaling. Bureau of Mines approved cap lights shall be acceptable for use in the tunnel heading) |
| 10 | General construction plant and shops (e.g., batch plants, screening plants, mechanical and electrical equipment rooms, carpenter shops, rigging lofts and active store rooms, mess halls, and indoor toilets and workrooms). |
| 30 | First aid stations, infirmaries, and offices. |

**Figure 25.1** OSHA's Table D-3, Minimum Illumination Intensities.

Chapter

# 26

# Hazardous Energy Control

## Quick Scan

1. Employees must be aware of the many types of energy (including electrical, mechanical, hydraulic, pneumatic, chemical, thermal, and radiation energy) that when mishandled can cause injury at the worksite.
2. Electricity is invisible and dangerous. Even common household electrical current can kill.
3. Set a rule prohibiting employees from cleaning, adjusting, or repairing electrical tools or components that are plugged in or energized. Establish a "zero energy" procedure for such work—which means unplugging or shutting power down so employees can't be accidentally shocked or electrocuted.
4. Employees must wear and use nonconductive personal protective equipment whenever the possibility exists of accidental contact with electrical components.
5. Power boxes, wiring, or even tools or equipment that are turned off but plugged in should not be handled by employees who are wet or who are standing on wet or damp surfaces.
6. Require that all worksite electrical systems be properly grounded and protected with ground-fault circuit interrupters. This must be part of an assured equipment grounding program planned and overseen by a competent person.
7. Follow manufacturer recommendations for extension cord usage. Use outdoor-rated cords for outdoor tasks, and heavy-duty cords for power tools.
8. Train employees to unplug and remove extension cords that aren't being used. Inspect tool cords and extension cords and plugs frequently.
9. Don't allow unqualified employees to work on, maintain, or install electrical equipment.

10. Employees must be aware of the location of overhead and underground power lines, and must ensure that equipment they are using does not come in contact with those power lines. Special care should be taken with cranes, backhoes, front-end loaders, dump trucks, and other mobile equipment, and while working with roofs, siding, or gutters.
11. Read and adhere to Subpart K—Electrical, 1926.400 through .449.

This chapter, for argument's sake, could be called the most important chapter in the book. "The prevention of unwanted and unexpected activation or energizing of equipment or power systems," as the more literate regulatory individuals among us may say, is absolutely necessary to maintain a safe worksite. And it is. Although that statement sounds impressive, it's really quite simple. It means to *make sure* that:

While an employee is

transporting,

working on,

maintaining,

or repairing something,

whatever he or she is working on or near cannot become unexpectedly energized or activated.

## Energy Sources

There are numerous sources of energy in the world. And, depending on the activities going on at your worksites, many different energy sources—and their associated hazards—can be present there:

- *Electrical* This includes alternating and direct current sources, as well as static electricity, or stored electrical energy devices such as batteries. Electricity is a major source of energy at the typical worksite.
- *Mechanical Potential* Mechanical potential includes sources of energy that derive their power from gravity (including items that could cause injury or damage by falling, slipping, or rolling) as well as from compression, such as springs or other stored mechanical forces. A rat trap with its jaws fully set is a good example of stored mechanical energy. A small sledgehammer placed on the top of a stepladder is an example of stored energy that is hazardous to humans. If an unsuspecting employee picks up the ladder, the hammer will probably topple off—perhaps into the employee's face.
- *Hydraulic* Hydraulic energy comes from fluids pressurized to perform work, such as those used to operate heavy mobile equipment such as backhoes and front-end loaders and dumping mechanisms on trucks. Equipment misuse or poor maintenance procedures can expose operators and other employees to hot, pressurized hydraulic fluids.

- *Pneumatic* Pneumatic energy comes from gaseous systems operating at positive (compressed) pressure or negative (vacuum) pressure. Most pneumatic systems are air-powered. Perhaps because "air-driven" sounds less threatening than electric or even hydraulic, employees are less likely to fear pneumatic systems—which, in turn, can lull individuals into a false sense of security.
- *Chemical* Chemical energy is released through simple contact with a chemical or chemicals, or results from the combination of different substances or gases.
- *Thermal* Thermal energy is heat that comes from electrical, combustion, chemical action, mechanical friction, or other sources.
- *Radiation* Radiation is energy consisting of alpha, beta, neutron, gamma, and x-rays. Although unlikely to be present at the typical construction site in dangerous forms, it does include nonionizing sources, including ultraviolet, infrared, microwave, and visible light. Lasers are classified here and can be found as components of surveying equipment.

## Electricity

Near the top of the list of dangerous worksite energy sources is electricity. Although we use electricity day in and day out, few individuals truly understand what electricity is and how it works. One problem with electricity is that it's invisible. Unless sparks are flying or smoke is coming from somewhere that it shouldn't, an electrically charged wire, component, or piece of equipment can appear to be harmless. And because electricity is such a common, everyday part of life, people tend to forget how dangerous it really is.

A common perception is that "normal" electricity, the kind available at standard 110-volt outlets, is not dangerous. In truth, at one time or another, most people have experienced a mild electrical shock and never considered it to be harmful or even dangerous. Nothing could be further from the truth. *All electricity should be considered dangerous.* It doesn't take much electricity to kill.

### What Is Electricity, Really?

Put simply, electricity is an energy source created by an imbalance of electrons in tiny building blocks of matter called atoms. As the electrons try to "correct" or "balance out" their unequal distribution, they're driven by a force or electrical pressure called "voltage" to move in what's known as an electrical current. A higher voltage can force more current through your body more easily than a lower voltage can. A higher voltage can cause violent muscular contractions, often so severe that the victim is thrown clear of the circuit. Although low voltage also results in muscular contractions, the effect is not so violent. Low voltage, however, often prevents the victim from freeing him or herself from the circuit. This is what makes contact with low voltage dangerous. In fact, 110 volts (common household electrical service) is the voltage that electrocutes most people today. A low voltage might appear to be harmless, but it can be deadly if a victim is exposed to it over a long period of time.

Current flow is the factor that causes injury in electrical shock. A person's resistance to current flow is found mainly in his or her skin resistance. Callous or dry skin has a fairly high resistance, but a sharp decrease in resistance occurs when the skin becomes moist or wet. An individual standing on a surface that won't conduct electricity, while making electrical contact with dry hands, may just feel the "buzz" of the electricity. On the other hand, that same individual—wet with perspiration or standing in water—may be killed by the vastly increased current conducted through his or her body. Plus, in general, the longer a current is allowed to flow through a body, the more damage done.

One common analogy used to describe the basic principles of electricity is water. Electricity travels through a system of wires much like water travels through pipes. One main difference, however, is that if a water pipe breaks, the water continues to flow and will spill out from the break. Not so with electricity. If a wire breaks, the flow of electricity will stop because electricity needs a complete circuit to flow. A circuit is simply a path that returns to its power source or point of origin, in a circular fashion. A broken wire, a loose connection, a switch in the off position, or a short-circuiting of the circuit by some other electrical conducting material can all cause the original circuit to be broken.

Metals such as copper and aluminum are used to transmit electricity because those materials are good conductors—electricity flows through them easily, and does not meet with much resistance. Water is another good conductor of electricity, and so is the human body, by virtue of its watery nature. Most nonmetals, such as rubber, plastic, glass, and wood, are poor conductors of electricity. That's why they're often referred to as insulators. An extension cord is a good example of both: a metal conductor (the copper wires) protected by an insulator (the rubber coating or exterior).

## Electrical hazards

There are three basic types of electrical hazards: fire, shock, and burns.

**Fire.** See Chap. 29, on fire prevention. In that chapter, the three essential ingredients of fires are discussed. First, for each fire there must be a source of fuel—something has to be burned or consumed to sustain or feed the fire. Second, there needs to be enough oxygen to keep the burning going. And third, there must be a source of heat or ignition to start the fire. Electricity is a potential source of dangerous heat. The delivery systems, such as the worksite's electrical drop, or generator delivery system are engineered to prevent electricity from building up enough heat to start fires. Controls and safeguards such as fuses, circuit breakers, ground-fault circuit interrupters (GFCIs), and insulators are designed to prevent electricity from being a heat source for fires. A system should shut itself down before temperatures reach a critical level.

Some electrically related fires can be prevented by the use of safe installation and operating practices. Fires have been associated with the use of electricity in the following appliances and installations, and others:

- Electric tools and equipment
- Electric wiring, installation
- Electric household appliances
- Electric lighting
- Electric welding and cutting equipment
- Electric heating equipment

Consider the following pointers for preventing electrical fires:

- Carefully inspect electrical systems, outlets, and cords.
- Follow good housekeeping practices; keep potential fuel sources away from electrical equipment.
- Perform maintenance on electrical equipment. Faithfully follow manufacturer's recommendations for component replacements, cleanings, filter changes, and all repairs.
- Do not overload circuits and outlets.
- Don't operate equipment that has greater energy demands than your electrical system is engineered to supply.

**Shocks and burns.** The two are often related. Accidental contact with electricity can cause a variety of health consequences, from minor tingling feelings to internal organ damage, burns, and even death. Factors that effect the seriousness of injury include:

- The type of circuit and voltage. A shock resulting from just enough power to cause a slight tingling or hair standing on end, or an involuntary movement could be harmless. Or it could cause you to lose your balance and fall, or it could—if the right path is followed through your body—affect a pacemaker or trigger a delicate heart condition. A "medium" shock may result in a loss of muscle control that could cause an inability of being able to let go or pull away from the source of shock.
- The longer the electricity is allowed to run through the body, the more harm will likely be done. Strong shocks can cause damage to internal organs, internal and external burns, cardiac arrest, and respiratory failure.
- The path that the current takes through the body makes a difference. It depends on the body's position at the time—what the hands and feet or other body parts are touching. The circuit always gets completed in the most efficient route available.

## Personal protective equipment

The most effective way to safeguard employees is to make sure they're thoroughly trained to avoid accidental contact with electricity. They need to de-energize a system being worked on or to just stay away from live electrical parts. If that isn't possible, the task must be performed by qualified electricians who are wearing personal protective gear and following established safety rules and procedures.

Nonconductive personal protective equipment should be chosen by employees working with or around electricity. Leather gauntlet gloves can be worn over rubber insulated gloves. Rubber is an insulating material, and leather is a nonconductive material that will prevent the rubber gloves from being punctured or ripped. Nonmetal hard hats can offer protection whenever there's a potential of your head coming into accidental contact with an energized source. Safety glasses with side shields should be worn when there's the possibility of sparks and small particles flying through the air.

Metal jewelry such as watches, rings, chains, and earrings should be avoided when working around exposed energized parts. The metal items could cause serious burns to the skin.

### Assured equipment grounding conductor program

The following excerpt is from OSHA publication #3007, *Ground-Fault Protection on Construction Sites:*

> The assured equipment grounding conductor program covers all cord sets, receptacles which are not part of the permanent wiring of the building or structure, and equipment connected by cord and plug which are available for use or used by employees. The requirements which the program must meet are stated in 29 CFR 1926.404(b)(1)(iii), but employers may provide additional tests or procedures. OSHA requires that a written description of the employer's assured equipment grounding conductor program, including the specific procedures adopted, be kept at the jobsite. This program should outline the employer's specific procedures for the required equipment inspections, tests, and test schedule.
>
> The required tests must be recorded, and the record maintained until replaced by a more current record. The written program description and the recorded tests must be made available, at the jobsite, to OSHA and to any affected employee upon request. The employer is required to designate one or more *competent persons* to implement the program.
>
> Electrical equipment noted in the assured equipment grounding conductor program must be visually inspected for damage or defects before each day's use. Any damaged or defective equipment must not be used by the employee until repaired.
>
> Two tests are required by OSHA. One is a continuity test to ensure that the equipment grounding conductor is electrically continuous. It must be performed on all cord sets, receptacles which are not part of the permanent wiring of the building or structure, and on cord- and plug-connected equipment which is required to be grounded. This test may be performed using a simple continuity tester, such as a lamp and battery, a bell and battery, an ohmmeter, or a receptacle tester.
>
> The other test must be performed on receptacles and plugs to ensure that the equipment grounding conductor is connected to its proper terminal. This test can be performed with the same equipment used in the first test.

These tests are required before first use, after any repairs, after damage is suspected to have occurred, and at three-month intervals. Cord sets and receptacles which are essentially fixed and not exposed to damage must be tested at regular intervals not to exceed six months. Any equipment which fails to pass the required tests shall not be made available or used by employees.

An outline of the equipment grounding conductor program is found in Fig. 26.1.

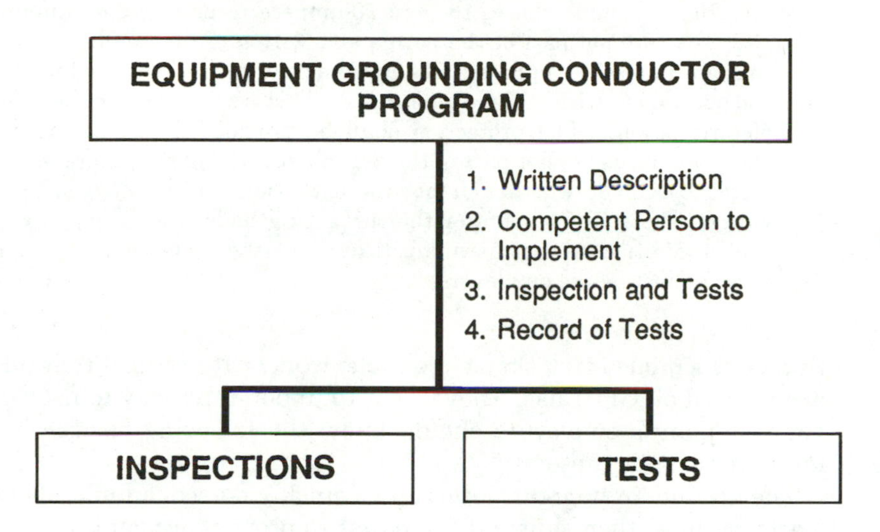


**INSPECTIONS**

Visual inspection of following:

1. cord sets
2. cap, plug, and receptacle of cord sets
3. equipment connected by cord and plug

Exceptions:

- receptacles and cord sets which are fixed and not exposed to damage

Frequency of Inspections:

- before each day's use

**TESTS**

Conduct tests for:

1. continuity of equipment of grounding conductor
2. proper terminal connection of equipment grounding conductor

Frequency of Tests:

- before first use
- after repair, and before placing back in service
- before use, after suspected damage
- every 3 months, except that cord sets and receptacles that are fixed and not exposed to damage must be tested at regular intervals not to exceed 6 months.

**Figure 26.1** OSHA Equipment Grounding Outline.

### Ground-fault circuit interrupters (GFCIs)

Ground-fault circuit interrupters (see Fig. 26.2) shall be used with all construction tools and equipment.

#### What OSHA says.

> The employer is required to provide approved ground-fault circuit interrupters for all 120-volt, single-phase, 15- and 20-ampere receptacle outlets on construction sites that are not part of the permanent wiring of the building or structure and that are in use by employees. If a receptacle or receptacles are installed as part of the permanent wiring of the building or structure and they are used for temporary electric power, GFCI protection shall be provided. Receptacles on the ends of extension cords are not part of the permanent wiring and, therefore, must be protected by GFCIs whether or not the extension cord is plugged into permanent wiring. These GFCIs monitor the current-to-the-load for leakage to ground, and are in addition to—not as a substitute for—the grounding requirements which must be met by the employer.

**How does a ground-fault circuit interrupter work?** It's critical that all employees are trained on GFCI use. They've got to understand why to use GFCIs. All of your company's employees should know the following basics about ground-fault circuit interrupters:

The ground-fault circuit interrupter quickly detects a fault, short-circuit, or electrical leak, then shuts off the power to prevent accidental shocks (see Fig. 26.3). A GFCI will help protect the employee from hazardous electrical shock that may be caused if his body becomes a path through which electricity travels to reach the ground. That could happen if he touches a power tool that is "live" because of a faulty mechanism, damp or worn insulation on the power cord, or other reason. The employee doesn't even have to be on the ground himself. He could be touching plumbing or other material that leads to the ground. When using the GFCI the employee may still feel a shock, but the GFCI is designed to cut off power quickly enough so a normal, healthy adult will not experience serious electrical injury.

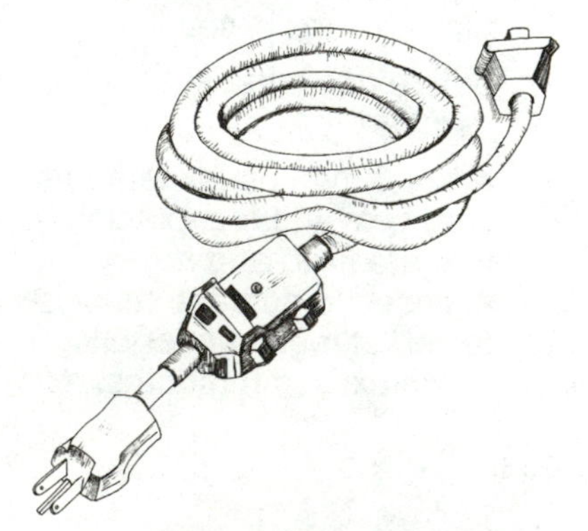

**Figure 26.2** A ground-fault circuit interrupter.

## Ground-Fault Circuit Interrupter

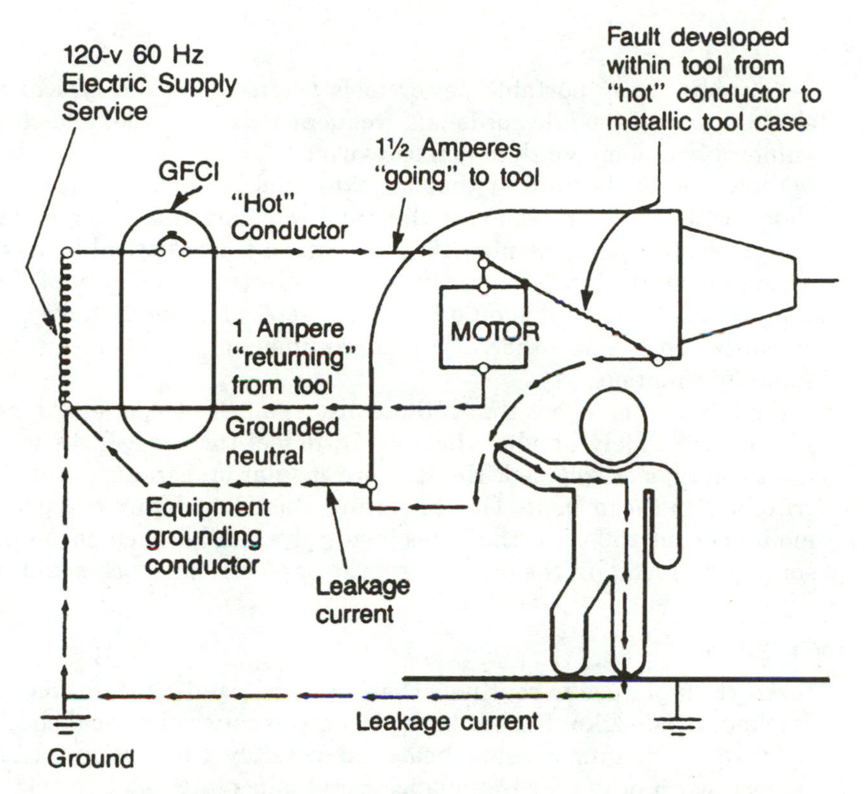


GFCI monitors the difference in current flowing into the "hot" and out to the grounded neutral conductors. The difference (1/2 ampere in this case) will flow back through any available path, such as the equipment grounding conductor, and through a person holding the tool, if the person is in contact with a grounded object.

**Figure 26.3** OSHA GFCI description.

Limitations of a GFCI include line-to-line shocks (the type received when touching metal inserted in both straight slots of a receptacle), current overloads, and line-to-neutral short-circuits.

To make sure that ground-fault interrupters will work when needed, GFCI units must be visually inspected and tested before *each use.*

1. Plug the GFCI cord set into a powered outlet. Plug a test lamp into the connector end of the cord set and turn the lamp *on. Push the test button* on the GFCI. This should result in the lamp going *off.* Caution—if test lamp remains lit, it means the GFCI unit is malfunctioning. *Do not use that GFCI cord set.*
2. If the GFCI tests okay, restore power by pushing the *reset* button and releasing it. The lamp should go *on* again. If the GFCI fails to reset properly, *do not use the GFCI cord set.*

3. If the *GFCI cord set trips by itself* at anytime during use, *reset* and perform test procedures 1 and 2 again.

### Extension cords

Because so many portable power tools are used on the typical construction site, the use of flexible cords is a frequent necessity. Flexible cords are more vulnerable to damage than is fixed wiring.

You've probably noticed that all extension cords are not alike. Some are short, some are long. Some are thin, and some are thick. Some are well made, others are flimsy. Remember that the sum of the electrical loads connected to an extension cord must never exceed the electrical capacity of the cord. How can you tell? The cord should have its electrical capacity marked in amperes or watts. Tools and other electrical equipment should be marked with the same information.

Explain to employees that if there are no markings on a cord, and they suspect it can't safely handle what's on it, to feel the cord. If it's warm or hot to the touch, it's overloaded. Heated cord insulation can eventually harden and crumble, or it can melt. The insulating plastic on many extension cords will melt at about 250° F. If the wires inside eventually touch each other, or touch something metal on the outside, sparks or electrical shock could result.

### Tool cords and plugs

Instruct employees to routinely check all cords and plugs for damage. A worn, cracked, twisted, knotted, or kinked cord can cause electrical shock if touched, with short circuits causing heat and possibly fires. Remind employees to replace worn or cracked appliance cords and plugs—not to just tape up the damaged parts.

Workers should also avoid wrapping cords around (or draping them over) hot furnace pipes, space heaters, or radiators. The heat may dry out or scorch the cord's insulation.

Newer plugs and electrical outlets have one metal prong that's wider than the other one, and one outlet slot that can accommodate the wider prong. This is called polarization. Polarized plugs help prevent electrical shock. The wider prong can only be inserted in the neutral side of the outlet. Electricity flows from the narrow-prong side, which is attached to the hot or "live" wire. The hot wire is the wire that's actually attached to whatever switch is being used for the electrical appliance or piece of equipment. At the same time, for example, a lamp plug is situated so the neutral wire "takes control" of the lamp to prevent someone from getting a shock while replacing a light bulb.

### Electrical work practices

The main causes of worksite electrical incidents that result in serious injuries or death are often listed as accidental contact with overhead power lines, improper lockout procedures, and attempting to work on live electrical components. While those reasons may be true, they're a little misleading. Consider the following list of *behavioral* reasons for electrical injuries at construction worksites:

- *Overfamiliarity with, and a lackadaisical attitude toward, electrical components.* Employees who have worked on and near electricity in the construction trades may never have had problems before, so they assume they can just sail along unscathed, and fail to take proper safeguards.
- *Lack of or inadequate electrical safety training.* If employees are not trained to respect and handle electricity correctly, they probably won't.
- *Nonqualified employees working near energized electrical circuits.* Qualified individuals are persons trained to work near exposed energized equipment, who can safely distinguish exposed live electrical parts from other electrical equipment, can test the voltage of live electrical parts, and are familiar with safe-work practices required to safely work on and near those charged components. Typically, qualified individuals are electricians, electrical engineers, and others who have comprehensive training and experience working with electricity. Other employees or persons *should not work on or near exposed energized equipment.* Pure and simple.
- *Working in unfamiliar situations or surroundings.* Employees thrust into new jobsites, when encouraged to begin work with little or no advanced planning and supervision, may attempt work activities that are unsafe, especially when electricity plays a role in the job.
- *Additional factors* leading to electrical injuries include: getting too close to energized circuits; accidentally making contact with tools, materials, or body parts; working in poorly illuminated areas; working in cluttered, disorganized environments; wearing or carrying conductive materials (such as jewelry or clipboards), and mental or physical fatigue.

### Safe electrical work practices

There are a number of safe work practices that OSHA, over the years, has identified as especially likely to be violated at construction worksites, with serious consequences to employees. They include:

- Prohibit work on new and existing electrical circuits until all power is shut off and a positive lockout/tagout system is in place.
- Keep all electrical tools and equipment in safe condition, and inspect them often for defects.
- Remove broken or damaged tools from the worksite.
- Protect all temporary power, including extension cords, with ground-fault circuit interrupters (GFCIs). Plug into a GFCI-protected temporary power pole, a GFCI-protected generator, or use a GFCI extension cord to protect against shocks.

- Locate and identify overhead electrical power lines. Make sure that ladders, scaffolds, equipment, or materials never come within 10 feet of typical residential electrical power lines.
- Equipment or circuits that are de-energized must be rendered inoperative and must have tags attached at all points where the equipment or circuits could be energized. 1926.417(b)
- Flexible cords must be connected to devices and fittings so that strain relief is provided that will prevent pull from being directly transmitted to joints or terminal screws. 1926.405(g)(2)(iv)
- Worn or frayed electrical cords or cables must not be used. Extension cords must not be fastened with staples, hung from nails, or suspended by wire. 1926.41(e)(1) and (2)
- In work areas where the exact location of underground electrical power lines is unknown, employees using jackhammers, bars, or other hand tools that may contact the lines must be protected by insulating gloves, aprons, or other protective clothing that will provide equivalent electrical protection. 1926.416(a) & .95(a)

### Lockout/tagout

Lockout/tagout is a way of handling electrical and other equipment that means the equipment has had all of its energy either released or blocked so the equipment is literally locked out and can't be restarted or released unintentionally.

It's not just turning off a switch. In the simplest sense, let's say you're working with a recessed electric light fixture. You need to do more than just turn off a wall switch. You need to access a circuit breaker box, turn off the circuit, close the box door (and prevent someone from opening it while you're on the second floor, and activating the circuit while your fingers are all over the electrical components), test the light to make sure it's off, or test the fixture for current.

*Lockout* means that a physical barrier has been installed to prevent restarting or reactivating some kind of energized component. There are many ways to lock out electrical components. Take a simple plug lockout, for example (see Fig. 26.4). Insert the plug, close the hinged box, and insert a padlock. There are similar lockout mechanisms for valves, handles, and other equipment controls.

*Tagout* means placing tags on equipment or machinery to notify workers not to start or operate the equipment. Tagout generally occurs right after lockout, and serves to notify others that someone is actively working on the equipment, and that it shouldn't be started or reactivated while the locks and tags are engaged. While an important part of the safety process, tags are not acceptable substitutes for locks. Tags must indicate the name of the employee who is locking the energy source out, must clearly display the danger/instructions involved, and must be durable enough to prevent accidental removal. Typical messages on lockout tags are "Do Not Start," "Do Not Energize," and "Do Not Operate."

Here's how it should go:

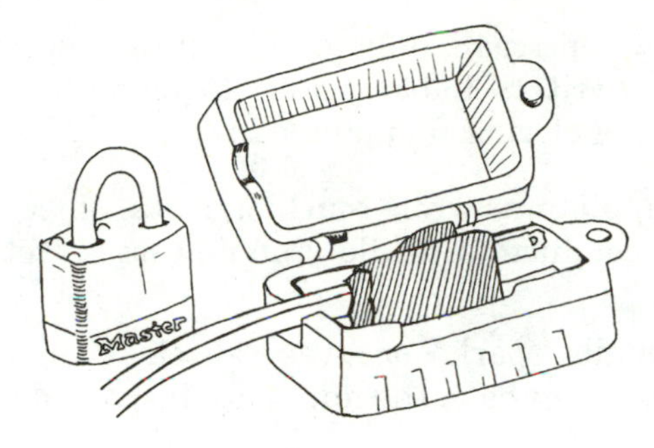


**Figure 26.4** Electric plug lockout device.

1. Shut off the tool, machinery, or equipment.
2. Release or physically restrain or block any stored energy.
3. Disconnect the equipment and electrical circuits from the power source(s).
4. Lock out the energy sources and operating controls with a lock that keeps the equipment in the "off" mode.
5. Place a "danger tag" with each lock. Tags should say that only authorized employees may remove the locks and tags and reconnect the power and the controls.
6. Test the circuit and equipment to make sure they can't be operated and that they're really de-energized.

Remember that the hazardous energy control and lockout/tagout principles are not solely for electrical equipment. The procedure applies to other energy sources as well, and should be used whenever the potential exists for unexpected equipment startups that could injure someone working with that equipment. Worksite examples of hazardous energy control include:

- Unplugging a power saw before changing the blade.
- Blocking up a backhoe front end while changing a wheel/tire.
- Removing the ignition keys from construction equipment when the equipment is not in use.
- Storing sheet materials such as plywood, flat.
- Keeping excavation spoils piles away from trench edges.
- Using toe boards on work platforms and scaffolding.
- Staging only the minimum amount of materials needed on scaffolds or other elevated areas.
- Not parking mobile equipment on steep grades.

### Electrical appliances and fixtures

All electrical appliances, including units wired directly into a household's electrical system, such as water heaters and furnaces, should display a seal from a certified testing laboratory. One commonly found testing laboratory symbol is "UL," representing Underwriters Laboratories. Beyond that, here are a number of safety tips for using electrical appliances:

- Never operate electrical appliances while touching metal objects (especially plumbing) or other good conductors, while standing on a wet surface, or while hands are wet.
- If an appliance smokes, smells, sparks, or gives a shock at any time, unplug it or turn off the electricity, then have the appliance inspected and repaired or replaced, if necessary.
- Keep electric motors free from buildups of dust, dirt, lint, and lubricating oil.
- Space heaters, toasters, irons, and other heat-producing electrical appliances and components require extra care to minimize fire hazards. Never place combustibles such as paper, drapes, or furniture near them. Keep heating appliances clean, in proper condition, and out of high-traffic areas. Unplug these appliances, if possible, after each use. Let them cool before storing them in a safe place. Avoid using lightweight extension cords with heat-producing appliances. If an extension cord must be temporarily used, select a heavy-duty model with enough capacity to handle the appliance, so the cord will not get warm to the touch.
- Space heaters should be equipped with a thermostat so they won't just run on and on by themselves, despite the room temperature. Another important feature is an automatic shut-off in case the unit is accidentally tipped over.
- Unless employees are skilled in electronic appliance repair, they must never attempt to troubleshoot or fix malfunctioning appliances themselves. Touching parts inside of a television set or other electronic unit could cause a severe shock, even when the power cord is unplugged. And an error in reassembly may result in radiation emissions that would otherwise not occur.
- Instruct employees to never operate an electronic appliance if liquid has been spilled into it. Rain or excessive moisture may cause electrical shorts which could result in fire or shock hazards. Such a unit must be unplugged, then inspected by a service technician before further use.
- Water and electricity can be a deadly combination. Plugged-in tools or appliances can cause shock and electrocution even if they're turned off, if they fall into a sink while employees are washing up. It's nice to listen to a radio or to have an electric clock in the bathroom or kitchen: just keep them away from water sources and out of the reach of wet hands. If an item that is plugged in falls into water, remind employees to never reach in for it. The power must be cut first—with the proper circuit breaker thrown or the correct fuse pulled. When certain the power is deactivated, the sink drain can be unplugged and the offending tool or appliance can be pulled out.
- Always unplug tools before cleaning them, removing parts, or repairing or adjusting components.

- Follow the manufacturer's instructions when using electrical fixtures. If a lamp comes with a label specifying that a 40-watt bulb is the maximum-rated bulb to use, don't try a 60-watt model. The label is saying that the 40-watt bulb runs at the maximum current the lamp can safely handle.
- Recessed lighting fixtures should be positioned within vented metal cylinders, cages, or protectors to shield surrounding insulation and construction materials from the fixture's heat. Some recessed lighting fixtures are available with a device that interrupts electrical power to the unit when it overheats for any reason. Inquire at building material stores and specialty lighting shops.
- Before beginning to work on any electrical fixture, shut off the electrical supply to that unit.
- When using power tools, the operator should dress properly. Rubber-soled shoes and rubber gloves should be worn when working in wet areas. Loose-fitting clothes that could become entangled in rotating equipment must be avoided. Rings, bracelets, wristwatches, and other jewelry must be removed before using power tools or equipment. Sawdust, shavings, rags, and other debris should be kept from piling up where they could create a fire hazard or cause tools to overheat by blocking proper ventilation.

### Outdoor electrical safety

These and similar guidelines should be followed for outdoor electrical use:

- Make sure outdoor electrical outlets are weatherproof and protected by ground-fault circuit interrupters.
- All outdoor lighting fixtures, extension cords, and other accessories must be designed for outdoor use.
- Avoid using electric landscaping or other power tools during wet conditions.
- Keep employees away from power lines, especially when they are working with ladders, dump trucks, gutters, antennas, or equipment that could come in contact with them. Touching power lines with metal tools or equipment is a major cause of electrocution associated with outdoor electrical shock. Victims fail to notice where the power line is, misjudge the distance to a line, or the tool/equipment is too heavy to be controlled and falls onto the line. Metal ladders are notorious for electric shock fatalities. Contact often occurs when the victim, while painting or doing roof or gutter work, fails to check the location of the power line in relation to the work area. Some people believe a covering on the power line will protect them, but coverings can't be trusted. A metal ladder scratching through a covering and touching the bare wire could easily cause a fatal shock. When employees must work with a ladder near power lines, they should use only nonconductive ladders made of fiberglass or wood.
- Underground wires must be located *before* digging begins. Contacting local electric utilities well in advance of the excavation or digging day will suffice.

One call will usually help flag where all underground lines are located, including cable television, electrical, natural gas, water, and phone lines.

### If someone at the worksite contacts electricity

If someone comes in contact with electricity, bystanders must be extremely careful with rescue attempts. Although every second of contact with the electrical source may cause more serious consequences, avoid actions which could result in a rescuer becoming injured as well.

First, the victim's contact with the source of current must be broken in the quickest and safest way possible (while this is occurring, someone else should be calling 911 for emergency help). Don't touch the victim or anything the victim is in contact with in a manner that could "freeze" you into the same position as the victim's. If the victim is in contact with a ladder or tool, those items could be electrically charged, too. Disconnect a tool plug or pull the main switch at the fuse or circuit box. If the source of current is from a utility power line, call 911 and the power company at once, because there is no safe way for you to shut off the current. Keep everyone away from the victim and from downed power lines until professionals arrive.

After you have broken the flow of current, check the victim for breathing and pulse, bleeding, and burns. Don't move a victim unless absolutely necessary. If you must move an injured person for safety reasons, EMS personnel advise to move the body lengthwise (not sideways) and head first, with the head and neck firmly supported.

Chapter

# 27

# Hazard Communication

## Quick Scan

1. A customized written hazard communication program must be prepared and maintained for and at all construction worksites having employees who are exposed or potentially exposed to hazardous chemicals.
2. The written Hazard Communication Standard (HCS) program and material safety data sheets (MSDS) must be readily accessible to all employees while they are in their work areas, and must be available at all times and on all shifts.
3. Employees must be trained at the time of initial exposure and whenever a new hazard is introduced into their worksite. Training may be hazard-specific or chemical-specific.
4. Training must include instruction on how to use the information on labels and MSDSs, including precautions, control devices, and personal protective equipment.
5. Purchased material containers must be labeled. In-plant containers must be labeled, tagged, or marked with the appropriate hazard warnings and the same identifying name as on the MSDS.
6. Employers must inform subcontractors of the content of their hazard communication program: that means that you show them yours, and they should reciprocate by showing you theirs.
7. Additional measures are required for multiemployer worksites to be in compliance. Good communications are needed.
8. Special standards (and safe handling practices) apply to various hazardous materials, including asbestos, lead, and crystalline silica.

## OSHA's Hazard Communication Standard Program

OSHA's Hazard Communication Standard (HCS) is based on a simple concept—that employees have both a need and a right to know the hazards and identities of the chemicals they are exposed to while working. They also need to know what protective measures are available to prevent adverse effects from occurring. The entire standard is long and somewhat technical, but it consists of fairly simple concepts.

As stated in 1926.59: Hazard communication.

> *Note:* The requirements applicable to construction work under this section are identical to those set forth at 1910.1200 of this chapter.

OSHA's final Hazard Communication Standard was published in *The Federal Register* on August 24, 1987. The HCS applies to all employers whose employees are exposed to hazardous chemicals in the workplace. Under the HCS, employers will gain knowledge that will help them provide and ensure a safer workplace, and employees will obtain the information they need to protect themselves from hazards.

### What is a hazardous chemical?

Many people envision hazardous chemicals as being horrible poisons, a few drops of which can cause serious or fatal injuries. Naturally, some hazardous chemicals are highly toxic. But the truth is that thousands of products that people use day after day contain chemicals that may be hazardous, more or less, depending on how they're used. Or abused. These include numerous products found on the shelves of your local hardware and grocery stores. Many of them are located in the cleaning supplies section.

Chemical products are generally not dangerous, unless they're used improperly, or are combined with other chemicals they shouldn't be mixed with, or come in contact with the body for unusually long periods.

That's why the Hazcom standard exists—to explain how chemical products can affect people, and how chemicals should be handled safely.

### Hazards from chemicals

There are a number of ways by which chemicals can harm the human body:

- Poisoning from ingestion
- Illness or injury (dizziness, nausea, headaches, unconsciousness, or even asphyxiation) from breathing or inhaling hazardous chemical fumes or vapors
- Burning of the skin or eyes from direct contact
- Skin rashes or diseases
- Burns or injury from fire or explosion
- Long-term damages to major organs, including pulmonary diseases and cancer

### The basic components of a Hazcom program

To meet OSHA requirements, employers must develop a formal hazard communication program. This needs to be prepared in writing and kept (along with appropriate MSDSs) available on the worksite, where it must be readily accessible to all employees at all times and on all shifts. To implement the program, employers must accomplish several key goals:

- Employers must explain what Hazcom is to their employees. Employees *must* be informed of the requirements of OSHA's Hazard Communication Standard.
- Employers must explain how their Hazcom program applies to the worksite, to employees, and to affected subcontractors. Additional communications and measures need to be made for multiemployer worksites to be in compliance.
- Employers must train employees how to recognize and understand the employer's communication of hazards, including MSDSs and labels, and how to protect themselves.
- Material safety data sheets need to be introduced with all products containing hazardous chemicals/materials.

OSHA has published an excellent pamphlet called *Hazard Communication Guidelines for Compliance.* Although the parts that follow here do not include every section, they will give you a representative sampling of the rule.

### Becoming familiar with the rule

OSHA has provided a simple summary of the HCS in a pamphlet entitled *Chemical Hazard Communication,* OSHA publication no. 3084. Some employers prefer to begin to become familiar with the rule's requirements by reading this pamphlet. A single, free copy may be obtained from your local OSHA office, or by calling the OSHA Publications Office at (202) 523-9667.

The standard is long, and some parts of it are technical, but the basic concepts are simple. In fact, the requirements reflect what many employers have been doing for years. You may find that you are already largely in compliance with many of the provisions and will simply have to modify your existing programs somewhat. If you are operating in an OSHA-approved state plan state, you must comply with the state's requirements, which may be different than those of the federal rule. Employers in state plan states should contact their state OSHA offices for more information regarding applicable requirements.

The HCS requires information to be prepared and transmitted regarding all hazardous chemicals. The HCS covers both physical hazards (such as flammability) and health hazards (such as irritation, lung damage, and cancer). Most chemicals used in the workplace have some hazard potential and thus will be covered by the rule.

One difference between this rule and many others adopted by OSHA is that this one is performance-oriented. That means that you have the flexibility to adapt the rule to the needs of your workplace, rather than having to follow specific, rigid requirements. It also means that you have to exercise more judgment to implement an appropriate and effective program.

The standard's design is simple. Chemical manufacturers and importers must evaluate the hazards of the chemicals they produce or import. Using that information, they must then prepare labels for containers, and more detailed technical bulletins called material safety data sheets (MSDSs).

Chemical manufacturers, importers, and distributors of hazardous chemicals are all required to provide the appropriate labels and MSDSs to the employers to whom they ship the chemicals. The information is to be provided automatically. Every container of hazardous chemicals that you receive must be labeled, tagged, or marked with the required information. Your suppliers must also send you a properly completed MSDS at the time of the first shipment of the chemical, and with the next shipment after the MSDS is updated with new and significant information about the hazards.

You can rely on the information received from your suppliers. You have no independent duty to analyze the chemical or evaluate its hazards.

Employers that use hazardous chemicals must have a program to ensure that the information is provided to exposed employees. *Use* means to package, handle, react, or transfer a chemical. This is an intentionally broad scope, and includes any situation where a chemical is present in such a way that employees may be exposed under normal conditions of use or in a foreseeable emergency.

The requirements of the rule that deal specifically with the hazard communication program are found in the Standard in paragraphs (e), written hazard communication programs; (f), labels and other forms of warning; (g), material safety data sheets; and (h), employee information and training. The requirements of these paragraphs should be the focus of your attention. Concentrate on becoming familiar with them, using paragraphs (b), scope and application, and (c), definitions, as references when needed to help explain the provisions.

### Identifying responsible staff

Hazard communication is going to be a continuing program in your facility. Compliance with the HCS is not a one-shot deal. In order to have a successful program, it will be necessary to assign responsibility for both the initial and ongoing activities that have to be undertaken to comply with the rule. In some cases, these activities may already be part of current job assignments. For example, site supervisors are frequently responsible for on-the-job training sessions. Early identification of the responsible employees, and involving them in the development of your plan of action, will result in a more effective program design. Evaluation of the effectiveness of your program will also be enhanced by involving affected employees in that phase as well.

For any safety and health program, success depends on commitment at every level of the organization. This is particularly true for hazard communication, where success requires a change in behavior. This will only occur if employers understand the program, become committed to its success, and are motivated by the people presenting the information to them.

### Identifying hazardous chemicals in the workplace

The standard requires a list of hazardous chemicals in the workplace as part of the written hazard communication program. The list will eventually serve as an inventory of everything for which an MSDS must be maintained. At this point, however, preparing the list will help you complete the rest of the program since it will give you some idea of the scope of the program required for compliance in your facility.

The best way to prepare a comprehensive list is to survey the workplace. A review of purchasing records may also help, and certainly employers should establish procedures to ensure that in the future purchasing procedures result in MSDSs being received before a material is used in the workplace.

The broadest possible perspective should be taken when doing the survey. Sometimes people think of chemicals as being only liquids in containers. The HCS covers chemicals in all physical forms—liquids, solids, gases, vapors, fumes, and mists—whether they are "contained" or not. The hazardous nature of the chemical and the potential for exposure are the factors which determine whether a chemical is covered. If it's not hazardous, it's not covered. If there is no potential for exposure (e.g., the chemical is inextricably bound and cannot be released), the rule does not cover the chemical.

Look around. Identify chemicals in containers, including pipes, but also think about chemicals generated in the work operations. For example, welding fumes, dusts, and exhaust fumes are all sources of chemical exposures. Read labels provided by the suppliers for hazard information. Make a list of all chemicals in the workplace that are potentially hazardous. For your own information and planning, you may also want to note on the list the location(s) of the products within the workplace, and an indication of the hazards as found on the label. This will help you as you prepare the rest of your program.

Once you have compiled as complete a list as possible of the potentially hazardous chemicals in the workplace, the next step is to determine if you have received MSDSs for all of them. Check your files against the inventory you have just compiled. If any are missing, contact your supplier and request one. It is a good idea to document these requests, either by a copy of a letter or a note regarding telephone conversations. If you have MSDSs for chemicals that are not in your list, figure out why. Maybe you don't use the chemical anymore. Or maybe you missed it in your survey. Some suppliers do provide MSDSs for products that are not hazardous. These do not have to be maintained by you.

You should not allow employees to use any chemicals for which you have not received an MSDS. The MSDS provides information you need to ensure proper protective measures are implemented *prior* to exposure.

## Preparing and Implementing a Hazard Communication Program

All workplaces where employees are exposed to hazardous chemicals must have a written plan which describes how the standard will be implemented in that facility. Preparation of a plan is not just a paper exercise—all of the elements must be implemented in the workplace in order to be in compliance with the rule. See paragraph (e) of the standard for the specific requirements regarding written hazard communication programs.

The plan does not have to be lengthy or complicated. It is intended to be a blueprint for implementation of your program—an assurance that all aspects of the requirements have been addressed.

Many trade associations and other professional groups have provided sample programs and other assistance materials to affect employers. These have been very helpful to many employers, because they tend to be tailored to the particular industry involved. You may wish to investigate whether your industry trade groups have developed such materials.

Although such general guidance may be helpful, you must remember that the written program has to reflect what *you* are doing in *your* workplace. Therefore, if you use a generic program, it must be adapted to address the facility it covers. For example, the written plan must list the chemicals present at the site, indicate who is to be responsible for the various aspects of the program in your facility, and indicate where written materials will be made available to employees.

If OSHA inspects your workplace for compliance with the HCS, the OSHA compliance officer will ask to see your written plan at the outset of the inspection. The written program must describe how the requirements for labels and other forms of warning, MSDSs, and employee information and training, are going to be met in your facility. The following paragraphs summarize the types of information compliance officers will be looking for to decide whether these elements of the hazard communication program have been properly addressed.

### Labels and other forms of warning

Containers of hazardous chemicals must be labeled, tagged, or marked with the identity of the material and appropriate hazard warnings. Chemical manufacturers, importers, and distributors are required by law to ensure that every container of hazardous chemicals they ship is properly labeled with such information and with the name and address of the producer or other responsible party. Employers purchasing chemicals can rely on the labels provided by their suppliers. If the material is subsequently transferred by the employer from a labeled container to another container, the employer will have to label that container unless it will be subject to the portable container exemption. See paragraph (f) for specific labeling requirements.

The primary information to be obtained from an OSHA-required label is an identity for the material, and appropriate hazard warnings. The identity is any term which appears on the label, the MSDS, and the list of chemicals, and thus links these three sources of information. The identity used by the supplier may be a common or trade name (e.g., "Black Magic Formula"), or a chemical name (e.g., 1,1,1,-trichloroethane). The hazard warning is a brief statement of the hazardous effects of the chemical (e.g., "flammable," "causes lung damage"). Labels frequently contain other information, such as precautionary measures (e.g., "do not use near open flame"). Labels must be legible and must be prominently displayed. There are no specific requirements for size, color, or any specified text.

With these requirements in mind, the compliance officer will be looking for the following types of information to ensure that labeling will be properly implemented in your facility:

- Designation of person(s) responsible for ensuring labeling of in-plant containers
- Designation of person(s) responsible for ensuring labeling of any shipped containers
- Description of labeling system(s) used
- Description of written alternatives to labeling of in-plant containers, if used
- Procedures to review and update label information when necessary

You need to make sure that every purchased container of hazardous chemicals is labeled. It is a continuing duty, that also applies to all in-plant containers, except those falling under the portable-container exemption (which essentially says that a material may be temporarily kept in personal containers if under the supervision of the same employee during one workshift, and if there won't be any leftover material in the same unlabeled container that someone else could mistakenly use.) It's important to designate someone to be responsible for ensuring that the labels are maintained as required on the containers at your worksites.

## Material Safety Data Sheets (MSDS)

Chemical manufacturers and importers are required to obtain or develop a material safety data sheet for each hazardous chemical they produce or import. Distributors are responsible for ensuring that their customers are provided a copy of these MSDSs. Employers must have an MSDS for each hazardous chemical which they use. Employers may rely on the information received from their suppliers. The specific requirements for material safety data sheets are in paragraph (g) of the Standard.

There is no specified format for the MSDS under the rule, although there are specific information requirements. OSHA has a nonmandatory format, OSHA Form 174, which may be used by chemical manufacturers and importers to comply with the rule. Older, shorter versions of the OSHA MSDS format, Form 20, were largely replaced after 1986 when hazard communication standards came into play. ANSI-approved MSDS formats are even more comprehensive.

The most complete material safety data sheets typically follow section numbers like these:

Section 1: PRODUCT IDENTIFICATION.

This is usually the identity or product name on the label. If the hazardous chemical is a single substance, it's the common and chemical name. If the material is a mixture, the chemical and common names of the ingredients will be listed. A brief description of the product may also be included, as may an emergency phone number for related information, so a user will not have to spend time looking through the sections in an emergency.

Section 2: COMPOSITION AND INFORMATION ON INGREDIENTS.

This section will list hazardous components and will provide exposure limits, including permissible exposure limits (PELs), short-term exposure limits (STELs), and ACGIH threshold limit values (TLVs).

Section 3: HAZARD IDENTIFICATION.

Gives an emergency overview, with major or minor hazards that may be involved with the material's handling and use. Included will be information on routes of exposure and health effects.

Section 4: FIRST-AID MEASURES.

Tells what to do if the material makes skin or eye contact, if it's swallowed, or if it's inhaled.

Section 5: FIRE-FIGHTING MEASURES.

Gives the flash point, autoignition temperature, flammable limits in air, fire-extinguishing procedures, and any special fire-fighting methods that may be required. Lists any unusual fire and explosion hazards, and also gives the NFPA health, flammability, reactivity, and other ratings.

Section 6: ACCIDENTAL RELEASE MEASURES.

Tells how to respond to accidental releases. How to approach and clean up spills and releases.

Section 7: HANDLING AND STORAGE.

Gives work, hygiene, storage, and handling practices, and protective practices during maintenance of contaminated equipment.

Section 8: EXPOSURE CONTROLS AND PERSONAL PROTECTION.

Reviews ventilation and engineering control information. Discusses respiratory protection, eye protection, and skin protection.

Section 9: PHYSICAL AND CHEMICAL PROPERTIES.

Lists information regarding vapor density, specific gravity, solubility in water, vapor pressure, appearance and color, and describes how employees can detect the substance.

Section 10: STABILITY AND REACTIVITY.

Discusses level of stability at room temperature, and conditions to avoid—such as exposure to heat, ignition sources, and materials with which the substance is incompatible. Reviews any hazardous polymerization or decomposition products.

Section 11: TOXICOLOGICAL INFORMATION.

Tells if the substance is a definite or suspected cancer agent. Lists toxicity data, medical conditions aggravated by exposure, and recommendations by physicians.

Section 12: ECOLOGICAL INFORMATION.

Tells if the material is environmentally stable, and gives the effect of the material on plants and animals, and the effect of the materials chemicals on aquatic life.

Section 13: DISPOSAL CONSIDERATIONS.

Explains how to prepare wastes for disposal, and supplies EPA waste numbers, if any.

Section 14: TRANSPORTATION INFORMATION.

Lists proper shipping name, hazard class number and description, UN identification number, DOT label(s) required, packaging group, emergency response guide number, and marine pollutant, if applicable.

Section 15: REGULATORY INFORMATION.

Reviews SARA reporting requirements, TSCA inventory status, California Proposition 65, CERCLA reportable quantities, and labeling—for precautionary statements, with signal word, such as "Caution," Target organs, such as "Eyes, or respiratory tract," and type of hazard.

Section 16: OTHER INFORMATION.

Tells who prepared the MSDS information, provides an address and phone number, notes whether the MSDS supersedes a previous one for the same substance, and gives the latest preparation date.

The MSDS must be in English. You are entitled to receive from your supplier a data sheet which includes all of the information required under the rule. If you do not receive one automatically, you should request one. If you receive one that is obviously inadequate, with, for example, blank spaces that are not completed, you should request an appropriately completed one. If your request for a data sheet or for a corrected data sheet does not produce the information needed, you should contact your local OSHA office for assistance in obtaining the MSDS. Realistically, what you'll eventually end up with will be a considerable number of different-looking material safety data sheets. Just make sure that the information you need is on each form.

The role of MSDSs under the rule is to provide detailed information on each hazardous chemical, including its potential hazardous effects, its physical and chemical characteristics, and recommendations for appropriate protective measures. This information should be useful to you as the employer responsible for designing protective programs, as well as to the workers. If you're not familiar with material data sheets and with chemical terminology, you may need to learn to use them yourself. The glossary of MSDS terms at the end of this chapter may be helpful in this regard. Generally speaking, most employers using hazardous chemicals will primarily be concerned with MSDS information regarding hazardous effects and recommended protective measures. Focus on the sections of the MSDS that are applicable to your situation.

MSDSs must be readily accessible to employees when they are in their work areas during their workshifts. This may be accomplished in many different ways. You must decide what is appropriate for your particular worksite. Some employers keep their MSDSs in a binder in a central location such as in a binder kept in a pickup truck or job trailer at a construction site (see Fig. 27.1). As long as employees can get the information when they need it, any reasonable approach may be used. The employees must have access to the MSDSs themselves—simply having a system where the information can be read to them over the phone is only permitted under the mobile worksite provision, paragraph (g)(9), when employees must travel between workplaces during the shift. In this situation, they have access to the MSDSs prior to leaving the primary worksite, and when they return, so the telephone system is simply an emergency arrangement.

When reviewing your program, the compliance officer will be looking for:

- Designation of person(s) responsible for obtaining and maintaining the MSDSs
- How such sheets are to be maintained in the workplace (i.e., in notebooks in the work area), and how employees can obtain access to them when they are in their work area during the workshift
- Procedures to follow when the MSDS is not received at the time of the first shipment

For employers using hazardous chemicals, the most important aspect of the written program in terms of MSDSs is to ensure that someone is responsible for obtaining and maintaining the MSDSs for every hazardous chemical in the workplace. The list of hazardous chemicals required to be maintained as part of the written program will serve as an inventory. As new chemicals are purchased, the list should be updated.

## Employee Information and Training

Each employee who may be exposed to hazardous chemicals when working must be provided information and be trained prior to initial assignment to work with a hazardous chemical, and whenever the hazard changes. *Exposure*

**Figure 27.1** MSDS Binder Collection.

or *exposed* under the rule means that "an employee is subjected to a hazardous chemical in the course of employment through any route of entry (inhalation, ingestion, skin contact or absorption) and includes potential (accidental or possible) exposure." Information and training may be organized to focus on individual chemicals, or to deal with categories of hazards (such as flammability or carcinogenicity). For example, if there are only a few chemicals in the workplace, you may wish to discuss each one individually. Where there are large numbers of chemicals, or the chemicals change frequently, you will probably want to train generally based on the hazard categories (i.e., flammable liquids, corrosive materials, carcinogens). Employees will have access to the substance-specific information on the labels and MSDSs.

Information and training are a critical part of the hazard communication program. Information regarding hazards and protective measures is provided to workers through written labels and material safety data sheets. Through effective training, workers will learn how to read and understand such information, determine how it can be obtained and used in their own worksites, and understand the risks of exposure to the chemicals as well as the ways to protect themselves. A properly conducted training program will ensure comprehension and understanding. It is not sufficient to either just read material to the workers, or simply hand them material to read. You want to create a climate where workers feel free to ask questions. This will help to ensure that the information is understood. You must always remember that the underlying purpose of the HCS is to reduce the incidence of chemical source illnesses and injuries.

In reviewing your written program with regard to information and training, the following items need to be considered:

- Designation of person(s) responsible for conducting training
- Format of the program to be used (audiovisuals, classroom instruction, etc.)
- Elements of the training program, which should be consistent with the elements in paragraph (h) of the HCS

- Procedures for training new employees when they are first assigned to work with a hazardous chemical, and for training employees when a new hazard is introduced into the workplace

The written program should provide enough details about the employer's plans in this area to assess whether or not a good-faith effort is being made to train employees. OSHA does not expect that every worker will be able to recite all of the information about each chemical in the workplace. In general, the most important aspects of training under the HCS are to ensure that employees are aware that they are exposed to hazardous chemicals, that they know how to read and use labels and MSDSs, and that, as a consequence of learning this information, they are following the appropriate protective measures established by the employer. OSHA compliance officers will be talking to employees to determine if they have received training, if they know they are exposed to hazardous chemicals, and if they know where to obtain substance-specific information on labels and MSDSs.

An employer can provide information and training to employees through any appropriate and protective means. Although there would always have to be some training on-site (such as informing employees of the location and availability of the written program and MSDSs), employee training may be satisfied in part by general training about the requirements of the HCS and about chemical hazards on the job which is provided by, for example, trade associations, unions, colleges, and professional schools. In addition, a worker's previous training, education, and experience may relieve the employer of some of the burdens of informing and training that worker. Regardless of the method relied upon, however, the employer is always ultimately responsible for ensuring that employees are adequately trained. If the compliance officer finds that the training is deficient, the employer will be cited for the deficiency regardless of who actually provided the training on behalf of the employer.

#### Training aids

The following items may be useful to your company to help with training on the Hazard Communication Standard (at this writing):

- *Highlights of OSHA's Hazard Communication Standard* (lecture slide set; price $60.00). Available from National Audiovisual Center, Customer Service Section, 8700 Edgeworth Drive, Capital Heights, MD 20743-3701, (800) 638-1300.
- OSHA-3104 *Hazard Communication—A Compliance Kit* (a reference guide to step-by-step requirements for compliance with the standard; price $18.00, GPO Order No. 929-022-000009). Available from Superintendent of Documents, U.S. Government Printing Office, Washington, D.C. 20402, (202) 783-3238.

### Other requirements

- Does a list of the hazardous chemicals exist in each work area or at a central location?
- Are methods the employer will use to inform employees of the hazards of non-routine tasks outlined?
- Are employees informed of the hazards associated with chemicals contained in unlabeled pipes in their work areas?
- On multiemployer worksites, has the initial employer provided other employers with information about labeling systems and precautionary measures where the other employers have employees exposed to the initial employer's chemicals?
- Is the written program made available to employees and their designated representatives?

If your program adequately addresses the means of communicating information to employees in your workplace, and provides answers to the basic questions outlined above, it will be found to be in compliance with the rule.

### Employer checklist for HCS compliance

- Obtain a copy of the rule.
- Read and understand the requirements.
- Assign responsibility for tasks.
- Prepare an inventory of chemicals.
- Ensure that containers are labeled.
- Obtain MSDS for each chemical.
- Prepare written program.
- Make MSDSs available to workers.
- Conduct training of workers.
- Establish procedures to maintain current program.
- Establish procedures to evaluate effectiveness.

## Asbestos

Asbestos is bad stuff. Its tiny particles can be inhaled and retained in the lungs indefinitely. The particles can become embedded in the tissues of the respiratory or digestive systems, where they may cause disabling or fatal diseases, including asbestosis (a condition similar to emphysema), lung cancer, mesothelioma (a cancerous tumor that spreads rapidly in the cells of membranes covering the lungs and body organs), and gastrointestinal cancer. Symptoms for some of these diseases generally do not appear until 20 or more years after the initial exposure. From my perspective, the long gestation period is one key reason why OSHA is so determined when it comes to asbestos. Remember the chapter on behavior? Consider that individuals, when left to their own means, may not be overly concerned about protecting themselves from hazards that are unlikely to affect them until two or more decades later.

Years ago asbestos was frequently used as insulation and as manufactured house siding and shingles. Children (myself included) used to peel threads of it from samples included in student mineral collections. Now you practically need a level-A suit just to get near it. In older homes, there could be asbestos in the attic, around the furnace or water heater, or in the siding. In commercial and industrial buildings, it could be anywhere.

Unfortunately, it's difficult to identify asbestos on sight. Even the trained eye will have problems distinguishing it from certain fiberglass and similar materials.

Although asbestos handling and removal are best left to companies that specialize in that field, your employees still need to be trained in asbestos-related hazards and safe work practices if there's a reasonable chance that they'll encounter it.

### What OSHA says

OSHA says a lot about asbestos. 1926.1101 (under Subpart Z—Toxic and Hazardous Substances) goes into page after page of details. It's one of the most comprehensive standards available.

In addition to the full standard, OSHA publication no. 3096, *Asbestos Standard for the Construction Industry,* is an excellent guide. It summarizes a recently established classification system for construction work, which categorizes asbestos hazards into four classes:

**Class I asbestos work.** This asbestos work is the most potentially hazardous class of asbestos jobs, and involves the removal of thermal system insulation and sprayed-on or troweled-on surfacing asbestos-containing materials or presumed asbestos-containing materials. Thermal system insulation includes asbestos-containing materials applied to pipes, boilers, tanks, ducts, or other structural components to prevent heat loss or gain. Surfacing materials include decorative plaster on ceilings, acoustical asbestos-containing materials on decking, or fireproofing on structural members. It's highly unlikely that readers of this book will be looking for guidance on Level I asbestos work—that work is best left to companies specializing in environmental services.

**Class II asbestos work.** This includes the removal of other types of asbestos-containing materials that are not thermal system insulation—such as resilient flooring and roofing materials containing asbestos. Examples of Class II work include removal of floor or ceiling tiles, siding, roofing, or transite panels.

**Class III asbestos work.** Class III asbestos work includes repair and maintenance activities where asbestos-containing or presumed-asbestos-containing materials are disturbed.

**Class IV asbestos work.** These operations include custodial activities where employees clean up asbestos-containing waste and debris. This includes dusting contaminated surfaces, vacuuming contaminated carpets, mopping floors, and cleaning up asbestos-containing or presumed asbestos-containing materials from thermal system insulation.

The asbestos standard for the construction industry (29 CFR 1926.1101) regulates asbestos exposure for the following activities:

- Demolishing or salvaging structures where asbestos is present
- Removing or encapsulating asbestos-containing materials
- Constructing, altering, repairing, maintaining, or renovating asbestos-containing structures or substrates
- Installing asbestos-containing products
- Cleaning up asbestos spills/emergencies
- Transporting, disposing, storing, containing, and housekeeping involving asbestos or asbestos-containing products on a construction site

There are specific standards for the four individual classes of asbestos work. These deal with:

- Permissible exposure limits.
- Exposure assessments and monitoring, with initial exposure assessments, negative exposure assessments, exposure monitoring, and periodic or additional monitoring.
- Medical surveillance for employees who work a certain number of days per year with asbestos, or who wear negative-pressure respirators.
- Record-keeping, keying in product data information, monitoring records, medical surveillance records, and training records.
- "Competent Person" requirements. On all construction sites with asbestos operations, employers must name a "competent person" who is qualified and authorized to ensure worker safety and health, and to do worksite inspections.
- Regulated areas need to be marked off where employees work with asbestos. These must include any adjacent areas where debris and waste from asbestos work accumulates or where airborne concentrations of asbestos exceed or can possibly exceed the PEL. Posted asbestos danger signs must be highly visible and readable, and contain the following information: DANGER. ASBESTOS. CANCER AND LUNG DISEASE HAZARD. AUTHORIZED PERSONNEL ONLY. RESPIRATORY AND PROTECTIVE CLOTHING ARE REQUIRED IN THIS AREA.

Employers must supply a respirator to all persons entering regulated areas. Employees must not eat, drink, smoke, chew tobacco or gum, or apply cosmetics in regulated areas.

An employer performing work in a regulated area must inform other employers on-site of the:

*Nature of the work*

*Regulated area requirements*

*Measures taken to protect on-site employees*

The contractor creating or controlling the source of asbestos contamination must abate the hazards. All employers with employees working near regulated areas must assess each day the enclosure's integrity, or the effectiveness of control methods to prevent airborne asbestos from migrating.

- Communication of Hazards include notifying all concerned, putting up signs, using labels, training all employees (with varying degrees of detail, depending on the asbestos class work involved).
- Respiratory protection must be provided to workers on:

  All Class I asbestos jobs

  All Class II work where an asbestos-containing material is not removed substantially intact

  All Class II and III work not using wet methods

  All Class II and III work without a negative exposure assessment

  All Class III jobs where thermal system insulation or surfacing asbestos-containing or presumed asbestos-containing material is cut, abraded, or broken

  All Class IV work within a regulated area where respirators are required

  All work where employees are exposed above the PEL or STEL

  All emergencies
- Protective clothing, including coveralls or similar whole-body clothing, head coverings, gloves, and foot coverings must be provided by the employer and is required for:

  Any employee exposed to airborne asbestos exceeding the PEL or STEL

  Work without a negative exposure assessment

  Any employee performing Class I work involving the removal of over 25 linear or 10 square feet of thermal insulation or surfacing asbestos-containing or presumed asbestos-containing materials
- Hygiene facilities, with a decontamination area consisting of an equipment room, shower area, and clean room in a series (next to each other), and having access to a HEPA vacuum (high efficiency particulate air).
- Housekeeping procedures require that asbestos waste, scrap, debris, bags, containers, equipment, and contaminated clothing consigned for disposal

must be collected and disposed of in sealed, labeled, impermeable bags or other closed, labeled impermeable containers. Employees must use HEPA-filtered vacuuming equipment and must empty it so as to minimize asbestos reentry into the workplace.

All vinyl and asphalt flooring material must remain intact unless the building owner demonstrates that the flooring does not contain asbestos. Sanding flooring material is prohibited. Employees stripping finishes must use wet methods and low-abrasion pads at speeds lower than 300 revolutions per minute. Burnishing or dry-buffing may be done only on flooring with enough finish that the pad cannot contact the flooring material. Employees must not dust, sweep, or vacuum without a HEPA filter in any area containing thermal system insulation or surfacing material or visibly deteriorated asbestos-containing materials. Employees must promptly clean and dispose of dust and debris in leakproof containers.

## Lead

### What OSHA says

29 CFR Part 1926.62 covers the construction lead standard. It is a long and technical standard. The standard describes procedures required for monitoring exposure of employees to lead, and protecting them from lead-based paint residues and airborne dusts. It applies to new construction, alterations and renovations, and repairs. Typical trades involved include painters, wall-covering contractors, plumbers, and in some cases, carpenters, but the standard does not exclude anyone who works on materials containing lead where the lead may become airborne or subject to inhalation or ingestion. The Environmental Protection Agency has also published a regulation that addresses training and certification requirements for specific categories of individuals working on abatement projects in residential housing (or in facilities where children six years old or younger are present). But this comes into play only when the work is true abatement, with the intention of permanently eliminating the hazard of exposure to lead-based paint. If it's not that—if, for instance, the work entails disturbing paint during routine preparation for repainting—then it's considered renovation or remodeling, and does not call for the requirements for trained and certified personnel.

The OSHA standard, however, always applies when there is known to be lead on a worksite. OSHA expects the employer to determine whether lead exists on a worksite, because the employer is responsible for recognizing all of the hazards to which the employees can be exposed. If you subcontract painting work, make sure that the painting company understands and abides by OSHA's lead-in-construction standard. Its plan must include acceptable procedures for disposing of lead-contaminated debris, in line with local landfill restrictions and permitted uses.

Consider that buildings and homes constructed before 1980 may contain walls, moldings, trim, door jambs, and windows that are coated with lead-based paint.

**Working with Lead in the Construction Industry**

U.S. Department of Labor
Occupational Safety and Health Administration

U.S. Department of Health and Human Services
National Institute for Occupational Safety and Health

April 1991

OSHA 3126

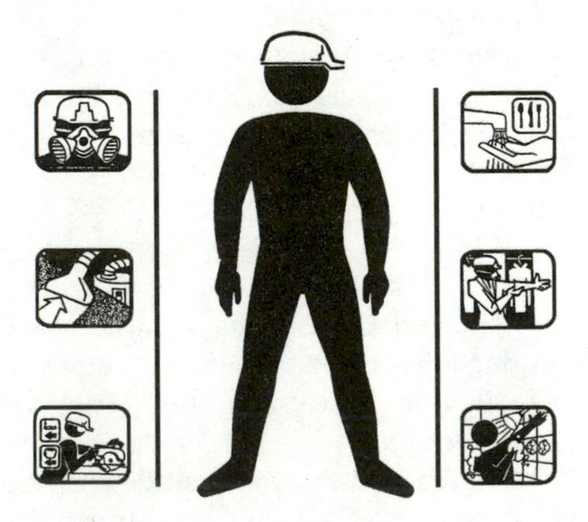

**Figure 27.2** OSHA Lead Booklet #3126.

OSHA has several special pamphlets, including OSHA no. 3126, *Working with Lead in the Construction Industry* (see Fig. 27.2) and OSHA no. 3142, *Lead in Construction,* that are available at area offices or from their publications office.

## Crystalline Silica

Crystalline silica is another name for quartz. It's a basic component of sand and gravel. In dust form, silica can be inhaled or ingested. Overexposure to dust that contains microscopic particles of crystalline silica can cause scar tissue to form in the lungs, which reduces their ability to extract oxygen from the air and may eventually cause a disabling, nonreversible, and sometimes fatal lung disease called silicosis.

There are three recognized types of silicosis, depending upon amounts of exposure to airborne concentrations of silica. Chronic silicosis usually occurs after 10 or more years of continued exposure. Accelerated silicosis, which results from higher exposure levels, can develop symptoms within 5 to 10 years. Acute silicosis occurs where exposures are the highest and can cause symptoms to develop within a few weeks or up to 5 years.

### Exposure activities

Construction workers can be exposed to crystalline silica dust in a number of activities:

- Sandblasting to remove paint and rust from stone buildings, metal bridges, tanks, and other surfaces
- Jackhammering
- Rock/well drilling
- Concrete mixing
- Concrete drilling
- Brick and block cutting and sawing
- Stonecutting (sawing, abrasive blasting, chipping, grinding)

### Limiting employee silica exposure

Limiting workers' exposure to silica dust can be done in ways similar to protecting employees from lead and asbestos:

- Installing and maintaining engineering controls to eliminate or reduce the amount of silica in the air and the buildup of dust on equipment and surfaces. This could involve exhaust ventilation and dust-collection systems, water sprays, and enclosed cabs.
- When possible, substitute less hazardous materials that don't contain silica. Use nonsilica blasting agents, for instance.
- Supply vacuums with high-efficiency particulate air (HEPA) filters, and instruct employees to vacuum, hose down, or wet-sweep work areas, instead of dry-sweeping.
- Educate workers about the health effects, engineering controls, and work practices that reduce dust, and train them in the proper choice and fitting of respirators. Make sure that they know which activities involve silica exposures.

NIOSH has published three alerts on silicosis: *Preventing Silicosis and Deaths from Sandblasting, Preventing Silicosis and Deaths from Rock Drilling,* and *Preventing Silicosis and Deaths in Construction Workers,* plus another one called *A Guide to Working Safely with Silica.* For a free copy, call (800) 35-NIOSH.

## Glossary

The following glossary of terms commonly found in material safety data sheets, provided courtesy of OSHA's Construction Safety and Health Outreach Program will help you decipher unfamiliar terms that you or your company's employees may run across while reading material safety data sheets.

**ACGIH** American Conference of Governmental Industrial Hygienists, which develops and publishes recommended occupational exposure limits for hundreds of chemical substances and physical agents. See **TLV.**

**acid** Any chemical with a low pH that in water solution can burn the skin or eyes. Acids turn litmus paper red and have pH values of 0 to 6.

**action level** Term used by OSHA and NIOSH to express the level of toxicant which requires medical surveillance, usually one half of the permissible exposure limit (PEL).

**activated charcoal** Charcoal is an amorphous form of carbon formed by burning wood, nutshells, animal bones, and other carbonaceous materials. Charcoal becomes activated by heating it with steam to 800–900°C. During this treatment, a porous, submicroscopic internal structure is formed, giving the charcoal an extensive internal surface area. Activated charcoal is commonly used as a gas- or vapor-adsorbent component in air-purifying respirators and as a solid-sorbent material in air-sampling devices.

**acute effect** Adverse effect on a human or animal, with rapidly developing severe symptoms leading quickly to a crisis. Also see **chronic effect.**

**adsorption** The condensation of gases, liquids, or dissolved substances on the surfaces of solids.

**AIHA** American Industrial Hygiene Association.

**air** The mixture of gases that surrounds the Earth. Its major components are as follows: 78.08 percent nitrogen, 20.95 percent oxygen, 0.03 percent carbon dioxide, and 0.93 percent argon. The level of water vapor (humidity) varies.

**air-line respirator** A respirator that is connected to a compressed breathing air source by a hose of small inside diameter. The air is delivered continuously or intermittently in a sufficient volume to meet the wearer's breathing requirements.

**air-purifying respirator** A respirator that uses chemicals to remove specific gases and vapors from the air or that uses a mechanical filter to remove particulate matter. An air-purifying respirator must only be used when there is sufficient oxygen to sustain life and the air contaminant level is below the concentration limits of the device.

**alkali** Any chemical with a high pH that in water solution is irritating or caustic to the skin. Strong alkalies in solution are corrosive to the skin and mucous membranes. Example: sodium hydroxide, also referred to as caustic soda or lye. Alkalies turn litmus paper blue and have pH values from 8 to 14. Another term for alkali is base.

**allergy** An abnormal response by a hypersensitive person to chemical and physical stimuli. Allergic manifestations of major importance occur in about 10 percent of the population.

**ANSI** The American National Standards Institute is a voluntary privately funded membership organization that develops consensus standards nationally for a wide variety of devices and procedures.

**asphyxiant** A vapor or gas which can cause unconsciousness or death by suffocation (lack of oxygen). Asphyxiation is one of the principal potential hazards of working in confined spaces.

**ASTM** American Society for Testing and Materials.

**atmosphere-supplying respirator** A respirator that provides breathing air from a source independent of the surrounding atmosphere. There are two types: air-line and self-contained breathing apparatus.

**atmospheric pressure** The pressure exerted in all directions by the atmosphere. At sea level, mean atmospheric pressure is 29.92 inches Hg, 14.7 psi, or 407 inches w.g.

**base** A compound that reacts with an acid to form a salt. Another term for *alkali.*

**benign** Not malignant. A benign tumor is one which does not metastasize or invade tissue. Benign tumors may still be lethal, due to pressure on vital organs.

**biohazard** A combination of the words *biological* and *hazard.* Biohazards are organisms or products of organisms that present a risk to humans.

**boiling point** The temperature at which the vapor pressure of a liquid equals atmospheric pressure.

**carbon monoxide** A colorless, odorless toxic gas produced by any process that involves the incomplete combustion of carbon-containing substances. It is emitted through the exhaust of gasoline-powered vehicles.

**carcinogen** A substance or agent capable of causing or producing cancer in mammals, including humans. A chemical is considered to be a carcinogen if: (1) it has been evaluated by the International Agency for Research on Cancer (IARC) and found to be a carcinogen or potential carcinogen; or (2) it is listed as a carcinogen or potential carcinogen in the current edition of the *Annual Report on Carcinogens* published by the National Toxicology Program (NTP); or (3) it is regulated by OSHA as a carcinogen.

**CAS** Chemical Abstracts Service is an organization under the American Chemical Society. CAS abstracts and indexes chemical literature from all over the world in *Chemical Abstracts.* CAS Numbers are used to identify specific chemicals or mixtures.

**ceiling limit (C)** An airborne concentration of a toxic substance in the work environment, which for reasons of health and safety should never be exceeded.

**CERCLA** Comprehensive Environmental Response, Compensation and Liability Act of 1980. Commonly known as "Superfund" (U.S. EPA).

**CFR** Code of Federal Regulations. A collection of the regulations that have been promulgated under United States law.

**Chemical cartridge respirator** A respirator that uses various chemical substances to purify inhaled air of certain gases and vapors. This type of respirator is effective for concentrations no more than ten times the threshold limit value (TLV) of the contaminant, if the contaminant has warning properties (odor or irritation) below the TLV.

**CHEMTREC** Chemical Transportation Emergency Center. This public service of the Chemical Manufacturers Association provides immediate advice for hazardous materials emergencies. CHEMTREC has a 24-hour toll-free telephone number—(800) 424-9300—to help those responding to chemical transportation emergencies.

**chronic effect** An adverse effect on a human or animal body, with symptoms which develop slowly over a long period of time or which recur frequently. Also see **acute effect.**

**combustible liquid** Combustible liquids are those having a flash point at or above 37.8°C (100°F).

**concentration** The amount of a given substance in a stated unit of measure. Common methods of stating concentration are percent by weight or by volume, weight per unit volume, normality, etc.

**corrosive** A substance that causes visible destruction or permanent changes in human skin tissue at the site of contact.

**cutaneous** Pertaining to or affecting the skin.

**degrees Celsius (Centigrade)** The temperature on a scale in which the freezing point of water is 0°C and the boiling point is 100°C. To convert to degrees F, use the following formula: degrees F = (degrees C × 1.8) + 32.

**degrees Fahrenheit** The temperature on a scale in which the boiling point of water is 212°F and the freezing point is 32°F.

**density** The mass per unit volume of a substance. For example, lead is much more dense than aluminum.

**dermatitis** Inflammation of the skin from any cause.

**dermatosis** A broader term than dermatitis; it includes any cutaneous abnormality, and thus encompasses folliculitis, acne, pigmentary changes, and nodules and tumors.

**DOL** U.S. Department of Labor.

**dose-response relationship** Correlation between the amount of exposure to an agent or toxic chemical and the resulting effect on the body.

**DOT** U.S. Department of Transportation.

**dusts** Fine solid particles generated by handling, crushing, grinding, rapid impact, detonation, and decrepidation of organic or inorganic materials, such as rock, ore, metal, coal, wood and grain. Dusts do not tend to flocculate, except under electrostatic forces; they do not diffuse in air but settle under the influence of gravity.

**dyspnea** Shortness of breath, difficult or labored breathing.

**EPA** U.S. Environmental Protection Agency.

**evaporation** The process by which a liquid is changed into the vapor state.

**evaporation rate** The ratio of the time required to evaporate a measured volume of a liquid to the time required to evaporate the same volume of a reference liquid (butyl acetate, ethyl ether) under ideal test conditions. The higher the ratio, the slower the evaporation rate. The evaporation rate can be useful in evaluating the health and fire hazards of a material.

**Federal Register** U.S. government publication containing new documents officially promulgated under the law, whose validity depends upon such publication. It is published on each day following a government working day. It is, in effect, the daily supplement to the Code of Federal Regulations (CFR).

**fire point** The lowest temperature at which a material can evolve vapors fast enough to support continuous combustion.

**first aid** Emergency measures to be taken when a person is suffering from overexposure to a hazardous material, before regular medical help can be obtained.

**flammable limits** Flammables have a minimum concentration below which propagation of flame does not occur on contact with a source of ignition. This is known as the lower flammable explosive limit (LEL). There is also a maximum concentration of vapor or gas in air above which propagation of flame does not occur. This is known as the upper flammable explosive limit (UEL). These units are expressed in percent of gas or vapor in air by volume.

**flammable liquid** Any liquid having a flash point below 37.8°C (100°F), except any mixture having components with flashpoints of 100°F or higher, the total of which make up 99 percent or more of the total volume of the mixture.

**flammable range** The difference between the lower and upper flammable limits, expressed in terms of percentage of vapor or gas in air by volume. Also often referred to as the "explosive range."

**flash point** The minimum temperature at which a liquid gives off vapor within a test vessel in sufficient concentration to form an ignitable mixture with air near the surface of the liquid. Two tests are used—open cup and closed cup.

**fume** Airborne particulate formed by the evaporation of solid materials (e.g., metal fume emitted during welding). Particles are usually less than one micron in diameter.

**gauge pressure** Pressure measured with respect to atmospheric pressure.

**gas** A state of matter in which the material has very low density and viscosity; can expand and contract greatly in response to changes in temperature and pressure; easily diffuses into other gases; and readily and uniformly distributes itself throughout any container.

**gram (g)** A metric unit of weight, equal to 0.035 ounce.

**HEPA filter** High-efficiency particulate air filter. A disposable, extended medium, dry-type filter with a particle removal efficiency of no less than 99.97 percent for 0.3-μm particles.

**IARC** International Agency for Research on Cancer.

**IDLH** Immediately dangerous to life and health. An atmospheric concentration of any toxic, corrosive, or asphyxiant substance that poses an immediate threat to life, or that would cause irreversible or delayed adverse health effects, or that would interfere with an individual's ability to escape from a dangerous atmosphere.

**ignition source** Anything that provides heat, spark, or flame sufficient to cause combustion/explosion.

**ignition temperature** The minimum temperature to initiate or cause self-sustained combustion in the absence of any source of ignition.

**impervious** A material that does not allow another substance to penetrate or pass through it. Frequently used to describe gloves.

**inches of mercury column** A unit used in measuring pressures. One inch of mercury column equals 0.491 psi.

**inches of water column** A unit used in measuring pressures. One inch of water column equals a pressure of 0.036 psi.

**incompatible** Materials which could cause dangerous reactions from direct contact with one another.

**ingestion** Taking into the body by way of the mouth.

**inhalation** Breathing of a substance in the form of a gas, vapor, fume, mist, or dust.

**insoluble** Incapable of being dissolved in a liquid.

**irritant** A noncorrosive chemical that is capable of causing a reversible inflammatory effect on living tissue by chemical action at the site of contact.

**latent period** The time that elapses between exposure to a harmful substance and the first manifestation of damage.

**LC-50** Lethal concentration that will kill 50 percent of the test animals within a specified time. See **LD-50.**

**LD-50** The dose required to produce the death in 50 percent of the exposed species within a specified time.

**liter (L)** A metric unit of capacity, equal to 1.057 quarts.

**lower explosive limit (LEL)** The lower limit of flammability of a gas or vapor at ordinary ambient temperatures expressed in percent of the gas or vapor in air by volume. This limit is assumed constant for temperatures up to 120°C (250°F). Above this, it should be decreased by a factor of 0.7 because explosion potential increases with higher temperatures.

**malignant** As applied to a tumor. Cancerous and capable of undergoing metastasis, or invasion of surrounding tissue.

**metastasis** Transfer of the causal agent (cell or microorganism) of a disease from a primary focus to a distant one through the blood or lymphatic vessels. Also, spread of malignancy from the site of the primary cancer to secondary sites.

**meter** A metric unit of length, equal to about 39 inches.

**micron (micrometer, μm)** A unit of length equal to one millionth of a meter, approximately 1/25,000 of an inch.

**milligram (mg)** A unit of weight in the metric system. One thousand milligrams equal one gram.

**milligrams per cubic meter ($mg/m^3$)** Unit used to measure air concentrations of dusts, gases, mists, and fumes.

**milliliter (mL)** A metric unit used to measure volume. One milliliter equals one cubic centimeter.

**millimeter of mercury (mmHg)** The unit of pressure equal to the pressure exerted by a column of liquid mercury one millimeter high at a standard temperature.

**mists** Suspended liquid droplets generated by condensation from the gaseous to the liquid state or by breaking up a liquid into a dispersed state, such as by splashing, foaming, or atomizing. Mist is formed when a finely divided liquid is suspended in air.

**MSDS** Material safety data sheet.

**MSHA** Mine Safety and Health Administration, U.S. Department of Labor.

**mucous membranes** Lining of the hollow organs of the body, notably the nose, mouth, stomach, intestines, bronchial tubes, and urinary tract.

**NFPA** The National Fire Protection Association is a voluntary membership organization whose aim is to promote and improve fire protection and prevention. The NFPA publishes 16 volumes of codes known as the National Fire Codes.

**NIOSH** The National Institute for Occupational Safety and Health is a federal agency. It conducts research on health and safety concerns, tests and certifies respirators, and trains occupational health and safety professionals.

**NTP** National Toxicology Program. The NTP publishes an annual report on carcinogens.

**nuisance dust** Dust that has little adverse effect on the lungs and does not produce significant organic disease or toxic effect when exposures are kept under reasonable control.

**OSHA** U.S. Occupational Safety and Health Administration, U.S. Department of Labor.

**oxidizer** A substance that gives up oxygen readily. Presence of an oxidizer increases the fire hazard.

**oxygen deficiency** That concentration of oxygen by volume below which atmosphere-supplying respiratory protection must be provided. It exists in atmospheres containing less than 19.5 percent of oxygen by volume.

**oxygen-enriched atmosphere** An atmosphere containing more than 23.5 percent of oxygen by volume.

**particulate matter** A suspension of fine solid or liquid particles in air, such as dust, fumes, mist, smoke, or sprays. Particulate matter suspended in air is commonly known as an aerosol.

**PEL** Permissible exposure limit. An exposure limit that is published and enforced by OSHA as a legal standard.

**personal protective equipment (PPE)** Devices worn by the worker to protect against hazards in the environment. Examples include respirators, gloves, and hearing protectors.

**pH** Means used to express the degree of acidity or alkalinity of a solution, with a pH value of 7 indicating neutrality.

**polymerization** A chemical reaction in which two or more small molecules (monomers) combine to form larger molecules (polymers) that contain repeating structural units of the original molecules. A hazardous polymerization is a chemical reaction of the above type that results in an uncontrolled release of energy.

**ppm** Parts per million parts of air by volume of vapor or gas or other contaminant. Used to measure air concentrations of vapors and gases.

**psi** Pounds per square inch (for MSDS purposes) is the pressure a material exerts on the walls of a confining vessel or enclosure. For technical accuracy, pressure must be expressed as psig (pounds per square inch gauge) or psia (pounds per square inch absolute; that is, gauge pressure plus sea-level atmospheric pressure, or psig plus approximately 14.7 pounds per square inch).

**RCRA** Resource Conservation and Recovery Act of 1976 (U.S. EPA).

**reactivity (chemical)** A substance's susceptibility to undergo a chemical reaction or change that may result in dangerous side effects, such as an explosion, burning, and corrosive or toxic emissions.

**respirable-size particulates** Particulates in the size range that permits them to penetrate deep into the lungs upon inhalation.

**respirator (approved)** A device that has met the requirements of 30 CFR Part 11, is designed to protect the wearer from inhalation of harmful atmospheres, and has been approved by the National Institute for Occupational Safety and Health (NIOSH) and the Mine Safety and Health Administration (MSHA).

**respiratory system** Consists of (in descending order) the nose, mouth, nasal passages, nasal pharynx, larynx, trachea, bronchi, bronchioles, air sacs (alveoli) of the lungs, and muscles of respiration.

**route of entry** The path by which chemicals can enter the body. There are three main routes of entry: inhalation, ingestion, and skin absorption.

**SARA** Superfund Amendments and Reauthorization Act of 1986 (U.S. EPA).

**SCBA** Self-contained breathing apparatus.

**sensitizer** A substance that on first exposure causes little or no reaction but which following repeated exposure may cause a marked response not necessarily limited to the contact site. Skin sensitization is the most common form of sensitization in the industrial setting.

**short-term exposure limit (STEL)** ACGIH-recommended exposure limit. Maximum concentration to which workers can be exposed for a short period of time (15 minutes) for only four times throughout the day, with at least one hour between exposures.

**"skin"** A notation (sometimes used with PEL or TLV exposure data) which indicates that the stated substance may be absorbed by the skin, mucous membranes, and eyes—either through airborne means or by direct contact—and that this additional exposure must be considered part of the total exposure to avoid exceeding the PEL or TLV for that substance.

**solubility in water** A term expressing the percentage of a material (by weight) that will dissolve in water at ambient temperature. Solubility information can be useful in determining spill cleanup methods and re-extinguishing agents and methods for a material.

**solvent** A substance, usually a liquid, in which other substances are dissolved. The most common solvent is water.

**sorbent** (1) A material that removes toxic gases and vapors from air inhaled through a canister or cartridge. (2) Material used to collect gases and vapors during air-sampling.

**specific gravity** The ratio of the mass of a unit volume of a substance to the mass of the same volume of a standard substance at a standard temperature. Water at 4°C (39.2°F) is the standard usually referred to for liquids; for gases, dry air (at the same temperature and pressure as the gas) is often taken as the standard substance. See **density.**

**stability** An expression of the ability of a material to remain unchanged. For MSDS purposes, a material is stable if it remains in the same form under expected and reasonable conditions of storage or use. Conditions which may cause instability (dangerous change) are stated. Examples are temperatures above 150°F, and shock from dropping.

**synergism** Cooperative action of substances whose total effect is greater than the sum of their separate effects.

**systemic** Spread throughout the body, affecting all body systems and organs, not localized in one spot or area.

**threshold** The lowest dose or exposure to a chemical at which a specific effect is observed.

**time-weighted average concentration (TWA)** Refers to concentrations of airborne toxic materials which have been weighted for a certain time duration, usually eight hours.

**threshold limit value (TLV)** A time-weighted average concentration under which most people can work consistently for eight hours a day, day after day, with no harmful effects. A table of these values and accompanying precautions is published annually by the American Conference of Governmental Industrial Hygienists.

**toxicity** A relative property of a chemical agent, that refers to a harmful effect on some biologic mechanism and the conditions under which this effect occurs.

**upper explosive limit (UEL)** The highest concentration (expressed in percent vapor or gas in the air by volume) of a substance that will burn or explode when an ignition source is present.

**vapor pressure** Pressure (measured in pounds per square inch absolute—psia) exerted by a vapor. If a vapor is kept in confinement over its liquid so that the vapor can accumulate above the liquid (the temperature being held constant), the vapor pressure approaches a fixed limit called the maximum (or saturated) vapor pressure, dependent only on the temperature and the liquid.

**vapors** The gaseous form of substances that at room temperature and pressure are normally in the solid or liquid state. The vapor can be changed back to the solid or liquid state either by increasing the pressure or decreasing the temperature alone. Vapors also diffuse. Evaporation is the process by which a liquid is changed into the vapor state and mixed with the surrounding air. Solvents with low boiling points will volatilize readily. Examples include benzene, methyl alcohol, mercury, and toluene.

**viscosity** The property of a fluid that resists internal flow by releasing counteracting forces.

WORLDWIDE LEADER ON-SITE EXPERT℠

## Material Safety Data Sheet Description and Explanation of Terms

*Emergency Telephone Number* (24 hours)
Medical (800) 462-5378 or (800) I-M-ALERT

## Overview

The Material Safety Data Sheet (MSDS) is the major media for transmitting health and safety information on chemical products. It is the communication tool to comply with OSHA's Hazard Communication Standard (29CFR 1910.1200) in providing hazard and safety information to our employees and our customers.

These OSHA regulations require chemical manufacturers to evaluate each chemical produced to determine if it is hazardous. The definition of "hazardous" is expanded from the usual "flammable, corrosive, oxidizer, explosive, toxic or highly toxic agents, carcinogen, etc." to include combustibles as well as irritants.

The regulations require chemical manufacturers and importers to prepare and distribute an MSDS for all hazardous chemicals and that each container of hazardous chemical leaving the workplace be labeled. In addition, the chemical manufacturer is to provide a copy of the MSDS to the purchaser at the time of shipment or before.

We have a computerized program to send MSDS's to each purchaser upon receipt of a **first order**. The MSDS will automatically be sent to the attention of the Purchasing Agent at the "ship to address." A **revised** MSDS will also be sent to that same address when a change has been made to the MSDS. Our system allows MSDS's to be sent to alternate locations and/or to additional recipients. You should contact your sales representative or Environmental Health and Safety at 630-305-1449 to customize your distribution requirements. MSDS's continue to be available through your Sales Representative.

Our MSDS complies with all of the requirements of the OSHA regulation. We also provide you with information regarding the safe handling of our product, recommended protection measures, toxicological data, and the status of our product under various federal, state, international, state right-to-know, and environmental and safety regulations. We have put all of this information, that is in compliance with the U.S. and European approved ANSI Z400.1, 1993 Standard, into one document for your convenience.

While the OSHA regulations require that an MSDS is prepared only for hazardous chemicals, as a commitment to product safety, we provide MSDS's on all our chemical products.

As additional commitment to the safe use of our products, we have characterized the human and environmental risk associated with our products so that our sales representatives and our customers can safely use our products.

This MSDS ***Description and Explanation of Terms*** brochure is designed to assist with your interpretation of the MSDS so that you may receive additional customer value from the document.

## Section 1 — Chemical Product Identification

This section identifies our product by Trade Name and/or Product Number. This is the same trade name or product number that will appear on the product container allowing you to match the product label with the MSDS.

We provide a generic chemical description of all major ingredients, both hazardous and nonhazardous. This gives your health and safety personnel information on the class of chemistry(s) in our product without compromising the proprietary nature of the formulation.

The third part of this section is the National Fire Protection Association (NFPA) 704M and Hazardous Materials Identification System (HMIS) rating designations. These popular rating systems are used to give you a quick summary of the hazards of the product regarding health, flammability, reactivity and other hazards. Based on HMIS definition, an organic product will always have a flammability rating of at least (1) with inorganics generally having a rating of (0).

## Section 2 — Composition/Ingredient Information

We have evaluated our formulation for hazardous properties and identify those chemical ingredients which cause or contribute to the hazard. As required by OSHA, the substances are identified if present in quantities of 1% or greater, or in the case of carcinogens, of 0.1% or greater, or if our hazard evaluation determines a hazard exists at lower concentrations. The hazardous ingredients are identified by specific chemical name and their CAS number (the Chemical Abstract Service number for that specific chemical).

**NALCO CHEMICAL COMPANY** One Nalco Center □ Naperville, Illinois 60563-1198 (630) 305-1000
**NALCO/EXXON ENERGY CHEMICALS, L.P.** P.O. Box 87 □ Sugar Land, Texas 77487-0087 (281) 263-7000

**Figure 27.3** NALCO MSDS Description & Explanation, Page 1.

To assist your industrial hygiene and safety personnel, we identify concentration ranges into which the exact percentage of the hazardous ingredient falls. This should enable your safety professional to evaluate the need for air sampling, employee monitoring, or other protective measures. Since our product formulation is proprietary, exact percentages will be given only when there is no trade secret concerns.

Where disclosure of specific chemical name and CAS number of a hazardous ingredient would release trade secrets, we have identified the chemical as "proprietary" as permitted by OSHA and State Right To Know regulations. In the event of an injury or accident, we will communicate the specific identity and concentration to health professionals who may have need for this information.

## Section 3 — Hazard Identification

An emergency overview statement is provided. This warning statement, which is also found on all product containers, provides basic hazard and proper handling practices including the use of protective equipment.

This section also provides the likely routes of exposure when handling the product. If the product does not have a volatile ingredient that can be an inhalation concern during use, the primary routes of exposure are eye and skin contact. The effects from single and repeated exposure are identified.

## Section 4 — First Aid Information

This section is designed to provide first aid information for the typical routes of exposure. The recommendations should be followed in all cases. If exposure causes unexpected or delayed effects, or severe reaction or injury, you should immediately consult a physician. Our **ALERT®** (Alert Link Emergency Response Team), **(800-462-5378)** or **(800-I-M-ALERT®)**, should be called by the attending physician or others. Our **ALERT®** system operates 24 hours/day, seven days/week and is staffed by trained professionals.

## Section 5 — Fire Fighting Measures

If the product exhibits flammable characteristics, information is provided on the recommended method for fighting fire. Unusual fire or explosion hazards are also given. OSHA 29CFR 1910.1200 considers products with flash points of less than 100 degrees Fahrenheit (F) as flammable materials. Chemicals with flash points between 100 degrees F and 200 degrees F are classified as combustible. The Department of Transportation (DOT) 49CFR 173.120 considers products with a flash point less than or equal to 141 degrees F to be flammable. Products of 142 degrees F to 200 degrees F are classified as combustible. The Resource Conservation and Recovery Act (RCRA) - 40CFR 261 subpart C and D define those chemicals with flash points of less than 140 degrees F as ignitable. These differences result from different requirements and definitions from different federal agencies.

## Section 6 — Accidental Release Measures

This section provides information on how to handle and clean up product spills. If reporting of the spill requires notification to the **National Response Center (800) 424-8802,** the requirements will be explained in Section 15 — Regulatory Information.

## Section 7 — Handling and Storage

Guidance on safe handling and storage practices is provided. General practices are provided. If the product necessitates special requirements, the information is provided.

## Section 8 — Exposure Controls/Personal Protection

Handling chemicals such as attaching feed pumps or transferring chemicals from one container to another constitutes the most likely exposure to operating personnel. Recommendations are provided to protect personnel handling product spills, on the type of ventilation needed, and on the protective equipment (respirator, gloves, goggles, etc.) that should be used. This is one of the most important sections of the MSDS and of the overall hazard communication program.

We have evaluated the recommended usage and its frequency for our product, delivery and storage, and intended application equipment to determine the degree of exposure to those handling the products. This exposure can be managed through the use of protective clothing and feeding equipment. This information should be reviewed and put in practice by operating personnel.

## Section 9 — Physical and Chemical Properties

To assess the physical hazards of our product, we perform appropriate tests using procedures recommended by the American Society for Testing and Materials (ASTM). Their procedure number is identified accordingly. The tests vary depending on the physical form and chemical nature of the product. These physical or chemical test results are one of the factors reviewed to determine the need and type of toxicological testing. The results are also used to identify hazardous physical properties which require labeling according to the Department of Transportation (DOT)

**Figure 27.4** NALCO MSDS Description & Explanation, Page 2.

regulations or for waste classification for disposal under the Resource Conservation and Recovery Act (RCRA).

## Section 10 — Stability and Reactivity

The potential for our product to aggressively react with other commonly found chemicals or to decompose represents a special hazard. Information is provided on possible interaction with other chemicals as well as reaction of our product to commonly encountered materials of construction used for chemical feed handling systems.

## Section 11 — Toxicological Information

Our health hazard evaluation for a product is based upon one or more of the following: results of toxicological tests conducted on a product, toxicological test results for a product ingredient(s), use of test results on a similar formulation or product and, use of information obtained in the open literature or supplier information for an ingredient(s).

In this Section, we present summaries of results of toxicity tests. In most cases, results are those from acute, single exposure tests. It should be remembered that the test procedures are quite stringent so that direct extrapolation of results to comparable human exposure must be viewed in that context.

The types of acute animal tests which are routinely conducted include: oral, dermal and inhalation lethality studies and eye and skin irritancy studies. The lethality studies involve administration of the chemical or formulation to groups of test animals at various graded dose levels and recording mortality. The mortality-dose response allows for the calculation of the $LD_{50}$ or $LC_{50}$ by appropriate statistical methods. The $LD_{50}$ is that dose (amount) of chemical or formulation usually expressed in grams or milligrams per kilogram of animal body weight (g/kg or mg/kg), which would produce death in one half of a group of animals administered the chemical or formulation. The $LC_{50}$ is equivalent to $LD_{50}$ except it uses concentration rather than dose and is expressed as parts per million (ppm), milligrams per liter (mg/l) or milligrams per cubic meter of air ($mg/M^3$). Oral and dermal tests use $LD_{50}$, while inhalation tests use $LC_{50}$. In both cases, the smaller the value the more "toxic" the chemical or formulation.

Eye and skin irritancy tests utilize weighted numerical scores to assess degree of injury or irritation. In many instances, such numerical scores are also given descriptive ratings such as mildly or severely irritating. Most grading systems are modeled after those described by Draize, *et al* in their original eye and skin irritation test procedures.

Results of skin sensitization tests conducted primarily on animals are presented. Human data is given if available. Generally, these test results will be for one or more chemicals in a formulation rather than the formulation itself.

When available and applicable, results of tests conducted to assess hazards other than lethality, will be provided under "other toxicity results" and "chronic studies." These types of tests include life-time cancer studies, reproduction tests, and tests designed to uncover birth defects (teratology studies). These tests are usually conducted on individual chemical(s) rather than on a formulated product.

Other short-term bioassays of changes to genetic cells are run with bacterial and other cells. While these tests identify genetic changes in tissue, the usefulness of the information as a prediction of a similar effect to humans continues to be a scientific uncertainty. If this data is available, it will be provided in this section.

Since OSHA has broadened the criteria for acute health hazards and since the numerical rating is not uniformly accepted by all governmental agencies and scientific bodies, we are including OSHA's definitions below:

***Highly toxic*** substance is one having:

1. An oral $LD_{50}$ of 50 mg/kg or less.
2. A dermal $LD_{50}$ of 200 mg/kg or less.
3. An inhalation $LC_{50}$ of 200 ppm or less of gas or vapor; or 2 mg/l or less of mist, fume or dust.

A ***toxic*** substance is one having:

1. An oral $LD_{50}$ between 50 and 500 mg/kg.
2. A dermal $LD_{50}$ between 200 and 1000 mg/kg.
3. An inhalation $LC_{50}$ between 200 ppm, 2,000 ppm of gas or vapors, or between 2 and 20 mg/l of mist, fume or dust.

A ***corrosive*** substance is one which causes third degree burns and scar tissue from 4-hour skin contact to rabbits.

A ***skin irritant*** is one which causes redness and swelling which does not persist and results in a numerical score of 5 out of 8 in greater than 50% of the animals tested.

An ***eye irritant*** — under 29CFR 1910.1200 an eye irritant is one, which at a minimum, results in a grade 2 redness and/or swelling of the conjunctiva in at least 4 of 6 test animals when tested by the methods described in 16CFR 1500.42 or other appropriate techniques. The maximum attainable score using the Draize procedure is 110 (80 for cornea, 10 for iris, and 20 for conjunctiva).

**Figure 27.5** NALCO MSDS Description & Explanation, Page 3.

Use of a finite irritation index to assess a chemical's potential as an eye irritant, i.e., x/110 cannot always be made because of inconsistencies between OSHA's definition and the standard Draize scoring technique. In some instances, an index as low as 2.7/110 is sufficient to warrant the eye irritation hazard statement, while in other instances an index of 6/110 would not. In cases of conflict such as this, we will point them out on the MSDS. This rating system tends to classify many substances as irritants which would not be so classified under other regulations.

## Section 12 — Ecological Information

This section contains information useful for assessing the environmental impact from a product or its ingredients. When available and where applicable, information on partition coefficients, Biochemical Oxygen Demand (BOD), and Chemical Oxygen Demand (COD) is presented.

Results of acute aquatic bioassays are presented. These bioassays are useful in assessing potential for adverse effects on aquatic invertebrates and vertebrates. Results are usually expressed as 48 or 96-hour $LC_{50}$ values in milligrams per liter water (mg/l) or parts per million (ppm). The $LC_{50}$ is the concentration which is lethal to 50% of a group of organisms exposed for the time period indicated. It is synonymous with the term $TL_{50}$ (the concentration which would result in the survival of 50% of a given test group). When applicable, a no-observed effect concentration is presented based upon lack of adverse effects and mortality.

Listed below are ratings we use as internal guidelines:

| 96-Hour $LC_{50}$ | Rating |
|---|---|
| <1.0 ppm | Extremely toxic |
| >1.0<5.0 | Highly toxic |
| >5<10.0 | Toxic |
| >10<100 | Moderately toxic |
| >100<1000 | Slightly toxic |
| >1000 | Essentially non-toxic |

In our effort to evaluate the environmental risk of our product, we have determined the environmental hazard and exposure of our product for its intended use. This information allows you to manage your exposures to achieve an acceptable level of risk.

## Section 13 — Disposal Considerations

The disposal of wastes generated at a facility is one of the biggest problems facing industry. This section describes how to contain and to dispose of our product if it is classified as waste. This section identifies those classifications, that would qualify as hazardous waste under the Resource Conservation and Recovery Act (RCRA).

Empty containers may contain residual product and should be treated in accordance with the label requirement unless the empty container has been properly reconditioned. By EPA Standards (RCRA - 40CFR 261.7), a container is considered to be "empty" when it contains: (1) no more than 1 inch (2.5 centimeters) of product, or (2) no more than 3% by weight of the total capacity of the container if the container is less than or equal to 110 gallons in size or (3) no more than 0.3% by weight if the container is greater than 110 gallons in size. Empty drums that formerly contained chemicals listed in 40CFR 261.33 (c) must be triple rinsed using a solvent capable of removing the commercial product to qualify as "empty." Quantities of chemical greater than those indicated above which remain in the container are considered "wastes" when disposing of the container and appropriate RCRA regulations will apply.

## Section 14 — Transportation Information

All hazardous chemicals are subject to regulation by the U. S. Department of Transportation (DOT). This section identifies the DOT proper shipping name and hazard class for the product, if any. The proper shipping name/hazard class may vary by packaging, properties, and mode(s) of transportation. Many times the name will be a generic name and not necessarily the exact chemical name identified in Section 2. DOT and OSHA hazard classifications are not always in agreement due to the differences in definitions.

## Section 15 — Regulatory Information

Today chemical products are regulated from the time they are manufactured, during use, should any environmental release occur, and when the material is finally ready for disposal. This section provides information on the status of our product under the various federal, state and international regulations that may govern its manufacture, use or disposal. Under the OSHA Hazard Communication Rule 29CFR 1910.1200, the reason for classifying the product as hazardous is as follows: (1) being combustible (flash point 100-200 degrees F), (2) being flammable (flash point less than 100 degrees F), (3) being a skin or an eye irritant, (4) having a chronic health hazard such as liver damage, nerve damage, etc., (5) listed on the National Toxicology Program (NTP) Annual Report on Carcinogens or found to be a potential carcinogen by the International Agency for Research on Cancer (IARC), (6) or OSHA having an established workplace exposure limit or recommended limits. A Threshold Limit Value (TLV) can be established by either OSHA [OSHA uses the term Permissible Exposure Limit (PEL)],

**Figure 27.6** NALCO MSDS Description & Explanation, Page 4.

the American Conference of Governmental Industrial Hygienists (ACGIH) or by the chemical manufacturer.

Three categories of TLV's are recognized: (1) the Threshold Limit Value-Time Weighted Average (TLV-TWA) — the time-weighted average concentration for a normal 8-hour workday and a 40-hour workweek, to which nearly all workers may be repeatedly exposed day after day, without adverse effect, (2) Threshold Limit Value-Short Term Exposure Limit (TLV-STEL) — the concentration that workers can be continuously exposed for a short period of time without suffering from a) irritation, b) chronic or irreversible tissue damage, or c) numbness of sufficient degree to increase the likelihood of accidental injury, impair self-rescue or materially reduce work efficiency, and provided that the daily TLV-TWA is not exceeded. A STEL is defined as a 15-minute time-weighted average exposure which should not be exceeded at any time during a workday even if the 8-hour time-weighted average is within the TLV and (3) Threshold Limit Value-Ceiling (TLV-C) – the concentration that should not be exceeded during any part of the working exposure.

We frequently receive questions on the status of our product under other federal, state and international environmental laws. For this reason, when applicable, we are providing information on product status under major laws.

1. ***OSHA Hazard Communication 29CFR 1910.1200*** — Regulated substances, the nature of their hazard and the regulatory requirements are provided. We identify those chemicals for which there is an established TLV or that appear on the NTP or IARC lists.

2. ***Other OSHA Regulations*** — OSHA has established specific regulations for various chemicals. If these regulations apply to our products, the regulation and its applicability is identified.

3. ***CERCLA/Superfund 40CFR 117.302*** — This Law requires the reporting of spills of certain chemicals when the quantity spilled exceeds certain specified amounts. If our product contains one of the specified chemicals, the quantity of the product, which must be spilled before the notification requirement is "triggered," is calculated and the chemical is identified. All products are reviewed for Section 313 40CFR 372 (List of Toxic Chemicals). If the product contains these substances, it will be identified with its CAS number and concentration range for reporting purposes.

4. ***Toxic Substances Control Act (TSCA)*** — Only substances that are included on the TSCA 8(b) Inventory list, have been exempted (e.g. for research and development only), or have been cleared through a TSCA premanufacturing notification (PMN), can be legally manufactured and used in the U.S.A. The TSCA status for every product is included.

5. ***If our product requires registration or governmental clearances*** for use in intended applications (examples, pesticides under FIFRA, food additives under FDA, drinking water additives, fuel additives under EPA, use in meat and poultry plants under USDA) the status under the appropriate law is indicated.

6. ***Resource Conservation and Recovery Act (RCRA)*** — Our product as sold is not a waste and therefore not covered by this Act. However, should someone decide to declare it a waste and discard it, then the formulation must be evaluated to determine how RCRA might define the waste. This information is provided for our product should it become "a waste." Please refer to comments in Section 13 — Disposal Considerations of this document regarding empty containers.

7. ***The Federal Clean Air and Water Acts 40CFR 60 and 61 and 40CFR 401.15 and 116*** contain sections which specifically list chemicals for which these regulations apply. If our product contains any of the chemicals listed under these sections, they will be identified. This will allow assessment of their impact, if any, on discharge or emission permits.

*State Regulations:*

We also get similar questions regarding the status of our products under state regulations including Right-to-Know laws. However, many states (such as Michigan or California's Proposition 65) list those materials which they consider hazardous or use specific criteria for listing chemicals. Examples of these criteria are the established TLV's by OSHA or ACGIH or the presence of the chemical on a list such as the National Toxicology Program (NTP) Annual Report or International Agency for Research on Cancer (IARC) list for suspect carcinogens.

***International Regulations:***

With the increase in global regulations, including Material Safety Data Sheets, product registrations and chemical substance notification of new substances, we have added the appropriate information. Specifically, this MSDS meets the requirements of Canada's Workplace Hazardous Material Information System (WHMIS).

**Figure 27.7** NALCO MSDS Description & Explanation, page 5.

## Section 16 — Other Information

Additional information not covered in other sections impacting health, safety, environmental or regulatory issues will be provided in this section.

## Section 17 — Risk Characterization

Our Product Stewardship process evaluates the risk of our product during its recommended use. Our evaluation includes both human and environmental risk. Risk characterization involves hazard and exposure considerations for the recommended use of our product. We define this as the product's general risk. Our human and environmental risk characterization results in a classification of high, moderate or low. Our goal is to either reduce the hazard and/or manage the product exposure to reduce human and environmental risk. The human hazard rating is found at the end of **Section 11 — Toxicological Information** with the human exposure rating found at the end of **Section 8 — Exposure Controls/ Personal Protection Equipment.** The environmental hazard and exposure rating is contained at the end of **Section 12 — Ecological Information**. The summation of the human and environmental risk is found on the MSDS in **Section 17 — Risk Characterization.**

Any use inconsistent with our recommendation may affect our risk characterization. Our sales representatives can assist you to determine if the product application is consistent with our recommendations. This information is offered to assist with your risk management practices.

This product material safety data sheet provides health, safety and regulatory information. The product is to be used in applications consistent with our product literature. Individuals handling this product should be informed of the recommended safety precautions and should have access to this information. For any other uses, exposures should be evaluated so that appropriate handling practices and training programs can be established to insure safe workplace operations. Please consult your local sales representative for any further information.

## Section 18 — References

These references along with this document serve as a brief description of our risk characterization process.


*Threshold Limit Values for Chemical Susbstances and Physical Agents and Biological Exposure Indices*, American Conference of Governmental Industrial Hygienists, OH.

*Hazardous Substances Data Bank*, National Library of Medicine, Bethesda, Maryland (CD-ROM version), Micromedex, Inc., Englewood, CO.

*IARC Monographs on the Evaluation of the Carcinogenic Risk of Chemicals to Man*, Geneva: World Health Organization, International Agency for Research on Cancer.

*Integrated Risk Information System*, U.S. Environmental Protection Agency, Washington, D.C. (CD-ROM version) Micromedex, Inc., Englewood, CO.

*Annual Report on Carcinogens*, National Toxicology Program, U.S. Department of Health and Human Services, Public Health Service.

*Title 29 Code of Federal Regulations*, Part 1910, Subpart Z, Toxic and Hazardous Substances, Occupational Safety and Health Administration (OSHA).

*Registry of Toxic Effects of Chemical Susbstances*, National Institute for Occupational Safety and Health, Cincinnati, Ohio (CD-ROM version), Micromedex, Inc., Englewood, CO.

*Shepard's Catalog of Teratogenic Agents* (CD-ROM version), Micromedex, Inc., Englewood, CO.

*Ariel Insight*™ (Databases of North American and Western European Regulations including Global Chemical Control Law Substances), Ariel Research Corporation, Bethesda, MD.

*The Teratogen Information System*, University of Washington, Seattle, Washington (CD-ROM version), Micromedex, Inc., Englewood, CO.

**Figure 27.8** NALCO MSDS Description & Explanation, Page 6.

Chapter

# 28

# Confined-Space Entry

## Quick Scan

1. Familiarize yourself with OSHA's General Industry Confined-Space Regulation.
2. Educate your employees regarding confined-space entries, training them to recognize hazards, how to take necessary precautions to avoid those hazards, and how to use required protective and emergency equipment.
3. All employees need to be able to identify confined spaces within their worksites and understand the entry hazards.
4. Plan entries into confined spaces in advance, involving all employees who will be working in or near them.

## Confined-Space Safety—A Major Employer Responsibility

For construction worksites, a confined space, as you might already have guessed, is a work setting or place with limited access. At the same time, it might be subject to the accumulation of toxic or flammable gases or vapors. It might lack enough oxygen to breathe in. It might have physical hazards such as the potential for soil cave-ins, electrical shock, explosion, falls, or objects falling from above. The point is that confined spaces, because of their "confinedness," may inherently be a lot more dangerous than other spaces.

Although there is not, at this writing, a detailed confined space standard in OSHA's Construction Safety and Health Regulations, there is one in OSHA's General Industry Regulation, at 1910.146, for permit-required confined spaces. The construction regulations also contain requirements that deal with confined space hazards in underground construction (Subpart S), underground electric transmission and distribution work (1926.956), excavations (Subpart P), and welding and cutting (Subpart J).

Additional information about how to avoid trouble in confined spaces may be found in the American National Standard ANSI Z117.1-1989, "Safety Requirements for Confined Spaces." That standard provides minimum safety requirements to be followed while entering, working in, and exiting from confined spaces.

### What OSHA says

Although the General Industry Confined Space Regulation does not apply to construction, you should be familiar enough with the regulation to identify confined-space hazards and to know how to avoid them. Plus if your employees ever work in manufacturing locations, the manufacturer's confined-space permit program will likely be in effect.

In short, you should train your employees to recognize confined spaces, teach them what potential hazards confined spaces may possess, and how to safely work in confined spaces—including what to do in an emergency.

Here's how the construction confined-space regulations read:

> 1926.21 Safety training and education
>
> (b)(6)(i) All employees required to enter into confined or enclosed spaces shall be instructed as to the nature of the hazards involved, the necessary precautions to be taken, and in the use of protective and emergency equipment required. The employer shall comply with any specific regulations that apply to work in dangerous or potentially dangerous areas.
>
> (ii) For purposes of paragraph (b)(6)(i) of this section, "confined or enclosed space" means any space having a limited means of egress, which is subject to the accumulation of toxic or flammable contaminants or has an oxygen deficient atmosphere. Confined or enclosed spaces include, but are not limited to, storage tanks, process vessels, bins, boilers, ventilation or exhaust ducts, sewers, underground utility vaults, tunnels, pipelines, and open top spaces more than four feet in depth such as pits, tubs, vaults, and vessels.

## Confined Spaces

Just what exactly is a confined space? Technically, a confined space must meet each of three criteria:

1. The space must be large enough to enable an employee to physically enter and perform work there.
2. The space must have restricted or limited means of entry or exit.
3. The space is not designed and constructed for continuous employee occupancy. A submarine meets the first two criteria, but fails the third, having been designed for continuous human occupancy.

The three characteristics noted above perform an important role in helping to technically define a confined space. But don't fail to see the forest for the trees here; just because one of the characteristics is missing doesn't mean that you shouldn't treat a particular situation with respect. There could be hazards associated with either of the other two characteristics, independently. Again, you want to be aware of whatever hazards are present, and factor a margin of safety into the work.

Some confined spaces are not easy to recognize. Trenches with open tops may contain hazardous gases which are heavier than air and consequently, remain trapped at the bottom of the trench, ready to affect employees working at the lowest reaches of the excavation. Again, we're not keying solely on mandatory regulations and rules. It's critical for employees to understand the theory behind confined space hazards.

For the purpose of categorizing confined spaces, there are two kinds: permit-required spaces and the less dangerous nonpermit confined spaces.

A permit-required space has one or more of the following characteristics:

- It contains or has the potential for containing a hazardous atmosphere that is low in oxygen, flammable, or toxic.
- It contains a material that has the potential for engulfing an employee or individual who enters the space.
- It has an internal configuration in which an entrant could be trapped or asphyxiated by inwardly converging walls or by a floor that slopes and tapers to a smaller cross-section.
- It contains any other recognized serious safety or health hazard.

Confined spaces encountered in residential construction include such installations as manholes and sewers, pipe assemblies, sumps, and transformers.

### Manholes and sewers

Manholes are, of course, means of entry to and exit from various vaults, tanks, pits, sewers, and similar spaces. A variety of hazards are associated with manholes, including oxygen-deficient atmospheres, explosive atmospheres, and toxic atmospheres, as well as physical hazards such as the potential for falls (covers can be left off, or never put on in the first place).

### Pipe assemblies

OSHA considers pipe assemblies, made of piping up to 36 inches in diameter, as constituting particularly dangerous and often unrecognized confined spaces. Potential hazards include oxygen-deficient atmospheres, welding fumes, or other gases. There's not much room to move about in pipe assemblies, and certainly it's very difficult to extricate any employee who may experience an injury or other problem within. Electrical shock is another hazard, due to ungrounded tools and equipment or inadequate extension or power cords. And heat within a pipe can cause heat illness.

### Sumps

Sumps are frequently needed to collect storm water. Employees entering sumps may encounter oxygen-deficient atmospheres, or atmospheres loaded with hydrogen sulfide or other gases from rotting vegetation. Due to the wet nature of a sump, electricity is a hazard here when power tools or portable lighting are used. Inadequate lighting can also create a hazardous situation.

### Electrical transformers

Electrical transformers may be located near the worksite. They often contain a nitrogen purge or dry air. Before they're opened, they have to be well-vented by pumping fresh air in. Electricians frequently enter these transformers through hatches on top for various reasons. Testing for oxygen deficiency and for toxic atmospheres is mandatory.

When certain confined spaces are linked together (in piping systems, for example), welding fumes or gases from another part of the system can easily migrate or spread to contaminate another part of the system, causing a hazardous atmosphere where work needs to be done. This poses a serious problem, especially if workers entering one space do not suspect or are not aware of hazardous materials leaking into their area.

## Confined-Space Hazards

The injuries historically associated with confined spaces within the construction industry are typically quite serious, with fatalities heading the list. In some circumstances, employees face several serious hazards at the same time: electrical shock and oxygen-deficient atmosphere, for instance. The hazards come from within the spaces themselves, and also from materials or work the employees introduce to perform the tasks.

### Oxygen-deficient atmosphere

Like electricity, an oxygen-deficient atmosphere is invisible. Most injuries occur to unsuspecting employees who are overcome before they know what hit them. If fellow employees go to rescue them, thinking there may be a medical illness problem, they're often overcome as well. A lack of sufficient oxygen causes the majority of confined-space injuries and fatalities, and studies have indicated that more than 60 percent of the fatalities occur among would-be rescuers of initial victims.

The typical air we breathe consists of approximately 21 percent oxygen, and most of the rest is nitrogen. For the typical person, a 19.5 percent oxygen level is the minimum amount of oxygen needed for normal functioning, and it's also the minimal percentage deemed acceptable for the atmosphere in a confined space. The following lower oxygen percentages all have harmful-to-fatal effects on individuals breathing them in:

- *15 to 19 percent oxygen.* Lessened ability to work strenuously. May affect coordination and may trigger early symptoms in employees with coronary, pulmonary, or circulatory problems.
- *12 to 14 percent oxygen.* Respiration labored in exertion, faster pulse, poor coordination, impeded perception and judgment.
- *10 to 12 percent oxygen.* Lips may turn bluish color, respiration faster, poor judgment. Exposure to atmospheres containing oxygen levels at or below 12 percent can cause unconsciousness so quickly that the individual cannot help him or herself.
- *8 to 10 percent oxygen.* Mental facilities at low point, fainting, unconsciousness, vomiting, nausea, blue lips, ashen face.
- *6 to 8 percent oxygen.* Four to five minutes, recovery with treatment. At six minutes, 50 percent fatal. At eight minutes, 100 percent fatal.
- *4 to 6 percent oxygen.* Coma in 40 seconds. Convulsions. Respiration ceases. Death.

### High-voltage shock

Electrical shock can result from equipment such as power tools, line cords, and extension lights. Shocks often happen because the company has not provided an approved grounding system or the protection afforded by ground-fault circuit interrupters or low-voltage systems, or because the employee had decided to work without them.

### Explosive or toxic gases, vapors, or fumes

Employees working in a catch basin, sump, or electrical vault may be exposed to the buildup of explosive gases such as those used for heating, including propane. An explosion in a confined space may present an even greater problem for employees because the explosive forces meet resistance within the space and can cause damaging forces that might otherwise be dissipated into an open atmosphere in directions away from individuals working there. Welding or any kind of sparking tools can ignite explosive gases or vapors within a confined space. Welding and soldering also produce toxic fumes which, when confined in a limited atmosphere, can be breathed in concentrated form. Unfortunately, every possible atmospheric hazard which may be encountered in a confined space cannot be listed.

Hydrogen sulfide and carbon monoxide are two of the most common toxic gases found in confined spaces.

Hydrogen sulfide is a dangerous gas that can be produced by rotting organic materials. It has an odor like rotten eggs, and it's easily detectable at low concentrations. One of its characteristics, though, is that it rapidly desensitizes a person's sense of smell by fooling the nerves and providing a false sense of security. High concentrations of hydrogen sulfide gas may overcome a worker with little or no warning. Hydrogen sulfide causes considerable eye and respiratory irritation after about an hour of exposure at concentrations of 200 to 300 parts per million (PPM), unconsciousness and death at between 30 and 60 minutes at 500 to 700 PPM, and unconsciousness and death in just a few minutes at 1000 or more PPM.

Carbon monoxide is an odorless, colorless, and invisible gas that's a product of combustion. In high concentrations it will also cause unconsciousness with little or no warning, and the affected employee would be unable to aid him or herself.

Methane is another explosive gas that can be produced by decomposing organic matter. While decomposing, organic matter will also use precious oxygen from the space.

It's always important to check for the presence of gases in confined spaces because of the possibility of leaking materials from storage tanks, natural gas lines, or underground tanks. All have found ways into sewer lines, catch basins, and manholes. Those gases can displace oxygen as well as posing explosive or toxic hazards threatening to life. Operating gasoline or diesel engines nearby can also fill a space with fumes and gases from incomplete combustion.

*Caution:* Never purge a confined space by using gas from a tank of compressed oxygen instead of air. Oxygen levels above the normal 21-percent level greatly increase the flammability range of combustible gases or material and cause them to burn violently.

Combustible gases and vapors have different explosive ranges. These ranges depend in part on the gases' lower explosive limits (LELs)—the lowest fuel-to-air ratio at which a particular gas can ignite. Concentrations below this limit are too lean to burn. The ranges are similarly dependent on the gases' upper explosive limits (UELs)—the highest fuel-to-air ratio at which a particular gas can be ignited. Above that concentration, the mixture is too rich to burn. A gas is only combustible between its LEL and UEL, but *any* concentration of a combustible gas should be a concern. Lean mixtures can accumulate in an area and reach a combustible level, or rich mixtures can be diluted with air to become combustible.

By using a monitoring instrument (see Fig. 28.1), you can tell in advance if there's an explosive atmosphere in a confined space. A gas monitor will test for LELs to tell if the space will need additional ventilation to rid the explosive gases and ensure a breathable atmosphere. Continuous monitoring will see that the air stays safe as the work proceeds.

Of course, it's critical that the employees counted upon to do the monitoring understand how to test and use the monitoring equipment. Typically, the manufacturer will supply a comprehensive manual, with clear instructions on how to calibrate and test the monitor before using it to check a confined space. It's important to "zero" the instrument in known fresh air before sampling for suspect gases or vapors. The checking for a proper zero indication should be done for combustible and toxic gases and for 20.9 percent oxygen in fresh air. Follow the manufacturers recommended calibration procedures and intervals. By checking the sensors against known concentrations of the gases you are trying to measure, you'll know the alarms are sounding when they should.

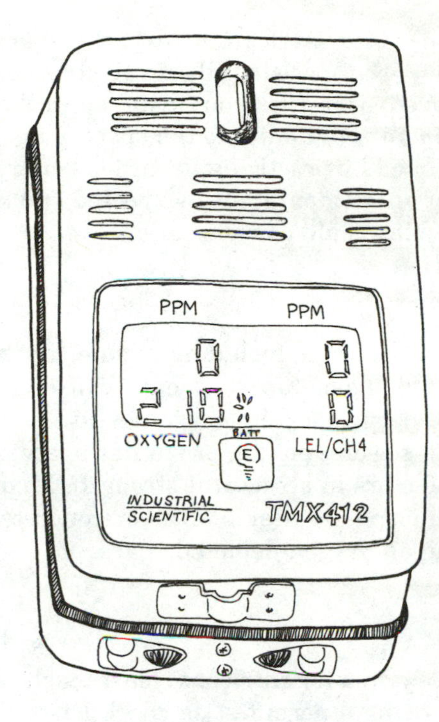


**Figure 28.1** Atmospheric monitor.

Employees expected to work in or around confined spaces should become familiar with all aspects of operation and any limitations or cautions. And follow all of the procedures for sampling techniques. The main one is to sample *at the entrance,* and then sample progressively into the space from the outside—*never* enter the confined space to take the first sample.

Some gases are heavier than air, and some are lighter. The lack of normal ventilation in a confined space allows gases to collect in layers, or at one level, depending on their vapor density (weight compared with air). So sample at several levels, and after the work begins sample frequently or continuously, because conditions can change. As work progresses, a once-safe atmosphere can become hazardous due to the work process going on within, or due to leaks, combustion, cleaning processes, changes in temperature, or other influencing factors.

When the potential hazards of a confined space are recognized, unseen atmospheric hazards can be avoided by preplanning work using your company entry permit as a guide, and by employing proper atmospheric testing and rescue procedures and precautions.

This information is not intended to give anyone a definitive guideline for working in confined spaces. It's meant to point out the atmospheric hazards that could be encountered, and the importance of atmospheric testing and monitoring.

## Mechanical hazards

Additional hazards encountered in confined spaces include material or equipment which has the potential for falling into the space from an overhead accessway, striking employees as they enter, work in, or exit the space. Vibration could cause the materials on top of the space to roll off and strike employees. If a manhole cover has been removed, or if it was not installed in the first place, materials, tools, or other equipment could fall into the space, causing injury.

These hazards must be controlled by practicing hazardous energy control—by eliminating the possibility of an accidental or unexpected starting or energizing of some form of energy which could cause injury.

### Employee-induced hazards

These can come from a variety of sources, including unsafe construction features and methods; physical arrangements that may cause unintentional worker contact with electrical energy sources; oxygen-deficient atmospheres created at the bottom of sumps, sewers, piping, or similar spaces; flammable atmospheres; or lack of safety factors in structural strength. Mixing a wrong mortar, or not checking the atmosphere after mortar or concrete has cured, could result in workers entering an oxygen-depleted space.

## Preplanning Confined-Space Entries

If you don't use a formal permit system for entries, at least teach your employees to preplan their entries by using a permit-style checklist so that nothing gets overlooked. Granted, most construction confined-space entries will not be facing the variety of hazards found within manufacturing industries, but at least by using a permit-style checklist, all possibilities will be addressed.

Such a checklist should consider:

1. Atmospheric testing and monitoring
2. Procedures
   - Initial plan
   - Standby person
   - Communications/observations
   - Rescue
   - The work
3. Preparation
   - Isolate/lockout/tagout
   - Purge and ventilate
   - Cleaning processes
   - Requirements for special equipment and tools
   - Labeling and posting
4. Safety equipment and clothing
   - Head protection
   - Hearing protection
   - Hand protection
   - Foot protection

- Body protection
- Respiratory protection
- Safety positioning belts
- Lifelines, harnesses

5. Rescue equipment

Indeed, the safety of your employees will depend on their knowledge and application of proper work procedures before the space is entered. Atmospheric testing and monitoring, as well as preplanning of the work and rescue procedures, are all critical aspects of job safety.

### Confined-space entry permits

A typical company preprinted confined-space entry permit or checklist contains a number of components:

1. The name of the space.
2. The permit's activation signature.
3. The permit's cancellation signature.
4. The scope of work to be done under the permit.
5. Hazards and exposure symptoms that are expected or that may be encountered.
6. Hazardous energy controls completed.
7. Personal protective equipment needed.
8. Method of communications used for emergency.
9. Subcontractor work/hazards.
10. Atmospheric monitoring.
11. Rescue/retrieval preparations.
12. Entry roster of entrants, observers, and entry supervisors.

Sample permits prepared by OSHA (listed as OSHA Appendix D-1 and Appendix D-2 in 1910.146) can be found in Figs. 28.2 and 28.3.

### General requirements

OSHA states that the employer must determine whether the worksite contains any permit-required confined spaces. And if the workplace does contain them, the employer must then inform employees of the existence of those spaces and their locations and the fact that they are dangerous. The following points will consequently apply:

- If the employer decides that employees will not enter permit-required spaces, the employer must positively ensure that they won't.

APPENDIX D TO § 1910.146—SAMPLE PERMITS

Appendix D - 1

Confined Space Entry Permit

Date & Time Issued: ____________

Job site/Space I.D.: ____________

Equipment to be worked on: ____________

Date and Time Expires: ____________

Job Supervisor ____________

Work to be performed: ____________

Stand-by personnel ____________ ____________ ____________

1. Atmospheric Checks: Time ______
   Oxygen ______ %
   Explosive ______ % L.F.L.
   Toxic ______ PPM

2. Tester's signature ____________

| 3. Source isolation (No Entry): | N/A | Yes | No |
|---|---|---|---|
| Pumps or lines blinded, | ( ) | ( ) | ( ) |
| disconnected, or blocked | ( ) | ( ) | ( ) |

| 4. Ventilation Modification: | N/A | Yes | No |
|---|---|---|---|
| Mechanical | ( ) | ( ) | ( ) |
| Natural Ventilation only | ( ) | ( ) | ( ) |

5. Atmospheric check after isolation and Ventilation:

| | | | |
|---|---|---|---|
| Oxygen ______ % | > | 19.5 | % |
| Explosive ______ % L.F.L. | < | 10 | % |
| Toxic ______ PPM | < | 10 | PPM $H_2S$ |

Time ______

Testers signature ______

6. Communication procedures: ____________

7. Rescue procedures: ____________

| 8. Entry, standby, and back up persons: | Yes | No |
|---|---|---|
| Successfully completed required training? | | |
| Is it current? | ( ) | ( ) |

| 9. Equipment: | N/A | Yes | No |
|---|---|---|---|
| Direct reading gas monitor - tested | ( ) | ( ) | ( ) |
| Safety harnesses and lifelines for entry and standby persons | ( ) | ( ) | ( ) |
| Hoisting equipment | ( ) | ( ) | ( ) |
| Powered communications | ( ) | ( ) | ( ) |
| SCBA's for entry and standby persons | ( ) | ( ) | ( ) |
| Protective Clothing | ( ) | ( ) | ( ) |
| All electric equipment listed Class I, Division I, Group D and Non-sparking tools | ( ) | ( ) | ( ) |

10. Periodic atmospheric tests:

| | | | |
|---|---|---|---|
| Oxygen ___ % | Time ___ | Oxygen ___ % | Time ___ |
| Oxygen ___ % | Time ___ | Oxygen ___ % | Time ___ |
| Explosive ___ % | Time ___ | Explosive ___ % | Time ___ |
| Explosive ___ % | Time ___ | Explosive ___ % | Time ___ |
| Toxic ___ % | Time ___ | Toxic ___ % | Time ___ |
| Toxic ___ % | Time ___ | Toxic ___ % | Time ___ |

We have reviewed the work authorized by this permit and the information contained here-in. Written instructions and safety procedures have been received and are understood. Entry cannot be approved if any squares are marked in the "No" column. This permit is not valid unless all appropriate items are completed.

Permit Prepared By: (Supervisor) ____________ ____________

Approved By: (Unit Supervisor) ____________ ____________

Reviewed By (Cs Operations Personnel): ____________ ____________

(printed name) (signature)

This permit to be kept at job site. Return job site copy to Safety Office following job completion.

Copies: White Original (Safety Office) Yellow (Unit Supervisor) Hard(Job site)

Figure 28.2 OSHA Sample Confined Space Permit (#1).

**Appendix D - 2**

**ENTRY PERMIT**

**PERMIT VALID FOR 8 HOURS ONLY. ALL PERMIT COPIES REMAIN AT SITE UNTIL JOB COMPLETED**

DATE: - - SITE LOCATION/DESCRIPTION ______________________

PURPOSE OF ENTRY ______________________

SUPERVISOR(S) in charge of crews Type of Crew Phone #

______________________

______________________

COMMUNICATION PROCEDURES ______________________

RESCUE PROCEDURES (PHONE NUMBERS AT BOTTOM) ______________________

______________________

*** BOLD DENOTES MINIMUM REQUIREMENTS TO BE COMPLETED AND REVIEWED PRIOR TO ENTRY***

| REQUIREMENTS COMPLETED | DATE | TIME | REQUIREMENTS COMPLETED | DATE | TIME |
|---|---|---|---|---|---|
| **Lock Out/De-energize/Try-out** | ____ | ____ | **Full Body Harness w/"D" ring** | ____ | ____ |
| **Line(s) Broken-Capped-Blank** | ____ | ____ | **Emergency Escape Retrieval Eq** | ____ | ____ |
| **Purge-Flush and Vent** | ____ | ____ | **Lifelines** | ____ | ____ |
| **Ventilation** | ____ | ____ | Fire Extinguishers | ____ | ____ |
| **Secure Area (Post and Flag)** | ____ | ____ | Lighting (Explosive Proof) | ____ | ____ |
| **Breathing Apparatus** | ____ | ____ | Protective Clothing | ____ | ____ |
| **Resuscitator - Inhalator** | ____ | ____ | Respirator(s) (Air Purifying) | ____ | ____ |
| **Standby Safety Personnel** | ____ | ____ | Burning and Welding Permit | ____ | ____ |

Note: Items that do not apply enter N/A in the blank.

** RECORD CONTINUOUS MONITORING RESULTS EVERY 2 HOURS **

| CONTINUOUS MONITORING** TEST(S) TO BE TAKEN | Permissible Entry Level | | | | | | | | | |
|---|---|---|---|---|---|---|---|---|---|---|
| **PERCENT OF OXYGEN** | **19.5% to 23.5%** | ____ | ____ | ____ | ____ | ____ | ____ | ____ | ____ | ____ |
| **LOWER FLAMMABLE LIMIT** | **Under 10%** | ____ | ____ | ____ | ____ | ____ | ____ | ____ | ____ | ____ |
| **CARBON MONOXIDE** | **+ 50 PPM** | ____ | ____ | ____ | ____ | ____ | ____ | ____ | ____ | ____ |
| Aromatic Hydrocarbon | + 1 PPM * 5PPM | ____ | ____ | ____ | ____ | ____ | ____ | ____ | ____ | ____ |
| Hydrogen Cyanide | (Skin) * 4PPM | ____ | ____ | ____ | ____ | ____ | ____ | ____ | ____ | ____ |
| Hydrogen Sulfide | +10 PPM *15PPM | ____ | ____ | ____ | ____ | ____ | ____ | ____ | ____ | ____ |
| Sulfur Dioxide | + 2 PPM * 5PPM | ____ | ____ | ____ | ____ | ____ | ____ | ____ | ____ | ____ |
| Ammonia | *35PPM | ____ | ____ | ____ | ____ | ____ | ____ | ____ | ____ | ____ |

* Short-term exposure limit:Employee can work in the area up to 15 minutes.

\+ 8 hr. Time Weighted Avg.:Employee can work in area 8 hrs (longer with appropriate respiratory protection).

REMARKS: ______________________

| GAS TESTER NAME & CHECK # | INSTRUMENT(S) USED | MODEL &/OR TYPE | SERIAL &/OR UNIT # |
|---|---|---|---|
| ____ | ____ | ____ | ____ |
| ____ | ____ | ____ | ____ |

SAFETY STANDBY PERSON IS REQUIRED FOR ALL CONFINED SPACE WORK

| SAFETY STANDBY PERSON(S) | CHECK # | CONFINED SPACE ENTRANT(S) | CHECK # | CONFINED SPACE ENTRANT(S) | CHECK # |
|---|---|---|---|---|---|
| ____ | ____ | ____ | ____ | ____ | ____ |
| ____ | ____ | ____ | ____ | ____ | ____ |

SUPERVISOR AUTHORIZATION - ALL CONDITIONS SATISFIED ____________ DEPARTMENT/PHONE ____________

AMBULANCE 2800 FIRE 2900 Safety 4901 Gas Coordinator 4529/5387

**Figure 28.3** OSHA Sample Confined Space Permit (#2).

- If the employer sends employees into permit-required spaces, the employer must develop and implement a written entry program.
- When changes in use or configuration of a nonpermit confined space develop that might increase the hazards to entrants, the employer must, if necessary, reclassify the space as needing a permit.
- On the other hand, a permit space may be reclassified as a nonpermit space:
  1. If there are no actual or potential atmospheric hazards and if all hazards within permit space are eliminated without entry, space may be reclassified for as long as the nonatmospheric hazards remain eliminated.
  2. If entry is required to eliminate hazards, it shall be according to regulations, and space may be reclassified as long as the hazards remain eliminated.
  3. The employer shall certify in writing that all hazards in permit space have been eliminated and make this document available to each entrant.
  4. If hazards arise in a declassified permit space, employees shall exit and the employer shall determine whether to reclassify the space.
- When your company arranges for a subcontractor to perform permit-required space entry work, you need to:
  1. Inform the subcontractor of permit space entry program.
  2. Apprise subcontractor of hazards of particular permit-required spaces and precautions and procedures that have been implemented for protection of employees in or near permit spaces.
  3. Coordinate entry operations with subcontractor when both will be working in or near permit-required spaces, and debrief subcontractor after entries.
- Subcontractors shall, in turn, inform you of the permit program to be followed and coordinate multiple entry operations.

## Permit-Required Space Entry Program

- Prevent unauthorized entry.
- Identify and evaluate hazards before entry.
- Establish safe practices, such as isolation, purging, inerting, ventilation, barricades, lockout/tagout, etc.
- Provide and maintain equipment necessary for safe entry, including testing and monitoring, ventilation, communications, personal protection, lighting, barriers, ingress and egress, and rescue.
- Test permit space and document results.
- Maintain acceptable conditions in permit space.
- Provide at least one attendant outside permit space for duration of entry operations.
- Identify duties of each employee and provide training.
- Implement proper procedures for rescue.

- Establish a written system for preparation, issuance, use, and cancellation of permits.
- Coordinate entry operations during multiple employer entries.
- Review entire entry program at least annually, unless previously reviewed at conclusion of a specific entry.

## Rescue Services

You can elect to use on-site or off-site team. If on-site:

- Must be properly trained in entry procedures, rescue procedures, and PPE requirements.
- Permit space rescues must be practiced at least annually from similarly configured spaces.
- Must be trained in basic first aid and CPR, and have at least one member currently certified in those procedures.

If you are using an off-site team or service, you must:

- Inform rescue service of hazards they may confront.
- Provide rescue service with access to all permit spaces so they can develop appropriate rescue plans and practice rescue operations.
- If injured entrant is exposed to substance with a required MSDS or similar document, it shall be made available to medical facility treating entrant.

## Nonentry Rescue

Retrieval systems or methods shall be used whenever entry is made, unless the retrieval equipment would increase overall risk of entry or would not be of value.

Each entrant shall use a chest or full body harness, with retrieval line attached at the center of back near shoulder level, or above head.

Wristlets may be used in lieu of the chest or full body harness if employer can show use of chest or body harness is infeasible or creates a greater hazard and that use of wristlets (or anklets) is safest and most effective alternative.

The other end of retrieval line shall be attached to a mechanical device or fixed point outside of permit space for immediate use. Mechanical device shall be used to retrieve personnel from vertical-type permit spaces more than 5 feet deep.

Relying on "911" for rescue services will not meet OSHA's rescue requirements for permit-required confined spaces, unless the specific requirements mentioned above are met before there is a need for rescue services, in paragraph K of 29 CFR 1910.146 (General Industry Standard).

Major types of individuals involved with confined spaces include the entrants (employees entering the spaces), the attendants (employees who remain outside, watching over the space while the entrants are inside), and entry supervisors (employees who make sure that the entrants and attendants know the hazards and are prepared to deal with emergencies such as nonentry rescue if needed).

### Confined-space entrants

Employee(s) who will be going into the excavation, trench, pipe, sewer, manhole, or other confined space are called entrants. You need to ensure that employees you will expect to enter confined spaces:

- Know the hazards that may be faced during entry, including the signs or symptoms, and effects of exposure.
- Are able to properly use all required equipment.
- Communicate with attendant as necessary to enable attendant to monitor status and alert entrants of need to evacuate.
- Alert attendant whenever any warning sign or symptom of exposure to a dangerous situation is detected.
- Leave the space as quickly as possible whenever an order to evacuate is given by the attendant or entry supervisor, or if the entrant recognizes any warning sign or symptom of exposure to a dangerous situation, or if the entrant detects a hazardous condition, or if an evacuation alarm is activated.

### Confined-space attendant

- Knows hazards that may be faced during an entry.
- Knows possible behavioral effects of hazards.
- Continuously maintains accurate count of entrants.
- Remains outside permit space during entry operations until relieved by another attendant.
- Communicates with entrants as necessary to monitor status and alert of need to evacuate space.
- Monitors activities inside and outside space to determine if safe for entrants to remain in space and orders evacuation when necessary.
- Summons rescue and emergency services when emergency exit from permit space is necessary.
- Takes the following actions when unauthorized persons approach or enter a permit space while entry is underway:

  Warns them to stay away, or advises them to exit immediately if they have entered.

  Informs authorized entrants and entry supervisor if unauthorized persons enter space.

- Performs nonentry rescues per employer's procedure.
- Performs no duties that might interfere with his or her primary duty to monitor and protect entrants who are inside the space.

### Entry supervisor

- Knows what hazards may be faced during entry.
- Verifies that acceptable conditions for entry exist.
- Stops the entry when the job is completed or a hazardous condition develops.
- Makes certain that rescue services are available.
- Prevents untrained or unauthorized persons from entering.

*Remember:* The time to clear up any questions about confined-space regulations, rules, or work practices *is during the entry's planning stage.* If you or your employees are unsure of how to proceed, stop and find out. Postpone the entry until later.

Chapter

# 29

# Fire Prevention

## Quick Scan

1. Train all employees on fire prevention techniques and on the use of fire extinguishers.
2. Establish a procedure for hot-works activities.
3. Store gasoline and other flammable liquids and gases in safety cans and containers outdoors or in an approved storage cabinet.
4. To prevent buildups of fumes and vapors that could cause explosions or fires, avoid the spraying of paint, solvents, or other types of flammable materials in poorly ventilated areas.
5. Follow good housekeeping procedures, with routine cleanup and removal of combustible debris from the worksite.

## What OSHA Says

1926.24 Fire protection and prevention

The employer shall be responsible for the development and maintenance of an effective fire protection and prevention program at the jobsite throughout all phases of the construction, repair, alteration, or demolition work. The employer shall ensure the availability of the fire protection and suppression equipment required by Subpart F of this part.

Subpart F includes standards:

1926.150—Fire Protection
- (a) General requirements
- (b) Water supply
- (c) Portable firefighting equipment
- (d) Fixed firefighting equipment, such as sprinkler systems
- (e) Fire alarm devices

1926.152—Flammable and Combustible Liquids

(a) General requirements
(b) Indoor storage of flammable and combustible liquids
(f) Handling liquids at point of final use
1926.153—Liquified Petroleum Gas (LP-Gas) many specific standards

## Fire

Two basic forms of fire exist: *flaming* (including explosions), and *flameless surface* (including deep-seated glowing embers). Flaming is a direct burning of a gaseous or vaporized fuel. The rate of burning is high, and a high temperature is produced. The flameless-surface variation of fire is a surface fire caused by a situation in which combustion occurs from a combination of the three legs of the fire triangle (source of fuel, temperature, and oxygen), but where the fire has not yet reached a stage where flames begin to burn the vapors given off by the fuel. The important thing to remember here is that you may not always see a fire that's started. A fire can smolder within a pile of debris or within a wall for days before bursting into flames.

## Types of Fires

There are three basic types of fires to be concerned about: types A, B, and C. Type A fires involve the burning of wood, paper, cloth, trash, and other ordinary materials. Type B fires involve flammable liquids such as gasoline, oil, paint, and lubricants. And type C fires occur with live electrical equipment. Another type of fire, type D, involves exotic kinds of metals that can catch fire and burn. Because the most common fires are types A, B, and C, the best all-around emergency fire aids to have at the worksite are multipurpose dry chemical extinguishers.

## Housekeeping

Everyone knows the importance of good housekeeping, but occasionally it falls to the bottom of the list of things to do.

- Instruct employees to dispose of rubbish and trash regularly. Why let it accumulate? Again, the less debris that is strewn about the worksite, the better. Keep containers emptied. It's important not only from the standpoint of preventing a fire from starting, but also to keep one from rapidly spreading.
- Either use or dispose of partially emptied cans of paint, thinner, solvents, and other flammable liquids, especially if there are no plans for using the remnants in the near future. Good cans of these materials should be stored away from sources of heat. Some of these liquids may not be flammable, but their vapors can be.
- Spills should be cleaned immediately—especially sawdust, flammable liquids, oil, or grease. Work areas should be cleaned after each use. Oily rags that are allowed to sit around for weeks could start a fire all by themselves, through spontaneous combustion.

Space heaters can provide crew members with temporary relief from the cold, but they too can be dangerous.

## Hot Works

"Hot works" include all welding, soldering, burning, and grinding operations, and any similar activities that could cause sources of ignition. Precautions against starting accidental fires while hot works activities are in progress include:

- Inspecting the location where the hot works will take place to make sure no flammable or combustible materials are nearby.
- Floors and surroundings swept clean of sawdust and debris, and wet down if necessary.
- Combustibles that can't be moved should be covered to control sparks.
- Nearby floor and wall openings that lead to lower levels should be covered to control sparks.
- Fire extinguishers/water must be available at the hot works location.

## Portable Kerosene and Electric Heaters

When portable kerosene or electric heaters must be used, allow only those models certified as safe by a well-known testing laboratory. Gas heaters should also be approved by the American Gas Association (AGA).

Consider the following points for safe operation of portable heaters:

### Kerosene heaters

- If you use kerosene heaters, you introduce hazards related to fuel storage and handling. The wick will have to be inspected every week or so during the heating season, and cleaned if it's dirty. Take care to use the right fuel. Kerosene heaters require 1-K grade kerosene. When colored or cloudy kerosene is burned it will give off an odor, it will smoke, and it will increase pollution levels because the fuel's higher sulphur content will result in higher sulphur dioxide levels than is emitted by the recommended 1-K fuel. Don't overfill the tank, and never refill a heater while it is still hot or burning. Keep kerosene in well-marked safety cans designed for storing fuel. Avoid using containers marked "kerosene" for gasoline or other fuel or oil storage.
- Kerosene heaters have constant flames while burning. For that reason alone, they should not be operated in a room where solvents, aerosol sprays, gasoline, oil, or other types of vapor-producing flammables are stored. Adequate room ventilation is also a must.
- Kerosene heaters should not be moved while in operation. The carrying handle could be hot and could cause a burn, and just think what would happen if the unit is dropped along the way. Train employees to play it safe by extinguishing

the flame and letting the heater cool before transporting the heater somewhere else.

- *Never* use gasoline to power a kerosene heater. Even small amounts of gasoline mixed with kerosene can greatly increase the likelihood of explosion and fire.

  Employees need to review the operator's manual to make sure they remember all operating and safety features.

### Electric heaters

- Is the wiring adequate to supply safe electric power for an electric heater? Their demands can be quite heavy on a circuit that may already be pushing its limit.
- If possible, avoid using extension cords with electric heaters. If there's no other option, make sure that the cord is marked with a power rating at least as high as that of the heater itself.
- To extinguish electrical fires, the offending appliance or component should be unplugged, if possible, or the power shut off to the circuit it's on. Water must never be used to extinguish an electrical fire—a shock could result. Multipurpose ABC dry-chemical extinguishers should be used instead.

## Flammable Liquids

Flammable liquids are dangerous because they give off vapors that burn. Flammable liquids must never be used to clean equipment or floors; the vapors collect and can be set off by a single spark, a lighted cigarette, or even a pilot light from a gas appliance.

Gasoline is the main cause of worksite flammable-liquid burns. Its frequent use with vehicles, power tools, plus landscaping and construction equipment is the main hazard. Many of the accidents result from refueling equipment that is still hot enough to ignite gasoline vapors. Consider the following prevention pointers:

- Let gasoline-powered tool engines sit and cool a while before refilling their tanks. Always handle gasoline fueling outdoors if possible, where the vapors can easily disburse—instead of inside a closed garage or shed, where vapors can collect and create fire and explosion hazards.
- Store gasoline in approved safety cans, and whenever possible, keep it in an outside shed or separate storage building other than in the house or in an attached garage. The shed should not be airtight—it should have good ventilation.
- Other liquids emit vapors that can easily catch fire or explode, including kerosene, paints, thinners, strippers, adhesives, and acetone. Try to limit the amounts employees have on hand to only enough to do the jobs planned for.

## Fire Extinguisher Use

Fire extinguishers (see Fig. 29.1) look simple enough to use, but during a real emergency, things happen fast. Employees need instructions on how to use extinguishers *before* an extinguisher is required. They need to know which extinguishers to use on individual fire types. Here are some points to consider about fire extinguisher use.

- When deciding on which types of extinguishers to keep at the worksite, consider that multipurpose ABC dry chemical extinguishers will cover the possibility of all three basic types of fires:

  Class A fires (see Fig. 29.2) involve ordinary combustibles or fibrous material such as wood, paper, cloth, rubber, and some plastics.

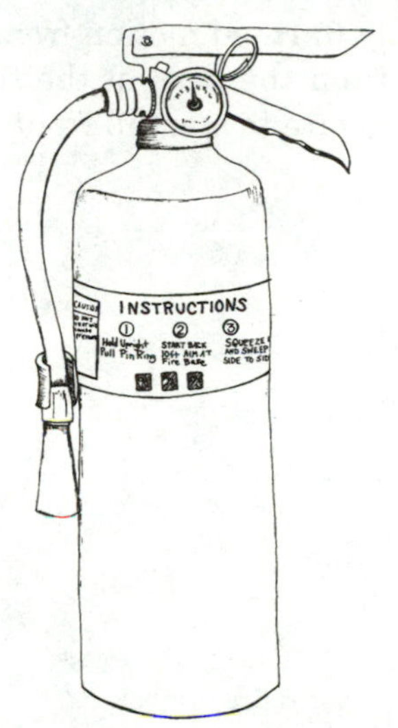


**Figure 29.1** Fire extinguisher.

**Figure 29.2** Class A fire symbol.

Class B fires (see Fig. 29.3) involve flammable or combustible liquids such as gasoline, kerosene, paint, solvents, and propane.

Class C fires (see Fig. 29.4) involve electricity, and include energized electrical equipment such as tools, appliances, panel boxes, and similar charged components. For more information on extinguisher selection, see OSHA's chart (Fig. 29.5).

- An extinguisher should be used only when everyone else is out of the area and the fire is still small or manageable. If any doubt exists as to whether the extinguisher will do the job, employees must get out and call the fire department.
- Train employees to never let a fire get between them and the only escape route.
- Supply hands-on fire extinguisher training to employees so you're sure they know how to operate extinguishers. Teach them to pull the pin, aim the extinguisher nozzle at the base of the flames, squeeze the handle, and use a sweeping motion vertically or horizontally in rapid motion from about six to eight feet away from the fire, covering from the base of the flames to just above the top or from side to side, making sure to start in front of the flames and to end just beyond the flames.

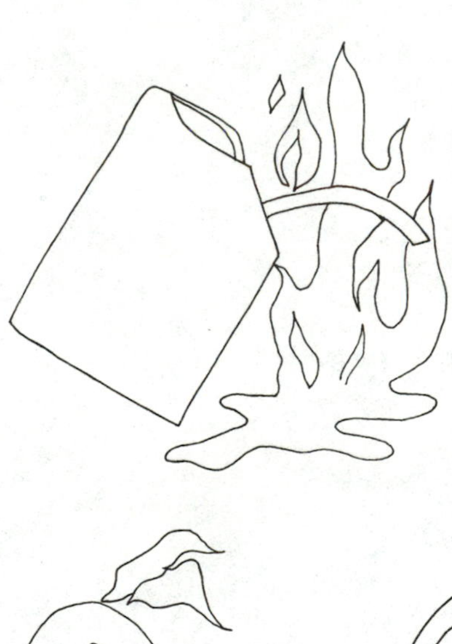

**Figure 29.3** Class B fire symbol.

**Figure 29.4** Class C fire symbol.

**FIRE EXTINGUISHERS DATA**

| | WATER TYPE | | | | FOAM | CARBON DIOXIDE | DRY CHEMICAL | | | |
|---|---|---|---|---|---|---|---|---|---|---|
| | | | | | | | Sodium or Potassium Bicarbonate | | Multi-Purpose ABC | |
| DEPARTMENT OF LABOR · UNITED STATES OF AMERICA | Stored Pressure | Cartridge Operated | Water Pump Tank | Soda Acid | FOAM | $CO_2$ | Cartridge Operated | Stored Pressure | Stored Pressure | Cartridge Operated |
| Class A Fires Wood, Paper, Trash, Having Glowing Embers | YES | YES | YES | YES | YES | NO (But will control small surface fires) | NO (But will control small surface fires) | NO (But will control small surface fires) | YES | YES |
| Class B Fires Flammable Liquids, Gasoline, Oil, Paints, Grease | NO | NO | NO | NO | YES | YES | YES | YES | YES | YES |
| Class C Fires Electrical Equipment | NO | NO | NO | NO | NO | YES | YES | YES | YES | YES |
| Class D Fires Combustible Metals | SPECIAL EXTINGUISHING AGENTS APPROVED BY RECOGNIZED TESTING LABORATORIES | | | | | | | | | |
| Method of Operation | Pull Pin Squeeze Handle | Turn Upside Down and Pump | Pump Handle | Turn Upside Down | Turn Upside Down | Pull Pin Squeeze Lever | Rupture Cartridge Squeeze Lever | Pull Pin Squeeze Handle | Pull Pin Squeeze Handle | Rupture Cartridge Squeeze Lever |
| Range | 30' - 40' | 30' - 40' | 30' - 40' | 30' - 40' | 30' - 40' | 3' - 8' | 5' - 20' | 5' - 20' | 5' - 20' | 5' - 20' |
| Maintenance | Check Air Pressure Gauge Monthly | Weigh Gas Cartridge - Add Water If Required Annually | Discharge And Fill With Water Annually | Discharge Annually - Recharge | Discharge Annually - Recharge | Weigh Semi-Annually | Weigh Gas Cartridge- Check Condition of Dry Chemical Annually | Check Pressure Gauge and Condition of Dry Chemical Annually | Check Pressure Gauge and Condition of Dry Chemical Annually | Weigh Gas Cartridge - Check Condition of Dry Chemical Annually |

**Figure 29.5** OSHA Fire Extinguisher Data Chart.

- Have fire extinguishers inspected at least every month. Keep records of the inspections by fastening a durable tag showing dates of inspections and any recharges that may have been necessary. Look at the pressure gauge to make sure the extinguisher hasn't developed a slow leak that has partially discharged the tank. Does the gauge still read in the safe zone? Is the extinguisher pin still in place? Lift the extinguisher off its bracket to check its overall condition and to make sure it is easy to remove in case of an emergency. Weigh carbon dioxide extinguishers every six months to determine whether the contents are leaking. Don't expel contents of regular or multipurpose dry chemical extinguishers in an effort to check them. Once any amount has been used, extinguishers should be recharged. If repairs or adjustments are needed, have them made by a professional fire equipment company technician.
- Distribute fire extinguishers to include at least one per floor. Place them near exits and in full view, so that they can be reached quickly and easily. Place extinguishers near any wood-burning appliances or portable heating units that are put into service.

## Fire Escape Planning

Despite the best prevention efforts, an accidental fire is still possible. Employees should discuss what to do in case of a fire—including how everyone will escape.

- Rooms should ideally have at least two escape routes.
- Windows should unlock and open easily.

- Escape ladders may be needed upstairs. Chain ladders are available through mail-order catalogs and in lock and security shops, hardware stores, or building supply centers.
- Remember that smoke and poisonous gases rise with hot air from the flames. Instruct employees to move along the floor where smoke and fume concentrations are the lightest. It's best to hold a wet cloth to the face and to take short breaths.
- In the event of a fire, interior doors should be closed to prevent the flames from spreading. Doors that feel unusually warm or hot to the touch should not be opened. If a door is not hot, the employee should brace him or herself against it, turn the face away, and open it carefully. Slam it shut again if any smoke or heat comes from the opening, then get via out a different route.
- Escape routes such as doors, windows, stairways, and hallways must always be kept clear.
- If a window that can't be opened must be broken, train employees to first make sure that all doors to the room are closed. (In general, open doors and windows will help to spread a fire.) Then stand to the window's side while swinging an object such as a chair or unplugged floor-lamp base toward the glass. Pieces of glass should be carefully removed from the frame, then a double-folded blanket or other thick cloth placed over the window sill.
- If employees can't escape through the window opening, they should signal for help by waving a white cloth or other light-colored article in the window.
- Instruct employees to jump only as a last resort, especially if they are trapped two or more stories above the ground. If it's their only option, they may hang by their hands from a window sill, then let go. Bending their knees will help cushion the landing.
- If an employee's clothes happen to catch fire, the proper response is for the victim to stop, drop to the ground with arms folded across his or her chest, and try to roll the flames out. If a blanket, heavy coat, or rug is available, use it to smother or retard the flames until someone can spray the victim with water.
- Have the worksite policed against careless smoking habits.

Chapter

# 30

# Tools and Equipment

## Quick Scan

1. Instruct employees on the safe use of power tools, hand tools, and other equipment they are expected to employ at the worksite, and provide refresher training when needed.
2. Discuss safe procedures for using tools and equipment at formal training sessions and shorter tool box meetings, using demonstrations and hands-on practice.
3. Maintain a collection of operating manuals for power tools and equipment for employees to use as training aids and reference materials.
4. Periodically audit tool and equipment usage at your worksites.
5. Review ergonomic tool designs with your employees so they know what new styles are available, and how the improved tools can reduce tool usage stresses.
6. Inspect the tools your employees use, even if the tools are their personal property, and insist that the tools be in good repair and properly maintained.

Power tools are no longer the Frankensteinish, huge, heavy, overengineered and unforgiving machines they used to be. They're safer. They come with clearly written operating instructions and manuals. They're less likely to throw their weight around and wrench or vibrate their user's hands and arms. They're not as bad as they used to be, but they're still dangerous. They carry risks of fire, shock, or electrocution. They create dust, shavings, or other small flying objects that can cause injury. Power tools cause numerous types of safety-related incidents, including eye injuries, head injuries, fractures, cuts and lacerations, contusions and abrasions, and strains and sprains.

Hand tools, because they seem so familiar and harmless to their users, also carry considerable risk when used improperly. Using a screwdriver as a chisel may cause the tip of the screwdriver to break and become an airborne missile.

If the wooden handle on a tool such as a hammer or axe is loose, splintered, or cracked, the head of the tool could fly off and strike the user or another employee. A wrench with sprung jaws might slip. Impact tools such as chisels, and wedges are unsafe if they have mushroomed heads; the heads could shatter on impact, sending sharp fragments flying.

Generators, compressors, welders, and similar pieces of equipment that are temporarily set up at the worksite must receive similar precautions.

## Power Tools

Your company is ultimately responsible for the safe condition of tools and equipment used by employees (even when the employees use their own tools). Employees also have a responsibility, too, for using and maintaining tools properly.

Power tools make great topics for tool box meetings. Here are some sample guidelines for review at your company's meetings. Add others to the list as needed.

1. *Maintain good housekeeping.* Cluttered work areas invite injury by making it difficult to walk while working, and may be responsible for employees assuming awkward and unsafe positions while using tools because of debris or bulky items in their way. This often results in backstrain from having to reach over or around materials while operating heavy drill motors or power saws.
2. *Be aware of environmental hazards.* Avoid exposing power tools to rain, even when they're being transported or stored. Don't use electrically powered tools in wet or damp environments. Keep the work area well illuminated, especially when working in areas where overhead lighting may not have been installed yet, such as in a basement or attic. Set up trouble lights or have someone hold a light over your work area while you're operating power tools. The same cautions apply, of course, when working with other equipment and hand tools.
3. *Watch out for flammables.* Because motors in most power tools normally spark, don't operate them near flammable liquids or in gaseous or explosive atmospheres. The sparks could ignite fumes. Instead, prepare the area by removing the flammable materials and/or by increasing ventilation.
4. *Protect against electric shock.* Prevent employees from making contact with grounded surfaces. Disconnect power tools when not in use, before servicing, and when changing accessories such as blades, bits, cutters, and grinding wheels. Never carry a tool by its cord, and never yank the cord to disconnect it from a receptacle. Electrical power tools must either be double-insulated or powered by an approved cord containing three wires, with a ground. Before using a power tool near components that are electrically charged, shut the current off to those components.
5. *Keep visitors a safe distance away.* All visitors should be kept out of the immediate area when power tools are being used. Why take a chance? A visitor may think nothing of walking up behind an employee who is using a loud power tool, and tapping the employee on the back to get his or her attention.

Naturally, a startled tool user may react instinctively by doing something dangerous. All employees must understand the importance of preventing visitors from barging in unannounced on worksite areas. Also, visitors have been injured by just picking up power tools—and turning the tools on accidentally. Don't let anyone but trained employees handle power tools, electrical cords, and other equipment.

6. *Put idle tools away.* When not in use, tools should be stored in dry and high or locked-up places. If they're not, they might be forgotten, and before you know it, some 12-year-old neighbor boy is playing around after hours with a power stapler. Make it a habit among your employees to put tools away unless the tools are being actively used.

7. *Don't force a tool.* A power tool, like any hand tool, will always do a better and safer job at the rate for which it was designed by the manufacturer. Applying additional pressure on a drill, saw, grinder, or other tool negates the human safety factors engineered into the tool's design. Too much pressure gone astray may not be able to be "caught" or controlled in time to prevent the operator's hands or other appendages from striking the piece being worked or some surrounding object. Too much pressure on a drill bit, saw blade, abrasive grinding or cutting wheel, or other tool will change its action and prevent it from operating cleanly. Use a steady pressure, without rocking or leaning into the tool.

8. *Use the right tool for the job.* You've probably heard this a million times. Don't try to force a small tool or attachment to do the job of a heavy-duty tool. On the other hand, carpet tacks shouldn't be driven with a sledge hammer, either. If necessary, buy or arrange access to several versions or sizes of the same tool.

Leave creativity for something other than tool usage; avoid using a tool for a purpose not intended by the manufacturer. Be assured that if a different job could be safely accomplished with a particular tool, the manufacturer would have already found out and advertised it. For example, using a circular saw for cutting tree limbs or logs would be asking for trouble. So would using wood bits for drilling holes in concrete blocks.

9. *Dress for success.* Please, do not wear loose clothing or jewelry. They can get caught in moving parts. Wear protective hair covering to contain long hair. The speed at which hands, arms, hair, and clothing can be drawn into moving parts is truly amazing. And once it happens, injuries will come fast and painfully, with dislocations, broken bones, scalpings, and other things described by emergency room personnel in awful-sounding medical terms. Rubber gloves and nonskid footwear are recommended when working outdoors, while vibration-resistant gloves can provide additional protection. For continuous work, use comfort grips or vibration-reducing gloves.

10. *Use safety eyewear.* Wear safety glasses with side shields, goggles, and face shields as required when using power tools. Also use hearing protection and a dust respirator, if needed.

11. *Respect the power cord.* Again, never carry a tool by its cord or yank it to disconnect from a receptacle. Keep the cord away from heat, oil, and

sharp edges. Remember to block doors or windows open if the cord must be run through those openings. Use caution when using the cord near sharp edges, abrasive surfaces, and in areas where pieces of mobile equipment are operating.

12. *Fasten down the work.* Whenever possible, use clamps or a vise to hold work. It's safer than using your hand, and it frees both hands to operate the tool. Be careful to keep clamped assemblies away from vibration and bumping by the tool. When the use of a vise or clamp is impractical, a tool holder, a temporary custom-made accessory, or some help from a fellow employee can be used instead.

13. *Avoid overreaching or being out of position.* Again, this is a sure way to cause strains, sprains, falls, lacerations—you name it. Keep proper footing and balance at all times. Start off slowly when engaging the tool with the work, making sure that the handle sits squarely within your hand or hands. If you need to change positions several times as the tool cuts or drills its way through the work, stop and do so. People have been injured because they stubbornly refuse to stop and reposition when straining through the last few inches of a cut.

14. *Care for your tools.* Take care of your tools and, as the saying goes, "they'll take care of you." Keep tools sharp and clean for better and safer performance. Carpenters, especially, require sharp tools to reduce the temptation of "bearing down," which can easily cause slipping and injuries. Follow instructions for lubrication and accessory changes. Keep tool handles dry, clean, and free from oil or grease. Inspect tool cords periodically and if damaged, replace them or have them repaired by an authorized service facility. If the only option appears to be welding or brazing a power tool to repair it, don't. Discard the tool instead. Any damaged tool, repairable or not, should be tagged out of service and removed from the worksite as soon as possible so no one else can accidentally use it.

Inspect tools for damaged parts, for improper alignment of moving parts, binding of moving parts, broken mountings, and for any other conditions that may affect the tool's operation. A guard or other part that is damaged should be repaired or replaced before a power tool is put back into service. Also have defective switches replaced; don't use a tool if its switch does not turn it on or off.

15. *Unplug or disconnect tools when not using them.* Think hazardous-energy control. You don't want any unexpected starting of a power tool while you're carrying it, servicing it, or when changing accessories such as hoses, blades, bits, or cutters.

16. *Remove adjusting keys and wrenches.* Although many power tools are "keyless" or "chuckless," many of them aren't, too. Form a habit of checking to ensure that keys and adjusting wrenches are removed from the tool before turning it on.

17. *Guard against unintentional starting.* Don't carry plugged-in tools with your finger on the switch. Be sure the switch is off when plugging in or activating

a tool. Inspect tool switch settings after they've been transported. Switches can be accidentally bumped "on" during travel.

18. *Outdoor-use extension cords.* When tool is used outdoors, use only extension cords designed and marked for outdoor use. And no cheating; residential construction not yet under roof—even if you ARE standing in the minimally framed living room—is still open to the weather and needing outdoor-use extensions.

19. *Stay sharp.* It's likely that your tools are. Watch what you're doing. Use common sense. Do not operate a power tool when you are tired.

## Tool Groups

### Saws

- Band saws must have guarded blades.
- Band saw wheels need to be completely enclosed.
- Circular saws must have a fixed guard over the top half of the blade and an adjustable guard over the bottom half.
- Radial saws used for ripping must have anti-kickback devices.
- Radial saws must return to start position.
- Radial saws must have a stop.

### Pneumatic or air tools

Pneumatic tools are powered by compressed air; they include chippers, drills, hammers, and sanders. There are several dangers encountered in the use of pneumatic tools. The main one is the danger of getting hit by one of the tool's attachments, or some kind of fastener the worker is using with the tool. Other considerations include:

- Pneumatic tools that shoot nails, rivets, or staples, and operate at more than 100 pounds per square inch, must be equipped with a special device to keep fasteners from being ejected unless the muzzle is pressed against the work surface.
- Eye protection is required and face protection is recommended for employees working with pneumatic tools.
- Noise is another hazard. Working with noisy tools such as jackhammers requires proper, effective use of ear protection.
- When using pneumatic tools, employees must check to ensure that the tool is fastened securely to the air hose by a positive means to prevent it from becoming accidentally disconnected. A short wire or positive locking device attaching the air hose to the tool will serve as an added safeguard. Radiator-type hose clamps should not be used to fasten hose ends to tools or air sources.
- In general, the same precautions that are recommended for electric cords should be taken with air hoses. An air hose is subject to the same kinds of damage, accidental striking, and tripping hazards.

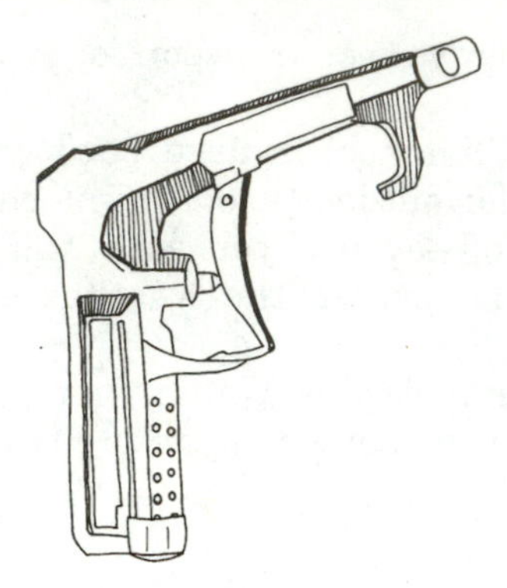

**Figure 30.1** 30-psi air gun.

- Compressed air must be at a pressure of less than 30 psi when used for cleaning. Special air blowguns are available that meet OSHA standards for 30 psi outlet pressure (See Fig. 30.1). Compressed air at any strength must never be directed at a person.

### Electrical tools

Employees using electric tools must be aware of several dangers; the most serious of these is the possibility of electrocution. Among the chief hazards of electric-powered tools are burns and slight shocks which can lead to injuries or even heart failure. Under certain conditions, even a small amount of current can result in fibrillation of the heart and eventual death. A shock can also cause the user to fall off a ladder or other elevated work surface.

- To protect the user from shock, tools must either have a three-wire cord with ground and be grounded, be double-insulated, or be powered by a low-voltage isolation transformer. Three-wire cords contain two current-carrying conductors and a grounding conductor. One end of the grounding conductor connects to the tool's metal housing. The other end is grounded through a prong on the plug. Whenever an adapter is used to accommodate a two-hole receptacle, the adapter wire must be attached to a known ground. The third prong should never be removed from the plug.

  Double insulation is more convenient. The user and the tools are protected in two ways: by normal insulation on the wires inside, and by a housing that cannot conduct electricity to the operator in the event of a malfunction.

### Electrical extension/flexible cords

- Flexible electrical cords shall be connected in such a manner so as to avoid the transmission of strain to joints, terminal screws, receptacles, and plug attachments.
- Electrical cords must not be fastened with staples or hung in a manner that could damage the jacket or insulation. Never wrap an extension cord around a fixed beam or other permanent structure for support. If cords are suspended, use only nonconductive materials such as plastic ties to secure them.

- Ground-fault circuit interrupters must be used with all electrical extension cords.
- Electrical cord and plug connected equipment must be visually inspected before each use for defects such as loose parts or damage to the jacket or insulation. Defective units must be removed from service immediately.
- Flexible extension cords connected to portable equipment must not be used for raising, lowering, supporting, or securing equipment; that is, the cord does not handle the load or weight of the load. A chain, cable, or rope must be used.
- Attachment plugs and receptacles shall not be connected or altered in any manner which could prevent proper continuity of the equipment grounding conductor.
- The employee's hands should not be wet when unplugging or plugging extension cords, flexible cords, or plug-connected equipment.
- Electrical cords and equipment must be protected from high traffic areas and sharp edges. Extension cords that cross pedestrian walkways must be secured so as not to present a tripping hazard.
- Employees must verify proper grounding and polarity of receptacles before using them.

### Powered abrasive wheel tools

Powered abrasive grinding, cutting, polishing, and wire buffing wheels create special safety problems because they can throw off flying fragments.

- Before an abrasive wheel is mounted, it should be inspected closely and sound or ring-tested to be sure that it's free from cracks or defects. To test a wheel, tap it gently with a light nonmetallic implement. If the wheel sounds cracked or dead, it could fly apart in operation and so must not be used. A sound, undamaged wheel will emit a clear metallic tone or "ring."
- To prevent the wheel from cracking, the user should be sure it fits freely on the spindle. The spindle nut must be tightened enough to hold the wheel in place, but not tight enough to distort the flange. Follow the manufacturer's recommendations.
- Due to the possibility of a wheel disintegrating or exploding during startup, the employee should never stand directly in front of the wheel as it accelerates to full operating speed.
- Portable grinding tools need to be equipped with safety guards to protect workers, not only from the moving wheel surface, but also from flying fragments in case of breakage.
- Always use eye protection with grinding tools.
- Turn off power when not in use.
- Never clamp a hand-held-type grinder in a vise.

### Power tools for metal construction

In the past five or so years, more and more metal building components have been employed at residential and commercial worksites. Carpenters and framers who used to rely mainly on circular saws and hammers are now learning to use powerful chop saws—large circular abrasive wheels driven by gutsy electric motors of 4 or more horsepower, corded screwdrivers, and rotary hammers. It's important to follow rules for electric tools, and to make sure that safety guards remain in place and are not temporarily removed by employees who feel that the job could go quicker if the wheel didn't have that "darn guard" in the way. Battery-powered tools are also being relied upon for use in difficult-to-reach places.

### Powder-actuated tools

Powder-actuated tools operate like loaded guns, and must be treated with the same respect and caution. They can't be used by just anyone. Operators are required by law to have formal training in their hazards and operation.

- Eye and face protection must be worn when using powder-actuated tools.
- Powder-actuated tools must never be used in explosive or flammable atmospheres.
- Before using the tool, the employee must inspect it to ensure that it is clean, that all moving parts work freely, and that the barrel is free from obstructions.
- Like any firearm, a powder-actuated tool should never be pointed at anyone.
- Powder-actuated tools must be used only by trained operators.
- All powder-actuated tools must be tested daily before each use.
- Powder-actuated tools shall not be loaded until just before use. A loaded tool must not be left unattended, especially where it would be available to unauthorized individuals.
- The tool operator must use the correct charge; an overcharged tool can drive a nail right through the floor.
- The operator's hands must be kept clear of the barrel end. To keep the tool from firing accidentally, two different motions are required for firing: one to bring the tool into position, and another to pull the trigger. These tools must not be able to operate unless and until they are pressed against the work surface with a force of at least five pounds greater than the total weight of the tool.
- If a powder-actuated tool misfires, the employee should wait at least 30 seconds before trying to fire again. If it still won't fire, the employee should wait another 30 seconds so that the faulty cartridge is less likely to explode, then carefully remove the load. The bad cartridge should be placed in water.
- The muzzle-end of the tool must have a protective shield or guard centered perpendicularly on the barrel to confine any flying fragments or particles

which might otherwise create a hazard when the tool is fired. The tool must be designed so it won't fire unless it has this kind of safety device.

- All powder-actuated tools must be designed for varying powder charges so that the user can select a powder level necessary to do the work without excessive force.
- If the tool develops a defect during use, it should be taken out of service immediately and kept out until properly repaired.
- When using powder-actuated tools to apply fasteners there are some precautions to consider. Fasteners must not be fired into material that would let them pass through to the other side. The fastener must not be driven into materials like brick or concrete any closer than three inches to an edge or corner. In steel, the fastener must not come any closer than a half-inch from a corner or edge. Fasteners must not be driven into very hard or brittle materials that might chip or splatter, or make the fastener ricochet.
- An alignment guide must be used when shooting a fastener into an existing hole. A fastener must not be driven into a spalled area caused by unsatisfactory fastening.

## Jacks

All jacks—lever and rachet jacks, screw jacks, and hydraulic jacks—must have a device which stops them from jacking up too high. Also, the manufacturer's load limit must be permanently marked in a prominent place on the jack and should not be exceeded.

- A jack should never be used to support a lifted load. Once the load has been lifted, it must immediately be blocked up.
- Use wooden blocking under the base if necessary to make the jack level and secure. If the lift surface is metal, place a one-inch-thick hardwood block (or equivalent) between it and the metal jack head to reduce the danger of slippage.
- To set up a jack, make sure that the base rests on a firm level surface, the jack is correctly centered, the jack head bears against a level surface, and the lift force is applied squarely.
- Proper maintenance of jacks is essential for safety. All jacks must be inspected before each use, and lubricated regularly. If a jack is subjected to an abnormal load or shock, it should be thoroughly examined to make sure it has not been damaged.

## Hand Tools

Hand tools are simple to use. No doubt about that. And because they are, individuals are lulled into a false sense of security with them. Employees can become overconfident and try to expand a tool's usefulness by attempting something the tool was not designed to accomplish, and often use the wrong tool for the job due to expediency. Using wrenches as hammers and hammers as striking tools for striking wrenches when working with stubborn nuts are typical examples.

Because we assume that everyone knows how to use hand tools safely, little training is done and that translates into a general lack of awareness of users. Among the factors leading to hand-tool safety incidents are:

- Use of incorrect tools for work being done. This includes the use of "cheap" or inexpensive tools that fail under normal worksite job conditions.
- Carelessness and improper procedures followed.
- Failure to keep tools in good condition and working order.
- Improper storage

Most hand-tool-related injuries can be avoided by paying attention to a few basic rules of hand-tool safety:

- Always protect the eyes from flying pieces and parts by wearing approved eye protection. Safety goggles should be in every tool box and worn for every project involving hand tools.
- Use the right tool for the job. Most tools are designed by the manufacturer to perform a specific function. Never substitute or use an inappropriate tool.
- Use the tools properly. Hammers, for instance, are designed to strike objects with the face, never the sides of the hammerhead.
- Service and replace tools regularly.
- All worn and damaged tools should be disposed of properly and replaced as soon as possible.
- Tools are not to be left on scaffolds, ladders, or overhead working spaces when not in use.
- Throwing tools from one location to another, from one employee to another, or dropping them to lower levels is not permitted.
- A considerable number of injuries with hand tools occur as they're being carried from place to place, or when they're inadvertently left somewhere and forgotten. Pointed tools and screwdrivers should never be carried loosely in an employee's pocket. Use a tool box or tool belt instead. That can also help prevent tools from being placed haphazardly while employees are working above ground level. Too often they're placed out of sight and forgotten until being kicked or dislodged down onto someone below.

**Hammers.** One of the most widely used hand tools, and also one of the most often abused. More than two dozen styles of hammers are available, with many more variations manufactured in sundry types, sizes, and configurations for very specific purposes.

- Claw hammers (see Fig. 30.2) are designed for driving and removing common and finishing nails only—not for driving hardened masonry nails into concrete, or to strike cold chisels, or to hammer other metal. Ball peen (see Fig. 30.3) or soft metal hammers of the proper size are designed for striking chisels and punches, and for riveting, shaping, and straightening unhardened metal. Numerous styles are available for concrete and

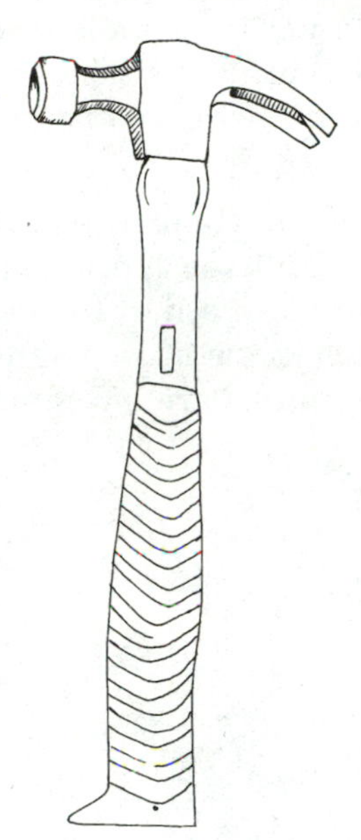

**Figure 30.2** Claw hammer.

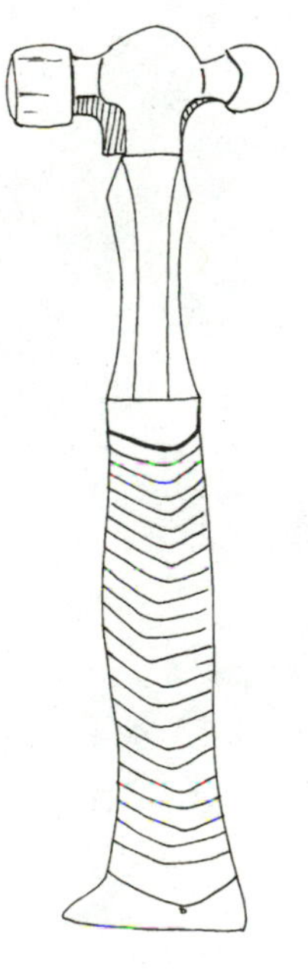

**Figure 30.3** Ball peen hammer.

other masonry work. Rubber mallets (see Fig. 30.4) can strike without marking.

A relatively recent development in hammers is the 19° offset handle (see Fig. 30.5), engineered to help reduce wrist stress and the likelihood of carpal tunnel syndrome. Because the hand has a maximum strength in its neutral (unbent) position, and blood supply to the hand is full and unrestricted in the neutral position, the wrist can remain straight while swinging the hammer (see Fig. 30.6), instead of being bent with a normal handle (see Fig. 30.7). When the hand and wrist move out of the neutral position, as they would while gripping a straight handle, the carpal tunnel is compressed, and grip strength drops off dramatically.

- When striking another tool, such as a chisel, punch, or wedge, the striking face of the proper-sized hammer will have a diameter approximately $^{3}/_{8}$ inch larger than the struck face of the tool. Never use a nail hammer to strike another tool. If chips, cracks, dents, or other signs of wear and stress are present, the tool should be discarded. Eye injuries are a common risk with

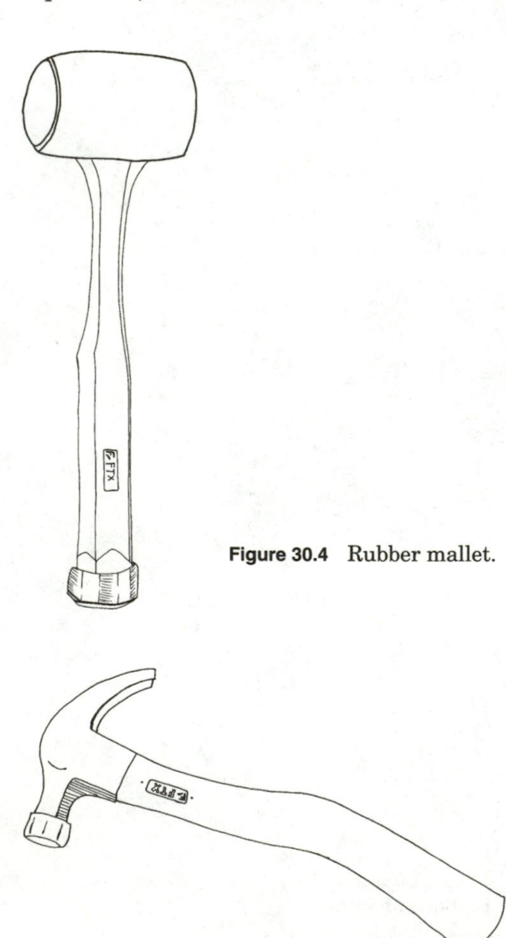

**Figure 30.4** Rubber mallet.

**Figure 30.5** Hammer with 19° handle.

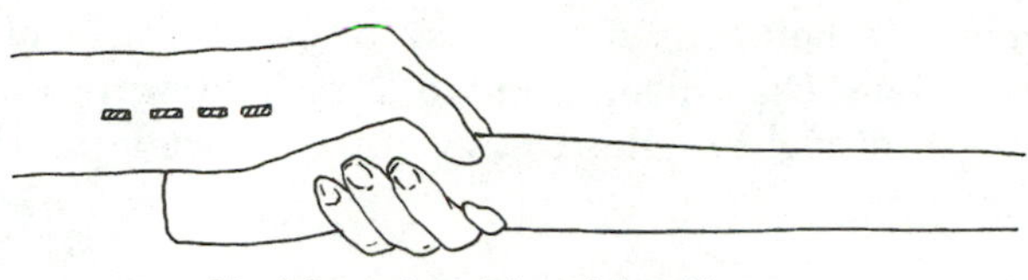

**Figure 30.6** Straight wrist with tool handle.

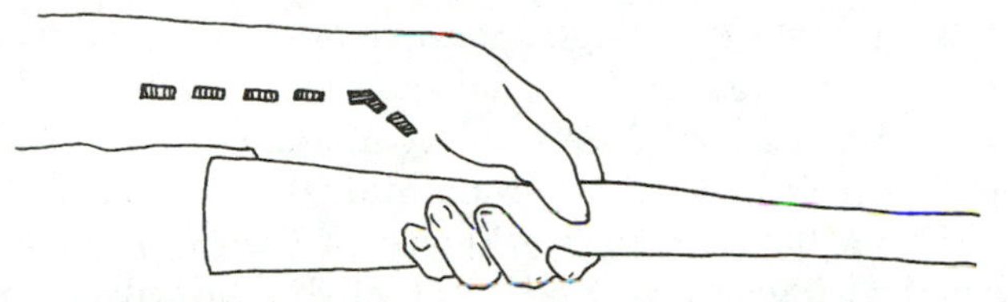

**Figure 30.7** Bent wrist with tool handle.

hammers, especially when driving hardened nails into concrete blocks and other masonry.

- Another hazard associated with hammers is the damage they can cause when dropped from above. Hammers can be left on scaffold planking, on ladders, or other platforms and places from where they can accidentally become airborne. A hammer falling even several feet can cause quite an injury to a finger, wrist, or face.
- Again, stress to your employees, when using hammers:

  Never use a hammer to strike another hammer.

  Avoid striking anything with the side of a hammer.

  Do something with the protruding ends of nails that are driven through the work; most often they're simply bent or driven over.

**Metalworking chisels.** Mushroom chisel heads can often produce flying chips or splinters that have an uncanny way of finding and burying themselves into unprotected eyes. Some chisel heads can be ground when mushrooming starts to develop, with a slight taper around the head to reduce the likelihood of further mushrooming. Chisel users—as well as other individuals working nearby—must wear eye protection. Goggles are recommended. Other chiseling pointers include:

- Cutting edges of chisels should always be kept sharp.
- The right-sized chisel, large enough for the job, is the only chisel that will perform correctly—and it must be used with an appropriately sized hammer.
- Small pieces to be chipped should be firmly clamped in a vise, with the chiseling being done toward the solid or stationary end of the vise, in a direction away from the user.

- If a sledgehammer is being used to strike a chisel, tongs or tool holders should be used to hold the chisel, and workers holding the chisels should position themselves at right angles to the employee wielding the sledge.

**Screwdrivers.** Screwdrivers are probably the most frequently used type of hand tool. They are also one of the most often misused tools. Numerous accidents are caused by misusing screwdrivers for prying, chiseling, scraping, scoring, and many other dangerous tasks. Although screwdrivers come in many sizes, shapes, and materials, they're all intended for one exact use: driving and withdrawing threaded fasteners. Too, they're often used by unskilled or semiskilled workers who do not keep their tools in a safe, well-maintained condition. A frequent cause of injuries comes from holding the workpiece in the palm of the hand while tightening or withdrawing screws, and a slip of the screwdriver results in the driver penetrating the individual's hand or wrist. That happens more than you'd like to think.

- A common screwdriver abuse is using a model that doesn't match or fit the screw. The result is usually a chewed up screw head, broken screwdriver, bloody knuckle, or damage to the work surface. For effective driving of screws, the screwdriver tip must be properly ground to fit the slot in the screw head, and it must be the correct size for the head.
- While screwdrivers with insulated handles should be used for electrical work, these insulated, plastic, or other coverings or handles are not for providing absolute protection against electrical shock, they're more for incidental contact or comfort only, and again, are *not* intended to provide total protection when working around electricity.

**Wrenches.** Wrenches provide the highest torque of any turning tool, but even they have strength limits. Wrenches should be the right size for the job and well maintained. If a wrench is too large or worn, it could slip and cause a whole series of events leading to a serious injury. It's dangerous to try to extend the handle of a wrench with the use of a cheater bar to add leverage: their use can result in excessive leverage which may result in cracks to the jaws of the tool, or a broken or bent handle. The safe and smart move is to go to the tool box and get another wrench with a longer handle.

- A common mistake with socket and other wrenches is using one whose head does not exactly fit the nut. If the opening is not the correct size for the fastener, the wrench is likely to damage the corners of the fastener, or to slip or break.
- A wrench handle should, if possible, be pulled (as opposed to pushed).
- The safest wrench is a box or socket type. Open end, flare nut, and adjustable wrenches are not as strong as the corresponding sizes of box or socket wrenches and are not intended for heavy loads, such as breaking loose frozen fasteners or final tightening.

- Adjustable wrenches should be tightly adjusted to the nut and pulled in such a manner that the force is on the side of the fixed jaw.

**Pliers.** Pliers are not wrenches. They have been designed and manufactured to perform specific functions—not one of which is to turn nuts or bolts. Pliers should not be used in place of a proper wrench.

**Pistol-grip handles.** Pistol-grip handles can be purchased separately to hold tools that are usually held with the more stressful pinch-style grips where the thumb and two fingers hold tools such as paint brushes and files (see Fig. 30.8). Check with your safety distributors for ergonomic tool and equipment catalogs for the latest and most user-friendly tools available.

**Clamps.** Clamps are designed to be tightened by using hand power alone (see Fig. 30.9). A wrench, pipe, hammer, or pliers shouldn't be used to gain extra tightening strength. Clamps should never be used to secure material in a vehicle. Vibration can loosen the tightening screw and create a hazard.

Clamps should never be used to secure material on which people will be standing or climbing nor to attach material to a lifting device. Never use a clamp as a lifting device.

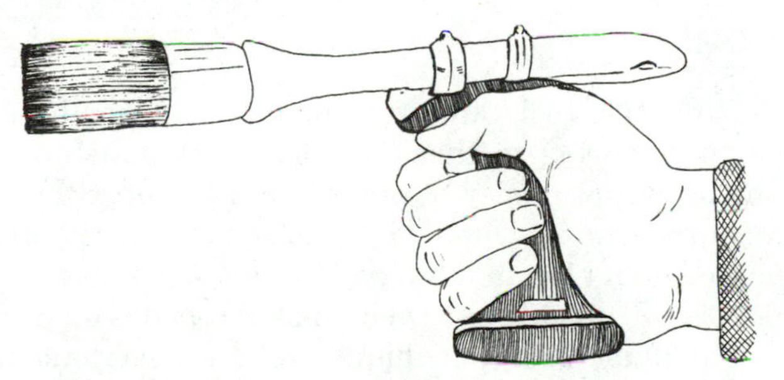

**Figure 30.8** Pistol-grip tool handle.

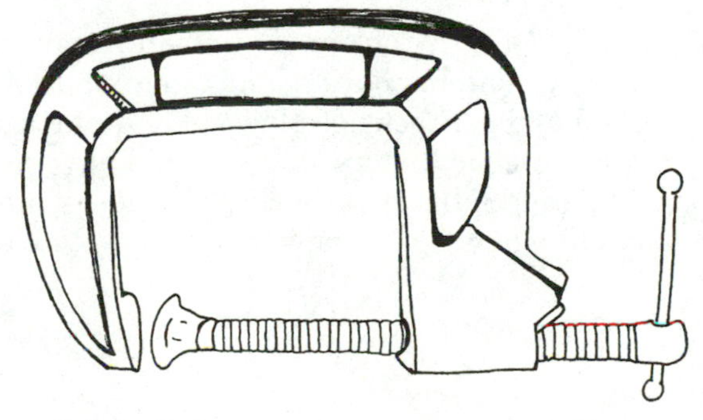

**Figure 30.9** C-clamp.

**Tool boxes, chests, and cabinets.** These may seem safe enough, but improper use can very easily lead to injury. First, they are designed and manufactured to hold tools, not to be stood upon or used as an anvil, sawhorse, or for other creative purpose. Using a tool box as an impromptu ladder to reach hard-to-reach items has been the cause of many an unfortunate tumble. Butt knives and other sharp tools should be sheathed or protected in some way before being placed in a tool-box drawer. Numerous workers have cut their hands on such items while rummaging through a drawer in search of another tool.

**Edged tools.** Many hand tools require a sharp cutting edge to enable employees to cut cleanly without the application of an inordinate amount of force. These include many wood-cutting tools such as knives, axes, picks, garden rakes, hoes, wood chisels, and planes. If cutting tools aren't maintained with a good cutting edge, employees tend to apply more pressure to the tools than the tools were originally designed for, causing slipping and injury because of the greater effort required to achieve the cutting action. Numerous sharp-edged-tool accidents come from failure of the tool owners to guard the cutting edge when the tools are transported, carried, or stored.

Sharpening hand tools can also be a source of injury if done incorrectly. When hand stones or steels are used, they should have a suitable guard to prevent the tool from slipping and cutting the hand. The stones or steels must be large enough for the blade being sharpened, to keep fingers far enough away from the edge.

**"Cheap" tools.** Inexpensive tools are often made from unsuitable materials, using incorrect manufacturing methods or poor workmanship. In any case, they're not worth the initial money saved. Hammers constructed from poor-quality steel can chip or shatter when a blow is struck. Poorly made chisels or punches will mushroom and fail in a hurry. Poor-quality knives, axes, shovels, and other bladed tools will quickly lose their edge. Wrenches made of unsuitable metal and inferior manufacturing techniques tend to open out under normal work conditions, and are likely to eventually cause hand injuries.

**Sheet materials.** While not really a tool or equipment, sheets of plywood, paneling, flooring or underlayment, insulation, and other kinds of material can be heavy and awkward to handle. Stacks of plywood and particle board have tipped over onto employees, crushed fingers and toes, and have caused all sorts of serious lifting injuries, especially with backs, shoulders, and arms. See Chap. 33, on material handling, for tips on how to properly store and transport sheet materials.

Chapter

# 31

# Stairways, Ladders, and Scaffolding

## Quick Scan

1. There are too many important points regarding stairways, ladders, and scaffolds to be listed as Quick Scans. Their individual sections should be reviewed in OSHA's construction regulations.

Stairs, ladders, and scaffolding have what in common? They're all things that your employees—if they're not careful—can fall from. And the falls from these common construction aids can range from dangerous to deadly.

## Stairways

Over the years, stairways have had their ups and (and mostly) downs with the construction industry. They're necessary evils needed for getting from one level to another. OSHA addresses stairways in 1926.1051 and .1052, and specifies numerous requirements, such as:

- Stairs must be provided where there's a greater elevation change than 19 inches and no ramp or ladder has been arranged for.
- Stairs must have tread and landings in the construction phase.
- Variations of riser height and tread depth must be less than $^1/_4$ inch so employees and others will not be faced with dimensions that could throw their balance off and cause slips and falls.
- Where doors open onto stairs, a platform is required so users do not immediately fall to the lower level.
- Stairs with metal pans must not be used until filled in with concrete, unless such use is limited to construction purposes.
- Stair rails 30 to 37 inches high are required when 4 or more risers are present.

- Spiral stairs must be offset, with an inside rail, to prevent walking where there's less than 6 inches of tread.
- Install permanent or temporary guardrails on stairs before allowing their use for general access between levels, to prevent someone from falling or stepping off edges.

## Ladders

Ladders are among the most common—and among the most dangerous—pieces of equipment at the typical worksite. Potential ladder hazards include: use of ladders that are broken or in poor condition; improper selection of ladders, such as using a ladder that's too short, too long, or made out of metal when set up near electrical hazards; and improper use of ladders. Careless use of ladders is a frequent cause of falls, injuries, and some fatalities at construction worksites. Fortunately, with a little understanding of safe ladder practices and regular inspections, ladders can provide a safe means for accessing higher and lower levels for employees.

### Ladder duty ratings

Make sure employees understand duty ratings. Ladders are ranked or rated based on their maximum-weight capacities: how much weight they can safely hold. Here's a simple chart that explains the duty rating system. Employees should understand the ratings—not so much because your company makes lower-duty ladders available at worksites, but because employees may otherwise purchase ladders that are too lightweight to provide safe at-home use.

| Category | Rating | Weight limit |
|---|---|---|
| Type IA | Heavy duty industrial | 300 lbs. |
| Type I | Heavy duty | 250 lbs. |
| Type II | Medium duty | 225 lbs. |
| Type III | Light duty | 200 lbs. |

To understand the duty rating system, employees have to understand what the weight capacity means. It means the employee's weight, plus the weight of his or her clothes and personal protective equipment, as well as the weight of any tools or materials the person will be working with while on the ladder. Longer ladders will not necessarily have greater weight capacities.

### Ladder types

The size and capacity is, of course, important when it comes to ladder selection, but so is the ladder type. Most ground- and first-level height tasks can be accomplished using a 6- or 8-foot stepladder. Higher-level tasks will require a larger stepladder or a straight or extension ladder. Perhaps several different ladders will be needed to complete certain jobs. A long straight ladder may place the employee too far away from the work location at midladder level, due to the amount of horizontal distance the bottom of the ladder must be positioned away from the building, so you may need a shorter straight ladder or a stepladder for that part of the job.

#### Straight or extension ladders

- Realize that the proper ratio of vertical height to the ladder support point to the horizontal distance of an extension ladder's feet away from the foundation is 4 to 1. That's also called the "one-quarter rule." This rule will result in a proper 75.5° ladder angle, which affords optimum balance and stability.
- An extension ladder must extend at least 3 feet past the roof line if its user will be climbing onto the roof (see Fig. 31.1). Never stand any higher than the third step from the top on a straight or extension ladder.

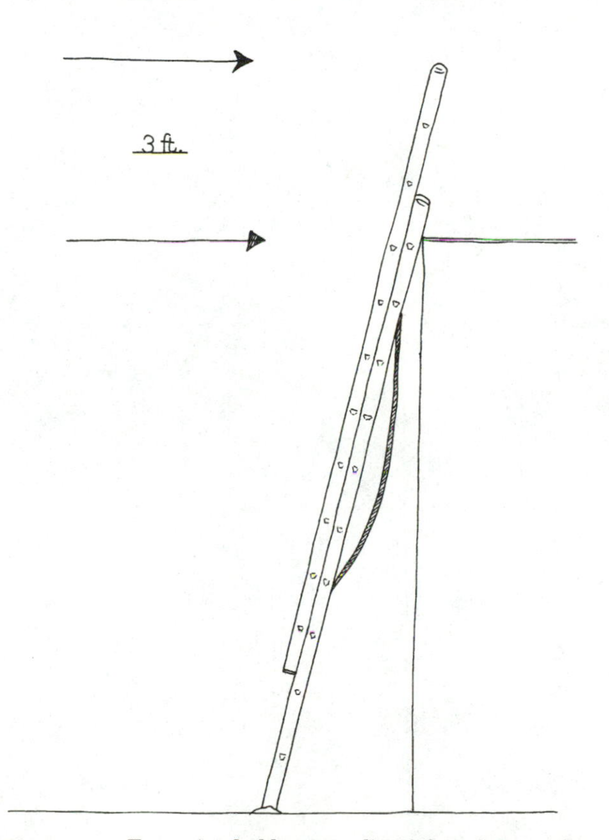


**Figure 31.1.** Extension ladder extending 3 feet above roofline.

- Short ladders should never be spliced together to provide long sections.
- The top section of an extension ladder is called the *fly* section, and the bottom part is called the *base* section. For safety's sake, the fly and base sections must overlap by at least $^1/_{12}$ of the total working height of the ladder (see Fig. 31.2).
- Employees must know how to set up a straight or extension ladder, whenever possible, if there's enough room, by first laying the ladder horizontally on the ground with its feet against the foundation of whatever they're using to rest the ladder against. Then raise or stand the ladder vertically by walking it up with their hands, rung by rung. Next, they should pull the base away from the foundation a few feet, then raise the fly section and lock the rungs. Make sure the ladder is positioned before it's extended. Pull the base of the ladder out until the 4:1 ratio is achieved, or an angle of about 75°.
- If a ladder must be placed on uneven ground, a leveler should be used—not boards, stones, bricks, or blocks. The feet of an extension ladder should be positioned flat on a hard surface, and in the spiked or vertical position on

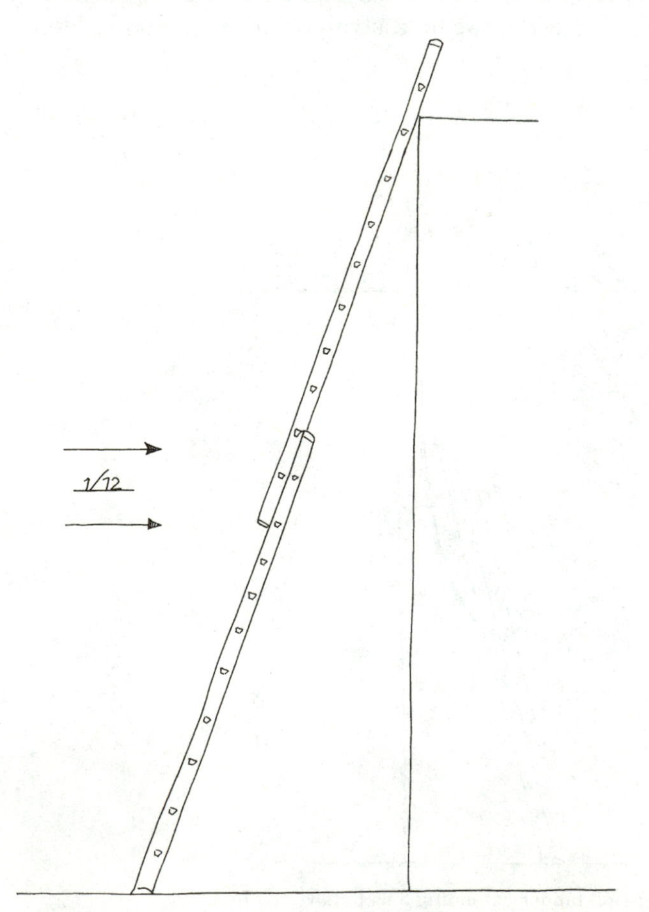


**Figure 31.2.** Extension ladder with minimum $^1/_{12}$ overlap.

grass or soft ground. Rubber feet are available for use on concrete or other potentially slippery surfaces, and "levelers" can be purchased to equalize the foot placement when a ladder must be set upon uneven ground. As an extra measure of caution, it's a good idea to have another employee hold a straight or extension ladder while an employee is doing work on it.

- Before active use, an extension ladder should be tied off at the bottom, middle, and top to prevent the ladder from moving or slipping.
- Never try to move a ladder while you're on it (as John Belushi did while peeking through second-floor sorority windows in the film *Animal House*).
- Never allow employees to climb too high on a ladder. That means not standing higher than the second step from the top of a stepladder, or the third step from the top of an extension ladder.
- If a ladder must be used in less-than-ideal conditions, at least one person should remain at the base, holding the ladder steady while the climber is aloft.
- When storing extension ladders, if hung horizontally they should be supported about every 6 feet. Be sure to avoid hanging and storing other items on or from a ladder. If stored on their side; vertically, support ladders at each end and in the middle.
- Recommend that two employees carry straight or extension ladders. When a single employee must move such a ladder, he or she should balance the center of the ladder on his or her shoulder, positioned so the front of the ladder is above head height, and the bottom end is positioned near the ground.

**Stepladders.** Stepladders are fairly simple to use, and many guidelines for straight and extension ladders apply here, too. Several others include:

- Step no higher than the second step from the top of a stepladder. Naturally, a stepladder should be opened up fully, with the upper "locking" arms engaged so the ladder cannot accidentally collapse and fall.
- When two hands are needed to perform work from a stepladder, *lightly* balance yourself, if needed, by touching your shins, knees, or legs against the next upper steps for a little extra stability.
- When a stepladder must be set up near a doorway, be sure to restrict access to the area so no one can accidentally open a door into or walk into the ladder.

#### General ladder use

- Inspect ladders before each use, and after they fall or when they are mishandled in any way. Are the steps secure and clean? Is anything broken or loose? Are the side rails free of cracks and splinters? Do the locks secure the base and fly sections of an extension ladder? Do the ropes for raising the fly section work properly? Immediately tag out, arrange to repair, or destroy and discard defective ladders; don't allow someone else to use them.

- Positioning ladders correctly is important. Ideally, you want the work to be in front of you, if possible, a comfortable reach away—not so far away that you have to stretch, and not so close that you'll have to lean back or twist, and certainly not directly above your head so you'll have to look and reach straight up.
- Avoid using metal ladders near electrical equipment. Serious injuries have occurred while aluminum ladders were being transported or set up and accidentally contacted wiring or malfunctioning electrical fixtures. It's best to bite the bullet and simply not use metal ladders. Wood ladders can be dangerous, too, if they contain metal reinforcements or if they're wet, or if they contain grease, oil, or dirt which may be conductive. Fiberglass ladders are the safest for the worksite.
- Wood ladders can be coated with clear protectants, but they should not be painted. The paint could cover up and prevent cracks or other defects from being seen.
- While climbing a ladder, maintain three points of contact (a hand and two feet, or two feet and one hand) with the ladder at all times. While working from a ladder, only when you're safely tied off can you work with both hands. Naturally, keep movements on a ladder slow and cautious.
- Allow only one person on a ladder at a time.
- Avoid carrying tools, materials, and other supplies up a ladder. Instead, have someone hand up the items, or pull them up with a tow line. Tools can be carried in a tool belt or pouch. Work with one hand on the straight or extension ladder, while keeping spare tools in the belt or pouch.
- Avoid using ladders as extra pieces of scaffolding—their weight capacity isn't accurate in a horizontal position.
- Ladders should not be placed in front of doors opening toward the ladder unless the door is locked, blocked, or otherwise guarded.
- Keep manufactured and job-made ladders in good condition and free of defects.
- Use ladders only for what they were made, not as platforms, runways, or scaffold planking.

**What OSHA says.** OSHA addresses ladder usage in 1926.1051 and .1053. These regulations stress that ladders shall be inspected by a competent person for visible defects on a periodic basis, and after any occurrence that could affect their safe use.

### Ladder and stairway training

OSHA specifies, in 1926.1060, training requirements for each employee using ladders and stairways:

> (a) The employer shall provide a training program for each employee using ladders and stairways, as necessary. The program shall enable each employee to recognize

hazards related to ladders and stairways, and shall train each employee in the procedures to be followed to minimize these hazards.

(a)(1) The employer shall ensure that each employee has been trained by a competent person in the following areas, as applicable:

(i) The nature of fall hazards in the work area;

(ii) The correct procedures for erecting, maintaining, and disassembling the fall protection systems to be used;

(iii) The proper construction, use, placement, and care in handling of all stairways and ladders;

(iv) The maximum intended load-carrying capacities of ladders used; and

(v) The standards contained in Subpart X.

(b) Retraining shall be provided for each employee as necessary so that the employee maintains the understanding and knowledge acquired through compliance with this section.

## Scaffolding

Scaffolding needs to be chosen and erected under the guidance of a trained, competent person. A major part of safe scaffold use involves how the units are set up in the job. Guidelines for scaffolding use include:

### Practice good housekeeping procedures

- Practice good housekeeping procedures on and near scaffolds. There's precious little walkway space on scaffolds, so keep those surfaces free of tools and debris.
- When you've got to stage materials and tools on a scaffold, keep them away from the edge of the platform.
- During winter and wet weather, ice, snow, mud, and slippery gobs of fresh mortar or concrete should be removed from the scaffold as soon as possible.
- Remember to remove belongings from the walkways before the scaffold is moved or altered from below.
- Avoid tossing or dropping items from the scaffold.
- Materials should be raised or lowered by a hoist or lift, with a tagline if the materials could swing and strike the scaffold.

### General considerations

- Provide ladders or stairs to get onto and off of scaffolds and work platforms safely.
- Erect scaffolds on firm and level foundations.
- Finished floors will normally provide a stable base and support a scaffolding load.
- Scaffold legs must be placed on firm footing and secured from moving or tipping, especially on dirt or similar surfaces.

- Erecting and dismantling scaffolds must be under the supervision of a competent person.
- The competent person must inspect scaffolds before each use.
- Don't use blocks, bricks, or pieces of lumber to level or stabilize the footings. Manufactured base plates or mud sills made of hardwood or the equivalent can be used.

#### Planking

- Fully plank or use manufactured decking to provide a full work platform on scaffolds. The platform decking and/or scaffold planks must be scaffold grade and not have any visible defects.
- Extend planks or decking material at least 6 inches over the edge, or cleat them to prevent movement. The work platform or planks must not extend more than 12 inches beyond the end supports to prevent tipping when stepping or working.
- Be sure that manufactured scaffold planks are the proper size and that the end hooks are attached to the scaffold frame.

#### Scaffold guardrails

- Use a standard guardrail on scaffold platforms that are more than 10 feet above the ground or floor surfaces. If guardrails are not practical, use other fall-protection devices such as safety harnesses and lanyards. (Safety belts were outlawed as fall-protection devices as of January 1, 1998, and can be used only as positioning devices.)
- Place the top rail 42 inches above the work platform or planking, with a midrail at half that height, at 21 inches.
- Install toe boards when other employees are working or accessing below the scaffold.

#### Scaffold design

Scaffolds must be designed by a qualified person and shall be constructed and used within those design specifications. A qualified individual is one who, by possession of a recognized degree, certificate, or professional standing, or who by extensive knowledge, training, and experience, has successfully demonstrated his or her ability to solve or resolve problems related to the subject matter, the work, or the project.

#### What OSHA says

On August 30, 1996, OSHA published their final rule for safety standards for scaffolds used in the construction industry. It's long and technically oriented and appears in Part 1926 as Subpart L, 1926:

.450 Scope, application and definitions
.451 General requirements
.452 Additional requirements applicable to specific types of scaffolds

Scaffold work is one of the most hazardous activities found on construction worksites. The following two examples (taken from OSHA's final rule on scaffolds) typify what can happen if scaffolding rules and regulations aren't strictly adhered to:

- Two employees were working on a pump jack scaffold doing roofing work. The scaffold became overloaded and broke. The employees fell 12 feet to the ground, resulting in one fatality and one serious injury.
- Two workers were erecting an aluminum pump jack scaffold. As they were raising the second aluminum pole, the pole apparently contacted an overhead power line. The pole being raised was 29 feet 10 inches long and the line was 28 feet 10 inches high. The line was approximately 11 feet from the house. One employee died and the other suffered severe burns and was hospitalized. The surviving employee thought they had enough room to work around the power lines, which were not de-energized or shielded.

# Chapter 32

# Signs, Signals, Barricades, and Tags

## Quick Scan

1. Train employees when and how to use barricades, signs, signals, and tags.
2. Enforce the active use of barricades, signs, signals, and tags throughout the worksite.

We live in a world that's practically dominated by signs, labels, and signals. Get in your car or truck and drive out on the open road. What do you see? Or walk past any store or business. Signs tell us what we can, and can't (or rather, shouldn't) do. Labels tell us what's inside a container or boundary, or instruct us how to operate something. Signals tell or show us which way someone is turning, or what to expect next. The same kinds of communications can be found on any worksite. You'll see signs and labels on mobile equipment, warning operators not to park without setting the emergency brake, or not to start moving until a certain air pressure builds up in the brake lines.

OSHA has established standardized types of warning communications through the use of certain colors and words in order to get peoples' attention and communicate quickly what kinds of hazards are present, or what to do to stay safe. By defining the meaning of warning words such as "Danger" and "Caution," OSHA is trying to make sure that workers won't misjudge the seriousness of a hazard.

A few definitions are in order so we're all on the same page.

- *Signs* are the warnings of hazards, temporarily or permanently affixed or placed, at locations where hazards exist.
- *Signals* are moving signs, provided by workers, such as flagmen, or by devices, such as flashing lights, to warn of possible or existing hazards. Horns and alarms can also be considered signals.
- *Barricades* are obstructions to deter the passage of persons or vehicles.

- *Tags* are temporary signs, usually attached to a piece of equipment or part of a structure, to warn of existing or immediate hazards.

Signs and symbols required by Subpart G (signs, signals & barricades) shall be visible at all times when work is being performed, and shall be removed or covered promptly when the hazards no longer exist.

## "Danger" Signs

Danger signs (see Fig. 32.1) shall be used only where an immediate hazard exists. Danger signs shall include red as the predominating color for the upper panel; black outline on the borders; and a white lower panel for additional sign wording. Danger signs identify something with a very high risk, and the word "Danger" always appears in white letters in a red oval on a black background. When employees see a danger sign they should know that there is some immediate danger in the area, and should take special precautions to prevent a serious accident, injury, or even death.

A few examples:

- Danger—Hard Hat Area [see Fig. 32.2]
- Danger—Construction Area, KEEP OUT [see Fig. 32.3]
- Danger—OPEN TRENCH [see Fig. 32.4]
- Danger—Corrosive Liquids, Wear Protective Equipment

## "Caution" Signs

Caution signs (see Fig. 32.5) should be used only to warn against potential hazards or to caution against unsafe practices. Caution signs have yellow as the predominating color; black upper panel and borders; yellow lettering of "caution" on the black panel; and the lower yellow panel for additional sign wording. Black lettering should be used for additional wording.

When you see a caution sign, you know that there are either potential hazards that call for you to take proper precautions or you're being cautioned against unsafe practices. Standard color of the background is yellow; and the panel, black with yellow letters. Any letters used against the yellow background should

**Figure 32.1** Danger sign design.

**Figure 32.2** Danger—Hard Hat Area.

**Figure 32.3** Danger—Construction Area, Keep Out.

**Figure 32.4** Danger—Open Trench.

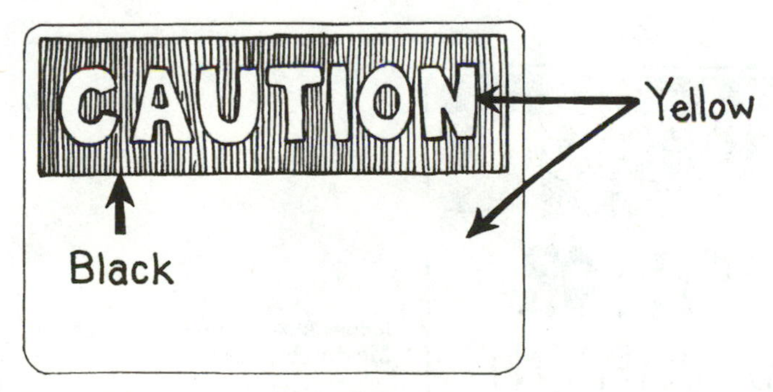


**Figure 32.5** Caution sign design.

be black. As with danger signs, a caution sign may have more information below the word "Caution" concerning the particular hazard, unsafe practice, or proper precautions.

For example:

- Construction Area—Caution—Watch Your Step [see Fig. 32.6]
- Caution—Watch for Moving Equipment [see Fig. 32.7]
- Caution—Automatic Equipment Will Start Without Warning

When you see a caution sign, the situation isn't as immediately dangerous as when you see a danger sign. But there's a definite risk, and employees or visitors should take precautions to avoid having anything serious develop.

## Safety Instruction Signs

Safety instruction signs, when used, should be white with a green upper panel with white letters to convey the principal message. Any additional wording on the sign should be black letters on the white background. Safety instruction signs deliver a useful safety instruction or suggestion.

For instance:

- Safety First—Wear Your Goggles While Sawing (see Fig. 32.8)
- Safety First—Ear Protection Required in This Area

**Figure 32.6** Caution—Watch Your Step.

**Figure 32.7** Caution—Watch for Moving Equipment.

**Figure 32.8** Safety message sign.

- Be Careful—Walk, Don't Run
- Safety First—Keep the Worksite Clean

## Temporary Guardrails and Barricades

Although temporary guardrails and barricades are often associated and used with methods of fall protection, they also can stand alone as important worksite safety aids.

- Temporary guardrails provide protection against the possibility of falling through an opening in a floor, roof, or wall, open-sided platforms or catwalks, stairways, or to prevent an employee from entering an area where other hazards may exist. They're usually constructed of two-by-four lumber, 42 inches high with a midrail at 21 inches, rail supports spaced at intervals no greater than 8 feet, and toeboards wherever the danger exists of items dropping onto employees who may be working below.
- Barricades can consist of items such as sawhorse-style plastic barricades (see Figs. 32.9 through 32.11), barricade lights (see Fig. 32.12), safety-orange cones (see Fig. 32.13), posts and chains or ropes (see Fig. 32.14), and yellow and black barricade caution tape (see Fig. 32.15), and red danger tape. Barricades are used to notify employees and prevent unauthorized access to areas where temporary hazards exist.

Yellow and black "Caution" tape is the most widely used barricade tape. It's frequently employed for barricading

- Trenches and other excavations
- Open manholes and sewers
- Places where employees are working overhead
- Areas where only designated employees are allowed to enter, such as confined space entry, asbestos removal, or around the working space of hydraulic hoes, cranes, or similar heavy mobile equipment
- Other temporary conditions that may pose hazards to employees

**Figure 32.9** Small plastic barricade.

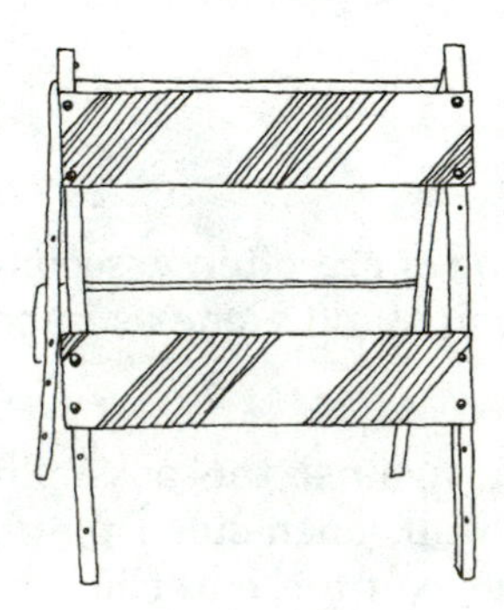

**Figure 32.10** Medium-size plastic barricade.

**Figure 32.11** Large plastic barricade.

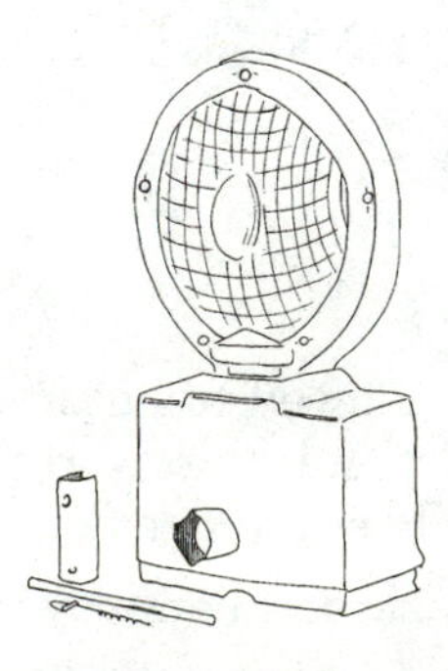

**Figure 32.12** Barricade light.

## Floor or Roof Opening Covers

Floor or roof openings should be blocked over with a cover of sufficient strength to withhold the weight of employees, tools, equipment, and whatever may be placed upon the cover. The cover must also be fastened or secured by some positive method to prevent it from being accidentally dislodged.

## Accident-Prevention Tags

Accident-prevention tags must be used as a temporary means of warning employees of an existing hazard, such as defective tools or equipment. They shall not be used in place of, or as a substitute for accident prevention signs.

Specifications for accident-prevention tags (see Fig. 32.16) include:

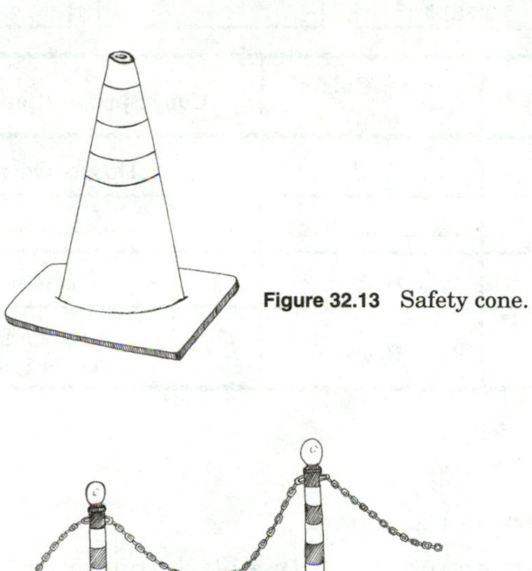

**Figure 32.13** Safety cone.

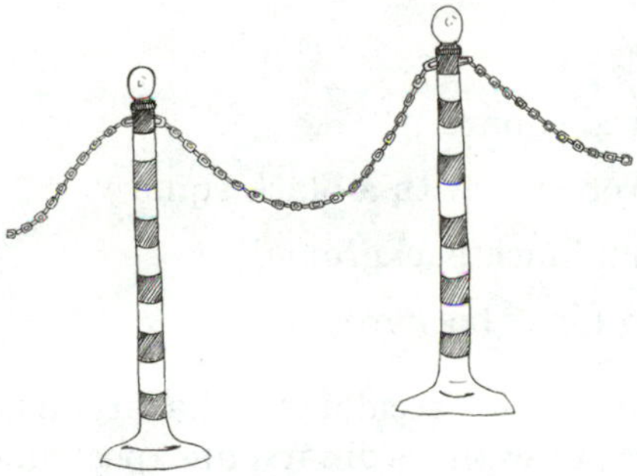

**Figure 32.14** Barricade posts and chain.

**Figure 32.15** Caution tape.

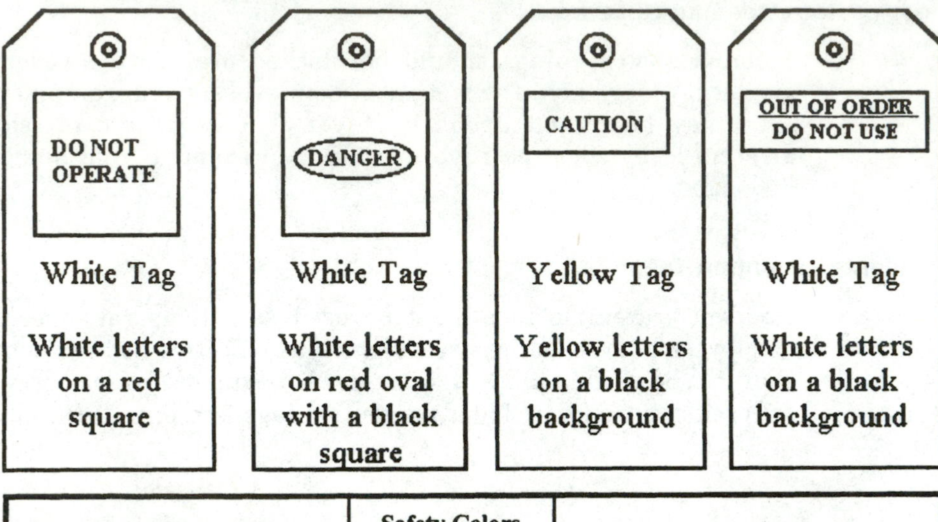


| Basic Stock (Background) | Safety Colors (ink) | Copy Specification (Letters) |
|---|---|---|
| White | Red | Do Not Operate |
| White | Black and Red | Danger |
| Yellow | Black | Caution |
| White | Black | Out of Order<br>Do not Use |

**Figure 32.16** OSHA tag specifications.

- White tag—white letters on red square
- White tag—white letters on red oval with a black square
- Yellow tag—yellow letters on a black background
- White tag—white letters on a black background

Tags, says OSHA, are used to prevent accidents in hazardous or potentially hazardous situations "which are out of the ordinary, unexpected, or not readily apparent." Tags shall be used until such time as the identified hazard is eliminated or the hazardous operation is completed. Tags give their safety alert with a signal word and a major message. Just as with safety signs, the signal word is the one that leaps out and gets your attention. It can be *Danger, Caution, Biological Hazard, Biohazard,* or the biohazard symbol. Tags can also include the term *Warning* for a situation whose hazard level is somewhere between *Danger* and *Caution.* In addition to the signal word, most tags contain other words and/or symbols that deliver what OSHA calls the tag's "major message." The major message explains the specific hazardous condition or the precaution. Examples might be:

- High Voltage
- Do Not Start This Machine
- Do Not Use
- Hands Off: Do Not Operate

Sometimes a tag may use a picture instead of, or in addition to, words. The pictures must be easy to understand at a glance. For instance, if the tag contains a message about wearing hard hats or hearing protection, there could be a picture of a head wearing a hard hat or earmuffs.

The main idea with these warning signs and barricades is because you can't possibly have someone standing by in every instance to warn employees, visitors, or others of hazards, the signs, tags, barricades, or signals will act as effective communications instead, and prevent individuals from inadvertently getting too close to places they should avoid.

Train employees on how to recognize hazards that should be safeguarded, then provide the signs, tags, barricades, and related materials, and see that they're used where and when required.

Chapter

# 33

# Material Handling

## Quick Scan

1. Unload and store materials, equipment, and tools out of the way, yet in the most convenient location possible, arranged in a secure fashion.
2. Keep aisles, passageways, and staircases clear for safe walking.
3. Combustibles and flammables must be stored so to minimize the potential of fire.
4. Plan difficult material or equipment moves in advance.
5. Use extra caution with materials such as sheet supplies, glass, metal banding, compressed gas cylinders, and heavy and bulky items.
6. Employees must be trained on the proper use of material-handling equipment such as hand trucks, wheelbarrows, temporary conveyors, hoists and chainfalls, power tailgates, loading dock components, and skid- and pallet-handling tools.
7. Employees need to understand how to perform manual lifting tasks in a safe way.
8. Train employees in the proper use of long-handled tools appropriate to their jobs, such as shovels, sledgehammers, log splitters, and rakes.

This chapter is mostly about common sense. Some topics that could have been discussed here as material-handling topics are found in other chapters, including the use of ladders and mobile equipment, and slips, trips, and falls, and housekeeping. But even with those topics set aside, there's still plenty of things left for a chapter on material handling. As such, this chapter is laid out as a kind of grab bag of points related to the storage, handling, and transportation of materials.

## Storing Materials

Where you put things can make a big difference in how efficient your operation is. Some people never learn. They unload two skids of bricks in the wrong spot, then have to transport the entire amount an extra 70 feet—by hand. Or they attempt to transport squares of shingles across a soft soil field—and the forklift sinks until it can't move. Instead of putting things helter-skelter across the worksite, try to use logic as to where materials, tools, equipment, and other items should be unloaded and stored for safety and convenience.

Especially watch the storage of materials in tiers: secure various layers to prevent falling. That includes wood and bricks, and skidloads of materials such as tiles, shingles, and even plumbing supplies. The unloading of building supplies can be one of the most dangerous tasks at the worksite. Never allow new employees to get assigned that work alone. Instead, someone with rigging and mobile equipment experience should supervise unloading and loading activities so that those tasks go smoothly and so the materials are stored properly.

Keep aisles and passageways—outside and inside—from being blocked by supplies. Naturally, stored materials must not block exits and emergency equipment.

Combustible/flammable materials should be stored in a manner that will minimize fire potential. They shouldn't be in the way of mobile equipment, or in a place where employees might smoke or perform any hot works. A fire extinguisher must be readily available in the area.

Scaffolds and work platforms must not be used to store or accumulate piles of material or debris. There should only be as much material or debris as can be used or generated by the immediate operations.

Used lumber, when stacked, should have nails removed first.

## Moving Materials

For complex or difficult material moves, do what professional movers are trained to do: they size things up first. They select the easiest and safest route. They measure dimensions, test weights, discuss options, and eventually arrive at what they feel is the best way. Sometimes they have to move certain items out of their walking path, temporarily. Whenever possible, they use a mechanical advantage: a handcart or dolly, power tailgate, wheelbarrow, or similar aid.

That's how your employees can do it too.

### General lifting

It's relatively simple to lift items without causing injury—if individuals use their heads, and their legs. Here's a lifting primer to share with your employees.

1. Planning the lift
   - Walk through the lift in your mind first. Is there any way to use mechanized lifting equipment?

- Can you break down the item to be moved into smaller, more manageable parts?
- Determine how to best grasp the item to lift and hold it. Check for nails and sharp edges. If the item is wet or muddy and slippery, clean it off.
- Select the straightest, clearest path, preferably flat, without stairs or grades, but with places to stop and rest as needed. If the shortest path has lots of stairs or uneven walking, take a longer route.
- Remove any tripping hazards from the path, and any obstacles which would prevent you from taking a safe position for a good grip.
- How far do you have to go while carrying the item?
- Consider where and how you'll place the item at the end of the lift.
- Should you get help?

2. Preparing for the lift
   - If the item becomes too heavy or clumsy to lift while you're first establishing a grip, don't attempt the lift.
   - Stand comfortably as close as you can to the load, with feet apart for balance.
   - If possible, squat to the load, keeping your back straight. Try to avoid bending.
   - Wear gloves that provide a good grip. Grasp the load firmly with your hands, with your fingers beneath the load if possible. Test it first to see that it's not too heavy.
   - Hold item as close to your body as you can, with the weight centered.
   - Lift gradually, using your leg muscles to rise to a standing position, with the legs providing most of the power, and doing most of the work. Not the back.

3. Carrying the item
   - Don't twist your body to change directions; use your feet instead. Twisting with a load puts enormous stress on the spine.
   - A major cause of accidents while carrying heavy or bulky items is when someone decides to change his or her grip in midstream. Never change your grip during a lift unless you can support the weight somehow during the grip change.
   - If you can't make it as far as you thought you could, stop, put down the item, and rest.

4. Placing the item
   - Remember to face the final resting spot for the item with your whole body—don't twist the item into place.
   - Reverse the lifting motion by bending your knees and squatting down with the item, keeping it close to your body, again without bending your back, if possible.
   - Don't forget where your fingers and toes are. Allow enough room to place the item so you can move all of your appendages out of the way. Put one corner of a box or similar item down first, so your fingers can be removed from beneath the item.
   - Test the item for stability before you leave it.

5. Other lifting situations
   - When lifting items from or to high places, use a ladder. Don't stretch from the ground.
   - When lowering an item from shoulder height, push against it first to test its weight and stability. Slide it as close to your body as you can, and hold the item close while lowering it.
   - Partner lifts can best be accomplished when two individuals who are about the same size pair up. Good communications are a must. Have lifting signals so you can both move in unison.
   - Be aware that rounded objects such as gas cylinders, drums, and small tanks can shift suddenly, as their contents may slosh back and forth or the rounded surfaces may begin to roll.

### Handling sheet material

As mentioned in the chapter on tools and equipment, sheet materials can be hazardous. Stacked sheet materials, as well as individual sheets, can cause crushing injuries or worse to hands and fingers, as well as back, shoulder, leg, and other strains and sprains. A thick four-by-eight-foot sheet of plywood, particle board, or drywall board is a heavy item—*especially* when carried vertically, perpendicular to a stiff wind.

When faced with storing and handling sheet materials, consider the following points:

1. If possible, store sheet materials flat. And keep the stacks low. Sheets stored vertically, leaning against a wall, building, or tree, occasionally fall over on employees and cause injuries.

2. If it's necessary to store sheet materials vertically, keep them in upright racks that are secure enough to handle the weight.

3. Sheet materials that may seem heavy and stable when you're putting them onto a roof or other exposed surface can nevertheless be picked up by a strong wind and blown off the edge of a roof, deck, or porch. Once airborne, individual sheets become fall-from-overhead (or from-the-side) hazards. The wind can also pick up individual sheets from an open storage stack like playing cards are lifted from the top of a deck. Similar precautions must be used when employees carry sheet materials during windy weather. They need to be aware of which direction the wind is blowing, especially when carrying plywood or other sheets to second-story and higher levels. It's easy to get "carried away" while hanging onto a piece of sheet material that is being blown by the wind.

4. When employees are opening banded stacks of sheet materials by cutting steel or plastic strapping or banding, instruct them to wear heavy gloves and a minimum of safety glasses with side shields or goggles. Use strap cutters (see Fig. 33.1), not hammers, pry bars, or big screwdrivers. Strapping injuries can range from nasty to severe. Banding that's too tightly strapped, or under a load, can spring out and snap at the face and eyes when snipped. Banding should be disposed of as soon as it comes off. Otherwise it can be a painful source of injuries. Watch for dangling ends of banding, or even very short, sharp ends that will easily cut through gloves or shirtsleeves.

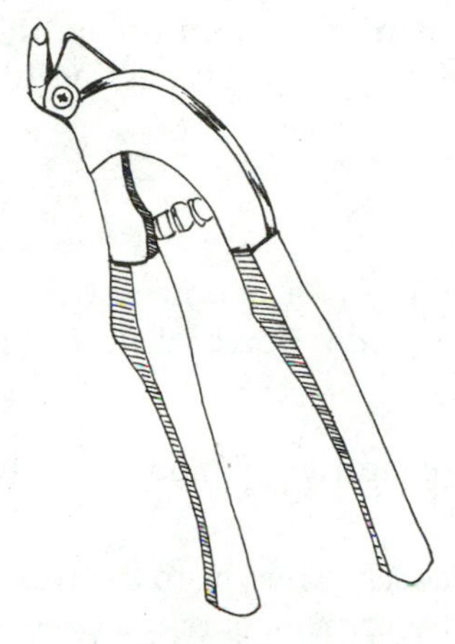

**Figure 33.1** Strap cutters.

5. When moving large quantities of sheet materials, it's safer to use a lift truck or other material-handling equipment. Sheet materials are too heavy for a single employee to manhandle. If full pieces need to be transported and no mechanized equipment is available, additional employees should be recruited to help.

6. Employees should wear safety-toed shoes or boots, eye protection, and gloves while handling sheet materials, to protect against dust, slivers, dropped pieces, and lacerations. If they may need to steady a sheet or sheets with their forearms, a long-sleeved work shirt or jacket should be worn.

7. Watch those fingers when carrying wide loads through doorways and other openings. Heavy sheets, pulled or pushed into fingers that are stopped against a door frame, can break bones. Keep fingers out of the way as sheet materials are maneuvered through doorways, stairways, other openings, and tight spaces. If need be, employees should put down the sheet, and reposition their hands. Movements should be coordinated with the lifting partners, so everyone knows what to expect.

### Handling bricks and blocks

Make a special effort to place skidloads of bricks or blocks on level ground. When set on muddy or rough terrain, the load could gradually shift—which could result in an unexpected spill. Bricks should be stacked no higher than 7 feet, and blocks, if stacked more than 6 feet high, must be tapered back at least a half-block per tier above the 6-foot level.

Handling masonry materials, of course, is heavy work. Extra attention should be given to safe lifting procedures, and supervisors should not expect employees to perform continuous lifting without frequent recovery times. Sturdy work gloves, of course, are a necessity.

### Handling lumber

Splinters, dust, chemicals, insects, sap, snakes, poison ivy or oak or sumac: these and other potential hazards are associated with lumber. There also are back injuries, wrist injuries, and shoulder injuries from unloading and stacking lumber, or from carrying it around the worksite. Handling lumber in the truss-state can be troublesome as well.

### Handling compressed gas cylinders

Each compressed gas cylinder (see Fig. 33.2) is a potential bomb waiting for just the right amount of mishandling to explode. Don't let that happen at your worksite.

- Accept only cylinders that are in good condition, capped, free from signs of corrosion, and legibly labeled.
- Don't allow cylinders to fall, or to strike each other, or to be transported loose in the back of a pickup truck. Never store or transport compressed gas cylinders in confined, unventilated spaces such as in cabinets, closets, or automobile trunks.
- Storage tanks containing flammable gases must be grounded to reduce the likelihood of explosion.
- Store cylinders in low-traffic, out-of-the-way places, away from mobile equipment operations. Outside storage locations are best. Keep the tanks upright and secured with a chain or strap so they can't fall (see Fig. 33.3).
- Cap compressed gas cylinders that aren't in use. Caps should only be hand-tight.

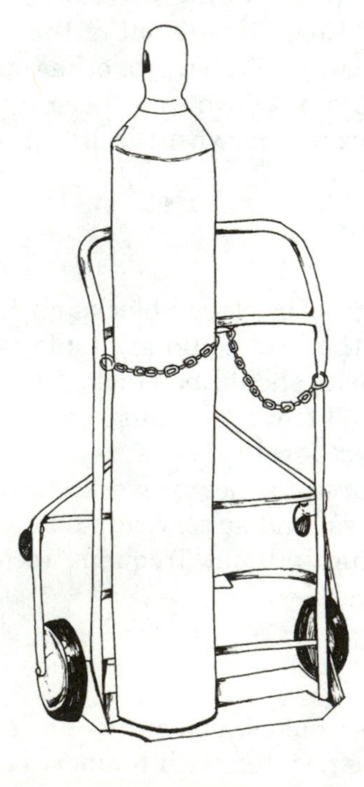

**Figure 33.2** Compressed gas cylinder.

**Figure 33.3** Compressed gas cylinder storage.

- Always consider cylinders as being full, even if you think or "know" they're empty.

### Carrying items up and down stairs

How many times a day do the employees at your worksite carry items up and down stairs? Consider these points:

- How adequate is the lighting? Is everything still under construction, including the light? If so, take extra time on the stairway. Make sure there's no box of nails, can of paint, misplaced claw hammer, or similar object on the stairs to trip over.
- Are handrails, even temporary ones, in place?
- Walk with your knees and feet pointing outward at an angle while descending stairs, instead of walking with feet and knees pointing straight ahead. Going straight ahead with feet and legs puts unnecessary strain on the knees.

### Hand trucks

Hand trucks are handy tools that can make light work out of moving heavy loads. But they've got to be used correctly for safe results. Hand trucks, dollies, and similar material-handling equipment—when not maintained or when in the hands of untrained employees—can be hazardous to the moving employees and to bystanders. The main hazards include: attempting to move an unbalanced load; losing the load and allowing it to fall on or against someone else or the dolly operator; overexertion, such as moving the dolly the wrong way, and working your muscles against (instead of with) the dolly's design; and smashing fingers against walls, door frames and other close-quarter fixtures and objects. Consider the following tips on the safe use of hand trucks:

- A hand truck must not fail while a load is being moved up a flight of stairs, or when a breakable item is being carried. So before a hand truck is used, it must be inspected for damages. Look for

    Cracks in the frame
    Loose, frozen, or damaged wheels
    Slippery surfaces, grease, oil

  and if any serious defects are discovered, the truck should be danger tagged and taken out of service.
- The load's center of gravity should be kept as low as practical. Heavier items should be on the bottom, with the load balanced over the axle while it's being moved. Bulky or awkward items should be secured to the cart before being moved. Gloves and safety shoes will help prevent injuries while working with a hand truck.
- Watch the fingers in door frames, on stairways, and when passing through other tight openings. If a mover can't see over the top of the load, he or she should use a spotter (someone to direct travel). Teach employees to lean in whatever direction they're traveling. When going downhill, they've got to resist the urge to keep the materials toward the top of the hill, and to back down. Instead, the load belongs in front of the employee as he walks downhill.
- The legs should be used to help shift the load from a standstill into a travel position, without jerking back with the arms and upper body. The knees should be kept bent, with the back straight.
- Employees must avoid pulling materials with a hand cart; push instead. Pulling a load will place too much weight on the mover's back. The load should be kept in front of the mover while it's being pushed into place. How else could it be maneuvered into a tight corner?
- If a hand truck with moving straps can be used, use it.

#### Loading docks

Watch weather conditions on a dock. Numerous accidental injuries occur when employees walk out the back of a materials supply warehouse and slip on ice- or snow-covered or wet metal dock plates.

Additional hazards are created by employees who insist on jumping off a loading dock.

Other dock-related hazards include:

- Not chocking trucks or trailers before loading or unloading.
- Accidentally stepping or backing off the dock.
- Getting impatient and attempting to load heavy items instead of waiting for the dock attendant.
- Overhead doors can be hazardous. They can seem to come down out of nowhere.
- Power tailgates must only be used by individuals trained in the power tailgate's safe operation. It's too easy to catch a finger, hand, or worse in the mechanisms.

### Skids and pallets

Skids and pallets seem harmless enough. Nevertheless, employees need heavy gloves when handling these materials, due to sharp splinters, nails, staples, unsecured banding, and similar laceration, puncture, and abrasion hazards.

Employees have been injured while walking on empty pallets, when their boots break through top boards.

Some pallets and skids are deceptively heavy and awkward to handle. When left on the ground, they're also low enough to blend into the terrain or to become camouflaged with weeds, so that they become serious tripping hazards. Best to stack them flat, not on end, and remove them from the worksite as soon as possible.

### Shoveling

A lot of jobs require materials such as sand, soil, mud, bark, gravel, stone, concrete, asphalt, and even snow to be shoveled occasionally. Shoveling causes a considerable number of strains, slips, and falls on worksites. Lifting and twisting is the most common way that employees get hurt, and unfortunately, that's the natural way that most people shovel. Twisting puts extra pressure on the back's discs. The further a person twists to one side, the more the pressure is applied to his or her discs. The discs have to support the upper body weight, the weight of the material on the end of the shovel—which is an arm's length away—plus the additional pressure from twisting.

For safe shoveling:

- Take a wide stance, with knees bent.
- Choke down on the shovel to keep the weight closer.
- *Pivot* instead of twisting with the shovel load. This point can't be stressed enough when it comes to shoveling.
- When shoveling to the left, the left foot should be forward, with the feet well separated. Load the shovel with a manageable amount of material, and pull the load in close. Keep the right foot planted. Move the left foot back and to the left, in the direction the material will be thrown. Throw.

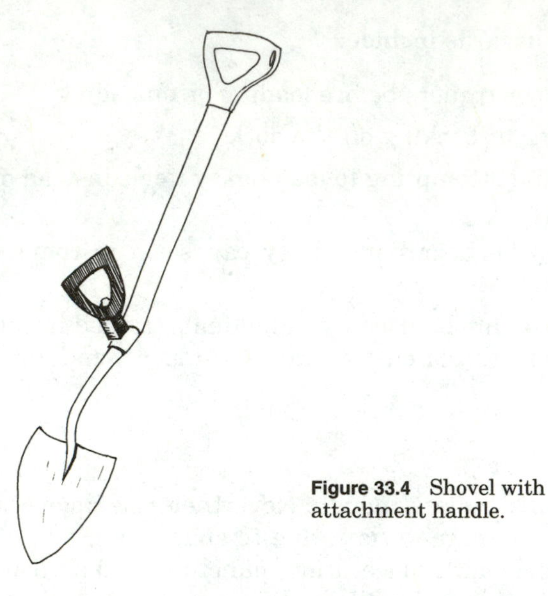

**Figure 33.4** Shovel with attachment handle.

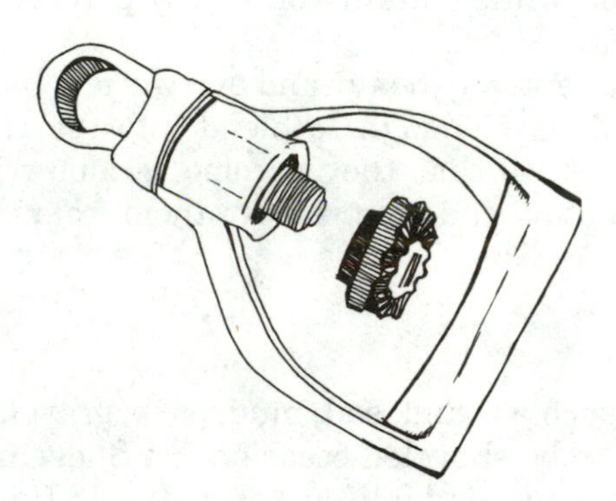

**Figure 33.5** Attachment handle.

- When shoveling to the right, reverse the process. Right foot forward. Left foot planted. Move right foot back and to the right. Throw.
- Also consider trying out some of the new ergonomically designed shovels (see Figs. 33.4 and 33.5) that feature offset handles and/or attachable handles that help to reduce the amount of bending that occurs while shoveling.

Chapter

# 34

# Mobile Equipment

**Quick Scan**

1. Allow only trained, authorized operators to run mobile equipment. Provide annual refresher safety training for operators, and awareness training for your other worksite employees.
2. Operators must perform safety and operating checklist inspections of equipment before they begin their day's work.
3. Mobile equipment must be outfitted with working safety components such as seatbelts, rollover and overhead protection, horn, backup alarms, lights, entry grab-handles, and fire extinguishers.
4. Prohibit employees and others from working alongside the business ends of heavy equipment, where operator error could prove fatal. If areas being worked on must be accessed by employees with hand or other tools, have the heavy equipment draw back temporarily while employees are working there.
5. Consider that falls from equipment cause about half of the injuries that occur to equipment operators. Most of these come from the operators jumping from the cab or body, not using a three-point system for climbing on or off of their vehicle, allowing steps and access surfaces to get muddy or slippery, and stepping down onto uneven surfaces or unstable objects on the ground.

Although varied types of heavy equipment are available for earthmoving, digging, loading, grading, and similar tasks at a worksite, most of them require similar precautions and share comparable safe operating procedures.

Some of the equipment found on many job sites include the miniloader or skid-steer loader, front-end loader, backhoe, hydraulic hoe, dump truck, mobile crane, aerial lift, and forklift. Lately the manufacturing trend seems to be introducing smaller units that, to old-timers, appear almost toylike. These new, smaller machines still have plenty of power, and they can access areas that larger equipment cannot. Despite their diminutive size, the new generations of digging and material-handling machines must still be treated with respect.

## General Guidelines for Heavy Equipment

The following guidelines apply to most of the individual mobile equipment machines found at residential construction and other worksites. They do not take the place of specific safe procedures issued by manufacturers or regulatory agencies for individual units, but can be considered of the type that will apply to such machinery. Always refer to the manufacturer operating manuals when developing company safety procedures for pieces of heavy mobile equipment being used at your worksites.

Much of this chapter is written as if it addresses equipment operators. It's written that way because it's meant to be shared directly with your operators.

## Fundamentals of Safe Operation

Basic rules:

- Know the capacity and operating characteristics of the machine so that it can be kept under control at all times.
- Preoperation checklists should be completed before work begins. The best checklists usually come from the machine's operator's manual.
- Never modify or remove any part of the machine except for servicing—and make sure that those components are reinstalled and inspected before operation resumes.
- Wear the seatbelt and any other applicable operator safety restraints.
- Keep others away from the operation—and that goes no matter how experienced the operator is. The working ends of heavy equipment are no places for hands, feet, and heads of fellow employees on the ground to be near.
- Look before backing up.
- Carry the load as low as safely possible.

You should have the original manufacturer's operating manuals available for every piece of equipment, and you should review it with the person who will be operating the machine. Don't rent a piece of equipment if the rental company cannot supply such a manual along with the unit. That means notifying the rental company long enough in advance that if they don't have a manual, they can get a replacement in time. They should keep the manufacturer's manuals inside the machine for reference.

### The operator's personal protective equipment

Suggested personal protection for an operator includes a hard hat, safety shoes, safety glasses and side shields at a minimum, hearing protection, reflective clothing, wet-weather gear (depending on the working conditions and if the unit has a cab), and perhaps a respirator. Avoid wearing loose clothing with flopping cuffs or dangling draw-cords. The same goes for jewelry, including wristwatches that can catch in moving parts. Make sure that the unit is supplied with a fire extinguisher and that a first-aid kit is nearby.

### The safe operator

Ask your operators: Are they really up to working today? Are they feeling sick or depressed? Are they capable of giving the big machines the total concentration and quick reaction necessary to operate them safely? Have they taken medication that could make them feel drowsy or stimulate them to where they might overreact? Do they know who and where to call for help in an emergency? Do they know where and how first aid is available? And where the nearest phone or two-way radio is?

- The operator and ground helpers must understand all signs, flags, and markings. They need to know what hand, flag, horn, whistle, siren, or bell signals mean, and they should discuss operator-to-ground helper/spotter communications before the job begins.
- Does the operator know how to operate all of the equipment on the machine? Does he or she understand the purposes of all the controls, gauges, and indicators? What about the rated load capacities, speed range, braking and steering characteristics, turning radius, and operating clearances? Do employees realize where main and secondary pinch points are? Pinch points are critical on machines such as miniloaders due to the small dimensions from seat to lifting arms and attachments. Everything happens right there in front of the operator.

### The safe machine

Manufacturers of loaders, backhoes, fork lifts and other machines provide operator manuals for their equipment which are usually quite complete and specific as to the preoperating inspection procedures as well as instructions for operating the machine and the handling of materials. Before beginning the day's work, employees must give their machine a thorough inspection for defects, such as leaks, worn parts, frayed hoses, and a cracked or damaged frame. Bypassing an inspection before they start to operate it may allow a machine to break down while working, which could result in a serious injury.

### Checking out the safety equipment

- Falling object protective structure (FOPS). It's not enough to just have rollover protection anymore. The operator must also be protected from objects the bucket or attachment could kick into the air—and items dropped or airborne from anywhere on the worksite.

- Rollover protective structure (ROPS)
- Seatbelt
- Operator seat bars
- Side shields, screens, or cab
- Grab handles
- Lift-arm restraining devices
- Lights
- Antiskid tread/steps
- Safety signs
- Horn
- Guards
- Back-up alarm
- Fire extinguisher
- First-aid kit

### Checking the machine

- Check for broken, missing, or damaged parts. Have necessary repairs made.
- Check the tires for cuts, bulges, and leaks. The tires on a miniloader, front-end loader, backhoe, or forklift are critical to correct and safe operation. Have badly worn or damaged tires replaced.
- Check the parking brake.
- Check the hydraulic system. Have leaks repaired. Remember that diesel fuel or hydraulic fluid under pressure can penetrate the skin or eyes and cause serious injury, blindness, or even death. And fluid leaks may not be visible. They may be in places difficult to see. Use a piece of cardboard or wood to find leaks that cannot be seen, but never use a bare hand to feel for leaks. Wear a face shield or safety goggles for eye protection. If any fluid is injected into the skin, it must be removed as soon as possible by a doctor familiar with that type of injury.
- Check the cooling system. If air-cooled, check for unobstructed air flow. If liquid-cooled, check coolant level at overflow tank, if provided. *Warning:* allow the radiator to cool before checking the fluid level.

### Clean the machine

- Always lower the lift arms or attachments and stop the engine before cleaning off the machine.

- Clean the windshield, lights, and safety signs.
- Make sure that the operator's area, steering levers, pedals, steps, and grab handles are clean. Oil, grease, mud, snow, ice, or debris in these areas can cause a slip or fall. Clean your boots of excess mud before getting on the machine. Remove all personal items or other objects from the operator's area. Secure these items in a tool box or remove from the machine. A pop bottle could roll under and get wedged beneath a pedal and cause it to be inoperative.

#### Use caution when fueling

Avoid filling a fuel tank with the engine running, while smoking, or when near an open flame. Never overfill the tank or spill fuel. If fuel is spilled, clean it up immediately. Use the correct type and grade of fuel. As added protection, before starting the flow of fuel, ground the fuel funnel or nozzle against the filler neck to prevent sparks, and be sure to replace the fuel-tank cap.

#### Know the work area

As much as possible, learn the lay of the land before operating. Check out the ground or floor. Inspect the surface over which you will travel. Look for holes, drop-offs, and obstacles. Look for rough spots or hidden obstacles on surfaces that could cause a collision or loss of control. Look for weak spots on docks, ramps, or floors. Look for oil spills, wet spots, slippery surfaces. Look for soft soil, deep mud, and standing water. Watch for anything that might make an operator lose control or cause the machine to tip over.

- Keep in mind that rain, snow, ice, loose gravel or stone, soft ground and similar conditions can change with weather conditions—which can affect the operating capabilities of a machine.
- Exercise caution when crossing ridges, side hills, logs, ditches, deep and loose gravel, and similar impediments to smooth travel.
- Never work a machine beneath overhangs, under energized power lines, or where there is a danger of a stockpile, rock, or earthen slide.
- When operating the machine inside a building, know what clearances you will encounter—overhead, in doorways, and in aisles, plus the weight limitations of floors and ramps.
- Plan the work. Know where to make pickups, lifts, and turns. Before raising a full loader bucket, know where to dump it, and always carry the load low. Check overhead. Check the clearances of doorways, canopies, and overheads. Know exactly how much clearance you have under power and telephone cables. Never, never approach power lines with any part of the machine unless all local, state/provincial, and federal (OSHA) required safety precautions have been taken. Electrocution can result from touching or being near a machine that is in contact with or near an electrical source.

- Check underground for all digging, drilling, and trenching operations. Know the location of underground cables, gas lines, and water and drain pipes.

#### Rules of the road

- Make sure that lights and warning signs are in place and visible.
- Make sure that a slow-moving vehicle emblem is installed and visible to any vehicle approaching from the rear.
- When traveling on public roads or streets, obey all local traffic regulations appropriate to the equipment classification. Find out if you must use an escort vehicle. Place the bucket in the transport position. Approach intersections with caution, and observe speed and traffic control signs. Avoid panic stops and sharp turns. If traffic backs up behind you, pull over and allow other vehicles to pass. Stop at all railroad crossings and look both ways before proceeding. Never park in traffic areas. If it is necessary to stop at night, pull off the road and set up flares or reflectors. When driving at night, use appropriate lights.

#### Mount safely

- Use a three-point mount or dismount, keeping two feet and a hand or two hands and a foot in contact with the machine steps and grab handles at all times.
- Face the machine.
- Never jump onto or off of the machine. Never attempt to mount or dismount from a moving machine.
- Going up the machine's ladder can be one of the most dangerous things an operator does all day. A significant number of injuries associated with front-end loaders have something to do with the ladder. The operator slips, falls, or jumps from the ladder and is injured. Operators shouldn't have a problem if they use the ladder properly. Again, this requires clean steps, no grease or minimal mud on shoes, and gloves to protect the hands. The gloves should be clean, too, to ensure a safe grip. Operators need to use *all* the steps, to use the handrails, and to look where they're going. They especially need to look where they're putting their feet. The ground is often rough, muddy, or full of ruts and debris.

  Carrying a lunch pail up an equipment ladder has caused numerous accidental injuries. To carry lunch boxes—or other items—up the ladder, the operator should always put the item a step or two above, climb a few steps, then move the lunch box or item into the cab ahead of his or her entrance.
- Do not use a steering or control lever as a handhold when you enter or leave the machine.

### Warn personnel before starting

Before starting, walk completely around the machine. Make sure no one is under the machine, on it, or close to it. Let other workers and bystanders know you are starting up, and don't start until everyone is clear of the machine.

### Starting the engine

Start the engine from the operator's seat only. Never attempt to start the engine by shorting across starter terminals. The machine could move uncontrollably and cause serious injury or death to anyone in its path.

Generally,

1. Sit in the operator's seat and adjust the seat so that you can operate the controls properly.
2. Fasten the seatbelt/operator restraint.
3. Familiarize yourself with the warning devices, gauges, and operating controls.
4. Lower the operator seat bars if so equipped.
5. Engage the parking brake and put all controls in the neutral/park position.
6. Clear the area of all persons.
7. Start the engine, following the instructions in the manufacturer's manual.
8. If it is necessary to run the engine or operate the machine within an enclosed area, be sure there is adequate ventilation. Exhaust fumes from gasoline, propane, or diesel can kill.
9. If jumper cables are used, follow the instructions in the manufacturer's manual. Jump-starting is a two-person operation. The operator must be in the operator's seat while jumping occurs so the machine will be under control when the engine starts. A battery explosion or a runaway machine could result from improper jump-starting procedures.
10. Ether cold-start fluid is highly flammable. Before using it, always read the instructions on the ether container and the instructions in the manufacturer's manual. Do not attempt to use ether if the engine is equipped with a glow-plug-type preheater or other intake-manifold-type preheater. Doing so could result in an explosion or fire, and severe injury or death.

### After starting the engine

Check gauges, instruments, and warning lights to assure that they are functioning and their readings are within the operating range.

### Run an operating check

- Don't use a machine that is not in proper operating condition. It's the operator's responsibility to check the condition of all systems, and to run the check in a safe area.
- Make sure that the engine is operating correctly.
- With the levers in neutral, test the engine speed control.
- Operate each control pedal or lever to make sure that all lift arm, bucket, and other attachment functions are correct. Be certain that you can control both speed and direction before moving. Operate the travel control levers to ensure correct operation in forward and reverse. Test steering—right and left—while moving slowly in a clear, safe area.

### Work safely

Remember these rules:

- Do not overload a bucket or attachments, or carry a load that could fall from the bucket or attachment.
- Carry a load high enough to ensure ground clearance, but low enough to maintain the stability of the equipment.
- Different attachments can change the weight distribution of the machine. They can also affect its stability and handling response.
- When changing buckets or installing attachments, make sure that all connectors are securely fastened. Tighten all bolts, nuts, and screws to torques recommended by the manufacturer. On quick-coupling units, test the attachment before actually using it. Sometimes the couplers could only partially engage—allowing for failures to occur when the actual work is being attempted.

### Remember the other person

- Never allow an untrained or unqualified person to operate your machine. Handled improperly, this machine can cause severe injury or death.
- Your loader, backhoe, or miniloader is a one-person machine. Never permit riders. And never use the bucket for a work platform or personnel carrier.
- Always look around before you back up or swing an attachment. Be sure that everyone is in the clear.
- Know the pinch points and rotating parts on the loader; awareness on your part can prevent accidents.
- Never overload a bucket with material that could injure someone if it spills. Keep a loaded bucket level as lift arms are moved and as the loader moves up or down on slopes and ramps.

- Never lift, swing, or move a load over anyone. Keep others away from your operation.
- Keep your body inside the operator's cab while operating a skid-steer loader. Never work with your arms, feet, or legs beyond the operator's compartment.
- Stay alert. Should something break, come loose, or fail to operate on your machine, stop work, lower the lift arms or attachment, stop the engine, and inspect the machine.

### Traveling on jobsite

- Take it slow and easy when traveling through congested areas. Use traffic courtesy.
- Give the right-of-way to loaded machines. Maintain a safe distance from other machines. Pass cautiously.
- Don't obstruct your vision when traveling or working. Carry buckets and other attachments low for maximum stability and visibility while traveling. Operate at speeds slow enough so that you have complete control at all times. Travel slowly over rough or slippery terrain and on hillsides.

### Operate safely

- Always have lift arms and attachments down when traveling or turning. Plan operation to load, unload, and turn on flat, level ground.
- Never ram a loader bucket into a material pile. Most loaders have more force at slow speeds. Also, when traveling over rough terrain, slow down to prevent losing control.
- Raise loads slowly at an even rate, and be ready to lower load quickly if the machine gets into an unstable situation.
- Avoid steep slopes or unstable surfaces. If you must drive on a slope, keep the load low and proceed with extreme caution. Do not drive across an excessively steep slope under any circumstances. Travel straight up and down the slope instead.
- Travel up- and down-slope with the heavy end of a loader pointed uphill. *Loaded* bucket...with bucket (and load) pointed uphill. *Empty* bucket...with bucket pointed downhill. Check loader manufacturer's recommendations. Most loaders are heavier on the rear axle when unloaded and heavier on the front axle when fully loaded. If you are working on a ramp or slope always have a flat, level turn-around area so that you can turn, load, and unload safely.
- The center of gravity of a skid-steer loader shifts as loads are lifted and lowered. Never attempt to make sharp turns or travel on steep slopes with a raised load.

### Watch out for hazardous working conditions

- Be alert for hazards. Know where you are at all times. Watch for branches, cables, or doorways.
- Use caution when working along docks, runways, banks, and slopes. Keep away from the edges of drop-offs.
- Use caution when working beneath an overhang. Never undercut a high bank.
- Never operate the machine too close to the edge of an overhang or gully. The edges could collapse, or a slide could occur, causing severe injury or death.
- Use caution in backfilling. Do not get too close to a trench wall. The combined weight of your equipment and the load could cause the trench wall to give way. Before backfilling, see the manufacturer's operator manual.
- When working under hazardous conditions, have a helper work with you to signal for dangers. Make certain he or she does not get too close to your loader.
- With machines employing buckets, be careful when handling materials such as rocks, gas cylinders, barrels, etc. Lifting too high and rolling bucket too far back could result in these materials falling into the operator's compartment.

### Exhaust gases can kill

Vent exhaust gases and assure a flow of fresh air when an internal combustion engine is used in a closed space. Exhaust fumes have deadly components which are odorless and invisible—and they can kill without warning.

### Loading/delivery of the machine

- Always wear your seatbelt/operator restraint when loading or unloading the machine.
- When transporting a loader or other machine, follow the manufacturer's recommended loading and unloading procedures.
- Several precautions are applicable to loading and unloading all machines:

1. Load and unload on a level surface.
2. Block the transport vehicle so it can't move.
3. Use ramps of adequate size and strength, low angle, and proper height.
4. Keep the trailer bed and ramps clear of clay, ice, oil, snow, and other materials which can become slippery.
5. Back the machine up the ramp onto the transport vehicle. (Unload by driving forward, down the ramp.)
6. Chain and block the machine securely for transport.

### Park safely

- Select a site having level ground whenever possible. If you must park on a slope or incline, position the machine at right angles to the slope, engage the parking brake, lower the bucket (and other attachments) to the ground, and block the wheels.
- Never leave a machine unattended with the engine running or the lift arms raised. If arms are left in a raised position, they must be restrained by lift-arm restraining devices.
- Whenever the operator leaves the machine, the lift arms should be lowered, with the bucket or attachments flat on the ground. Or lock the lift arms with any lift-arm restraints provided. Set the parking brake, turn off the engine, and cycle the hydraulic controls to relieve built-up pressure.

### Safe shutdown

- Detailed shutdown procedures can be found in the manufacturer's manual; they usually consist of:

1. Stop the machine.
2. Lower bucket and other attachments flat on the ground. A backhoe or loader bucket, or a forklift's forks or other attachments, should always be on the ground while the vehicle is parked. This gives additional braking support to the machine and will prevent the bucket, forks, or attachment from accidentally falling on someone if the hydraulics fail.
3. Position controls in neutral.
4. Engage parking brake.
5. Idle engine for short cool-down period.
6. Stop engine.
7. Cycle hydraulic controls to eliminate pressure.
8. Raise operator seat bar, if so equipped.
9. Check that controls are locked in neutral, if so equipped.
10. Unbuckle seatbelt/restraint.
11. Remove ignition key, and lock covers and closures.
12. Shut off master electric switch, if so equipped.
13. When you leave the machine, maintain three-point dismount, facing machine.
14. Be careful of slippery or uneven ground or slippery steps and grab handles.
15. Block wheels if on a slope or incline.

### Perform maintenance safely

- Do not perform any work on a machine unless you are authorized—and qualified—to do so.
- If you have been authorized to do maintenance, read the operator and service manuals. Study the instructions; check the lubrication charts. Examine all the instruction messages on the machine.
- Attach a *do not operate* tag or similar warning tag to the starter switch or steering lever before performing maintenance on the machine, and if possible, lock out the controls.
- If the engine should not be started, remove the ignition key.
- Never work on machinery with the engine running, unless so instructed by the manufacturer's operator's or service manuals, for specified service.

## Types of Mobile Equipment

### Miniloader or skid-steer loader

This relatively small piece of mobile equipment (see Fig. 34.1) comes in numerous forms and sizes, with many attachments: digging buckets, scooping and loading buckets, hole diggers, trench diggers, forks, and other attachments. Because of the their small size, miniloaders can be deceptive when it comes to safety. Certain employees look at them like large toys, but nothing could be further from the truth.

While the controls are generally simple to learn, and simple to handle, there is a definite skill to a smooth operation. Many operators tend to run them jerky,

**Figure 34.1** Skid-steer or miniloader.

**Figure 34.2** Front-end loader.

with quick, almost uncontrolled movements, constantly oversteering and spending considerable time correcting unwanted movements—to the point that a lot of wasted energy takes place. This is especially true when a unit is rented for a specific job, such as spreading stone in a basement, or landscaping.

It's critical that whoever runs a miniloader really knows how to operate it. Experienced, qualified equipment operators should be selected because miniloaders can be extremely hard to learn to control if the task will require fine maneuvering in close quarters. To be qualified, an operator must understand the written instructions supplied by the manufacturer, have practical training that includes actual operation of the machine, and know the safety rules and regulations for the jobsite. There's a lot to all of that. A fresh, unproven operator is not only dangerous to him or herself, but to bystanders and materials as well. An alternative is a different piece of equipment, perhaps a backhoe, or hiring a trained operator with the miniloader.

## Rubber-tired front-end loaders

Front-end loaders (see Fig. 34.2) are versatile machines that can be used for loading trucks, pushing material, grade work, and hauling materials. When fitted with quick-coupler systems and forks, they can act as substitute lift trucks—with a distinct advantage of being able to travel over extremely rough and muddy terrain. A disadvantage is that most of them are quite large and heavy, and can be difficult to fit into tight locations or to transport between worksites.

Rubber-tired front-end loaders (and forklift trucks) have two basic machine designs—the articulated machine and the rigid-frame machine. They're available in a wide range of sizes, with both two-wheel and four-wheel drive.

The frame of the articulated machine is constructed in two parts that are connected by a pivot. The machine pivots in the center to steer. Operator's controls may be located on the front or rear section. Front-end articulation improves vehicle maneuverability and front vision for the operator, but it also introduces a serious pinch-point, consisting of the center of the entire machine, to each side, formed when the machine turns sharply either way.

A rigid one-piece frame characterizes the rigid-frame machine. The steering of the machine is usually accomplished by rear wheels that pivot.

Allow *no riders* on a machine that's in operation. Also, a front-end loader is not designed as an elevated work platform. Such use has often proven fatal. Proponents argue for such a use if the machine can be fitted with a functional load-locking device or a device which prevents free and uncontrolled descent of the bucket or attachment, and if the loader can be blocked and secured to prevent rolling, and if the tilt mechanism is also disabled, and if the worker in the bucket or attachment wears a body harness with a lanyard, and...on and on. For my book, that's way too many "ands." If an elevated work platform is the only way to go, then arrange for one. Usually, though, safe alternatives can be planned—including contracting the job to someone who already owns the equipment needed.

As with any piece of heavy mobile equipment, it's impossible to cover every safety aspect involved with front-end loaders.

## Mobile cranes

Mobile cranes are used on residential construction sites most often to either unload or load materials or equipment being delivered to or from the site, or for lifting and placing structural members of whatever's being worked on. Because of their initial high cost and continuing expensive maintenance, most mobile crane services are contracted out or short-term rentals. Unless a crane is used frequently enough to pay for itself—and this is difficult for a typical home builder to do—the company does not have to train individuals to operate cranes, but does indeed need to instruct all employees on how to safely work with and around mobile cranes.

General guidelines for working with or around mobile cranes include:

- Never allow individuals to work beneath a suspended load.
- Use tag lines with suspended loads to help direct the placement of those loads.
- Know the basic hand signals of crane operation.
- Inspect all lifting slings, chokers, wire ropes, hooks, clamps, and related attachments and other rigging before each use. All hooks require safety latches. Damaged slings and other rigging needs to be destroyed or removed from the worksite as soon as possible.
- Edge pads or softeners must be used when lifting loads having sharp edges to prevent damage to the rigging, and to get a firmer grip on the load.
- Allow no employees to hitch rides on slings or other rigging attachments that are lifted off the ground.
- About 20 percent of all mobile crane injuries occur while people are climbing on or off the machine.

- The areas within the crane's swing radius, and the areas beneath the lifting action, must be barricaded to limit employee access and prevent accidental entry.
- Safe clearances must be maintained between all parts of the crane plus its payload, and energized power lines and equipment. If there's not proper clearance, the power lines will have to be either moved or shut down by authorized personnel—usually employees of the local electrical utility.
- Outriggers must be extended fully and set on firm ground or sturdily prepared pads or bases. The operating location needs to be far enough away from shoring, excavations, trenches, foundations, underground utilities, and any other area that could be affected by the weight and vibration action from the crane and its loads.
- Rated load capacities must be legibly displayed with paint, decals, or other conspicuous methods. The load capacity must never be exceeded.

### Aerial lifts

Aerial lifts (see Fig. 34.3) are really portable or movable scaffolds. They're incredibly convenient to have around the worksite, but can also be incredibly dangerous if recommended work practices are not followed. Employees need to be trained in the equipment's operating controls and capabilities, hazard recognition, safe work practices, and other factors critical to safe operation.

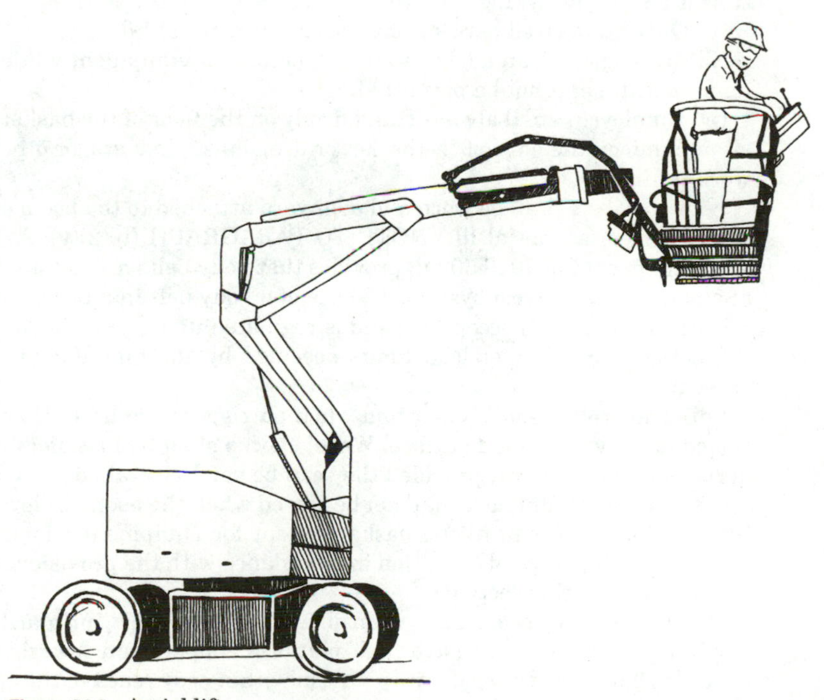

**Figure 34.3** Aerial lift.

The training must be provided (and verified for understanding, and documented) by a competent individual who should also plan and supervise the aerial lift activities.

There are numerous types of aerial lifts used in construction worksites. Each unit has unique characteristics that must be understood and trained on by all employees who will be working with or near that particular machine.

Serious hazards with aerial lift operations include:

- Operating too closely to overhead electrical lines and energized parts
- Overloading or attempting to overreach with the boom or basket
- Falling from the basket
- Improper setup on uneven ground without stable outrigging
- Traveling with the boom raised high in the air
- Using the aerial lift as a crane
- Failing to inspect the entire unit before each work job

**What OSHA says.** The OSHA standards regulating aerial lifts are found in Subpart L, the subpart on scaffolds. Here are some highlights of what OSHA has to say about them:

1926.453 (b)(2) Extensible and articulating boom platforms.

(i) Lift controls shall be tested each day prior to use to determine that such controls are in safe working condition.

(ii) Only authorized persons shall operate an aerial lift.

(iii) Belting off to an adjacent pole, structure, or equipment while working from an aerial lift shall not be permitted.

(iv) Employees shall always stand firmly on the floor of the basket, and shall not sit or climb on the edge of the basket or use planks, ladders, or other devices for a work position.

(v) A body belt shall be worn and a lanyard attached to the boom or basket when working from an aerial lift. NOTE TO PARAGRAPH (b)(2)(v): As of January 1, 1998, subpart M of 1926.502(d) provides that body belts are not acceptable as part of a personal fall arrest system. The use of a body belt in a tethering system or in a restraint system is acceptable and is regulated under 1926.502(e).

(vi) Boom and basket load limits specified by the manufacturer shall not be exceeded.

(vii) The brakes shall be set and when outriggers are used, they shall be positioned on pads or a solid surface. Wheel chocks shall be installed before using an aerial lift on an incline, provided they can be safely installed.

(viii) An aerial lift truck shall not be moved when the boom is elevated in a working position with men in the basket, except for equipment which is specifically designed for this type of operation in accordance with the provisions of paragraphs (a)(1) and (2) of this section.

(ix) Articulating boom and extensible boom platforms, primarily designed as personnel carriers, shall have both platform (upper) and lower controls. Upper controls shall be in or beside the platform within easy reach of the operator. Lower controls shall provide for overriding the upper controls. Controls shall be plainly marked as to their function. Lower level controls shall not be operated unless permission has been obtained from the employee in the lift, except in case of emergency.

(x) Climbers shall not be worn while performing work from an aerial lift.

(xi) The insulated portion of an aerial lift shall not be altered in any manner that might reduce its insulating value.

(xii) Before moving an aerial lift for travel, the boom(s) shall be inspected to see that it is properly cradled and outriggers are in stowed position except as provided in paragraph (b)(2)(viii) of this section.

Nonmandatory Appendix C of Scaffold Subpart L lists a number of national consensus ANSI/SIA standards for working with aerial lifts.

## Backhoes and excavators

Backhoes refer to traditional hoes, many of which have buckets and digging attachments at opposite ends of the machine. While they're not as large or powerful as most front-end loaders, mobile cranes, or excavators, they're just as dangerous if mishandled or not given their space. And with recent equipment developments, the differences between these machines are becoming more blurred all the time.

Major hazards with backhoes include:

- Operating too close to electrical lines.
- Digging into ground without checking for the presence of utility lines such as natural gas, electric, and water.
- Operating without seatbelts and rollover protection.
- Helpers getting too close to the digging or work end of the machine.
- Untrained operators running the hoe.
- Tipovers while unloading and loading the hoes from and onto trailers.

Excavator (see Fig. 34.4) hazards include, in addition to most of the backhoe hazards:

- The swing-area radius of the counterweight at the back of the machine poses a crushing hazard to individuals who inadvertently or otherwise approach the machine from behind or the sides.
- Materials dumped or falling from the excavator bucket can injure truck drivers or others standing too close to the bucket.
- The overall weight of the machine, and vibrations from its operations, can create extremely unstable earthen sides on trenches and other excavations.

## Forklifts

There are many specialized types and models of forklifts (also called highlifts or lift trucks) designed and manufactured for numerous purposes. Due to the rough terrain often found at residential and similar construction

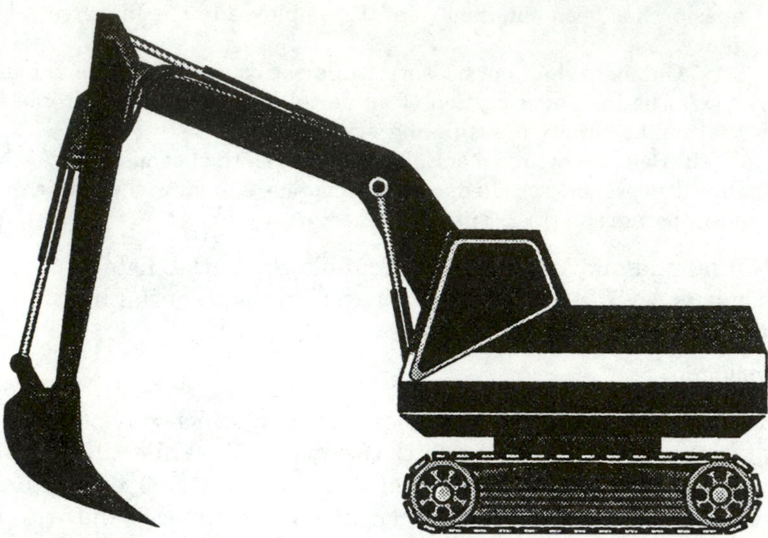

**Figure 34.4** Excavator.

sites, however, front-end loaders and backhoes are frequently outfitted with quick-change coupler systems that can in a few moments, safely drop a digging or loading bucket from the machine's boom or arms, and pick up an attachment set of forks. With forks, a loader or backhoe can perform most "occasional" lifting duties. But when a large number of lifts and transports are required—especially when offloaded from trailers or flatbeds—it's hard to beat the unloading and carrying capacity of lift trucks that are designed solely to do those tasks.

Like the other pieces of mobile equipment, forklifts should be outfitted with an overhead guard/cage, backup alarm, seatbelt, horn, lights, a model type and number/lifting capacity nameplate, and working controls and gauges.

Like other pieces of equipment, forklifts should be thoroughly inspected before each use—with an operator's checklist form.

Main hazards on construction sites include equipment operation by individuals who aren't properly trained, dropping loads that aren't safely rigged for transport across uneven ground, and crushing injuries to hands and feet when ground helpers work too closely to materials being unloaded or placed. Tipovers are also possible when heavy loads are carried by inexperienced operators or even experienced operators who are in a hurry.

### Dump trucks

Safety guidelines include:

- An operator waiting to get loaded by a front-end loader, backhoe, or excavator should remain in the cab, and not be wandering around behind or to the side of his or her vehicle. Too many times materials have struck an unsuspecting driver when they missed the dump bed or spilled over to the side or back.

- A dump truck's steps or ladder to the cab can get extremely muddy, icy, wet, and slippery. The ground a driver steps onto or from is also often muddy and slippery. These conditions make it extremely important for drivers to use three-point entries and exits from their vehicles.
- Dumping a load of stone exactly where it's wanted can often be complicated by overhead electrical lines, phone lines, cable television lines, and other utility installations. Remember that the bed of a tandem or triaxle dump truck can be raised awfully high, and make it a point to see that there's plenty of clearance before the dump occurs.
- Because backing up is such a large part of dump-truck operation, it's a true asset for worksite safety to have back-up alarms installed.
- Keep window glass and rear-view mirrors clean and frost-free, because visibility is such a critical part to safe dump-truck operations.

#### Other pieces of mobile equipment

- Concrete trucks with gates and chutes can cause strains, sprains, and pinched fingers and hands to employees attempting to direct the delivery of concrete, gravel, soil, or similar construction materials.
- Portable conveyor systems contain numerous strain, sprain, crushing, and pinch-point hazards.

Chapter

# 35

# Excavations, Trenches, and Landscaping

## Quick Scan

1. Worksites requiring excavations and trenches need the services of a competent person for supervising, inspecting, identifying, and correcting hazardous conditions and practices related to excavations and trenches.
2. The competent excavation and trench person needs to be knowledgeable in:
   - Soil type identification and analysis
   - The use of protective systems such as sloping, benching, shoring, bracing, underpinning, and shielding
   - OSHA's excavation and trenching standards
3. All affected employees need to be trained in safe work practices dealing with excavating and trenching operations.

## Excavations and Trenches

Excavating is a deceiving, dangerous business. It's deceiving because most employees will look at a 3-, 4-, 5-, and even 6-foot-deep trench as something they could easily climb out of if they had to. The truth is that even a shallow trench, if the sides collapse unexpectedly, can overcome and suffocate the strongest employee working within it. If you don't see it collapsing, you can't jump out of the way. It only takes a matter of minutes without oxygen for a fatality to occur. The fatality rate for trenching is about twice that for general construction work. Plus, in addition to cave-ins there are lots of other hazards associated with excavations and trenches.

Excavations and trenches are required for numerous reasons in the construction industry. Typical projects include building basements, installing foundations for buildings, and laying pipes for various drain, sewer, water, phone, electric, and gas lines.

### What OSHA says

OSHA's excavation standard can be found in 29 CFR Subpart P, 1926.650 through .652, which lays out basic safety measures for all open excavations, plus Appendices A through F, which cover soil classification, sloping and benching, timber shoring, aluminum hydraulic shoring, alternatives to timber shoring, and selection of protective systems.

OSHA's pamphlet on excavations, OSHA no. 2226, reviews important excavation and trenching standards and safe work practices. The section called "Planning for Safety" reads:

"Many on-the-job accidents are a direct result of inadequate initial planning. Correcting mistakes in shoring and/or sloping after work has begun slows down the operation, adds to the cost, and increases the possibility of an excavation failure. The contractor should build safety into the pre-bid planning in the same way all other pre-bid factors are considered."

It is a good idea for contractors to develop safety checklists before preparing a bid, to make certain there is adequate information about the job site and all needed items are on hand.

Those checklists should incorporate elements of the relevant OSHA standards as well as other information necessary for safe operations. Before preparing a bid, these specific site conditions should be taken into account:

- Traffic
- Proximity of structures and their conditions
- Soil
- Surface and ground water
- Water table
- Overhead and underground utilities
- Weather

These and other conditions can be determined by job-site studies, observations, test borings for soil type or conditions, and consultations with local officials and utility companies.

Before any excavation actually begins, the standard requires the employer to determine the estimated location of utility installations—including sewer, telephone, fuel, electric, and water lines—that may be encountered during digging. Also, before starting the excavation, the contractor must contact the utility companies or owners involved and inform them, within established or customary local response times, of the proposed work. The contractor must also ask the utility companies or owners to find the exact location of the underground installations. If they cannot respond within 24 hours (unless the period required by state or local law is longer), or if they cannot find the exact location of the utility installations, the contractor may proceed with caution. To find the exact location of underground installations, workers must use safe and acceptable means. If underground installations are exposed, OSHA regulations also require that they be "removed, protected, or properly supported."

### Excavation hazards

Excavation and trenching hazards include:

- Site-preparation and digging hazards, such as those encountered in felling trees and grading, accidentally digging into underground electric or gas lines, contacting overhead power lines, or undermining existing buildings or structures.
- Collapse of an excavation or trench, resulting from:

  1. *The intrinsic nature of the ground.* This refers to the mechanical stability of soil. As civil engineers tend to say, an unstable ground results from the soil's lack of strong cohesive properties and weak angles of internal friction. Layered, or nonhomogeneous ground can consist of different layers of soil which may enable cave-ins to be easily set off.

  2. *Overloading at the edge of the excavation or trench.* Material or equipment dangerously staged too close to an excavation's perimeter edge can cause cave-ins. Existing buildings or fixtures can also be weakened or undermined by an excavation.

  3. *The presence of water.* Increased humidity will usually result in reduced stability of excavation or trench walls. Surface water can run in, or water can permeate in from the ground. Freezing, thawing, and heavy rains can also have dramatic effects on soil stability. OSHA's excavation standard expressly prohibits employees from working in excavations and trenches where water has accumulated or is accumulating, unless adequate protection has been taken. If water-removal equipment such as trash pumps and siphons is used to control or prevent water from accumulating, the equipment and its operation must be monitored by a competent person to ensure proper use.

     OSHA standards also insist that diversion ditches, dikes, or other suitable means are used to prevent surface water from entering an excavation.

  4. *The presence of mobile equipment and machinery.* The use of earthmoving and lifting equipment can place excess loads along the edges of an excavation. Vehicles traveling near the edge of excavations are too common on construction sites, sometimes because there's not much clearance available between existing buildings, trees, or other obstructions. There can be difficulty in maneuvering heavy equipment due to lack of soil stability near the excavation. Other related hazards include noise, exhaust gases, and vibration from the equipment and machinery.

- Materials and objects falling or rolling into the excavation. Potential hazards include spoil heaps, stacks of material, large rocks or soil clumps, and equipment or machinery. Materials or equipment that could fall or roll into an excavation should be kept at least 2 feet from the edge, or should be kept from falling in through the use of retaining devices.
- Employees falling into the excavation. This hazard is increased if safe means of access, including ladders or ramps, is not provided.
- Hazardous atmospheres. Hazardous atmospheres may include oxygen deficiency, carbon monoxide, or any other hazardous atmosphere that could reasonably be expected to exist at the particular worksite.

- Mobile equipment hazards. Operator hazards include machine tipover, reckless or out-of-control machine operation, digging or placing items where they shouldn't be, falls from equipment, and accidental contact with electrical lines.
- Ground worker/helper hazards. These include sprained ankles from walking on rough surfaces and ruts, strains from digging and lifting, being struck by objects being swung overhead or dropped, overexertion, being run over by mobile equipment, and during extreme weather, heat, or cold exposure.

## Excavating and Trenching Operations

### Site preparation and digging

Find the location of all underground utilities by contacting the local utility locating service before digging. This can't be stressed enough. Even if you obtain a set of prints on your own, let the utility companies do the locating groundwork. They'd rather take the time—to prevent an unexpected utility outage and possible injury. And if any uncertainties crop up (they've been known to happen!), the experts will be right there to sort things through. Typical utility installations include sewer, water, natural gas and other fuel, electric, telephone, and television lines.

Remove or support any fixtures or obstacles on the surface that could present hazards to the excavation. Is a utility pole only inches away from where a trench will pass? The pole may need to be reinforced with additional guy wires, or it may have to be temporarily removed—after its lines are deactivated, if need be. Is a large pine tree's root ball protruding into an open trench—and does that leave the tree unsupported on one side, and in danger of tipping over? The tree will have to be removed or shored up so that it won't fall.

Keep drivers in their cabs and workers away from where soil or debris is being loaded onto dump trucks. If employees are working near public roadways or exposed to other traffic, provide them with warning vests having reflective surfaces or other bright materials such as fluorescent orange or safety yellow.

### Employee access and warning systems

Excavations and trenches that are 4 or more feet deep, need a stairway, ladder, ramp, or other safe means of employee access. This means of access must be within 25 feet of lateral travel for any individual worker inside the excavation. Structural ramps constructed solely for employee entrance and exit have to be designed by a competent person. Ramps designed for equipment access must be designed by a competent person qualified in structural design, and be constructed as the design prescribes.

If employees will have to cross a trench, they shouldn't be leaping from one side to the other. Instead, walkways or bridges with standard guardrails must be provided.

Put an effective warning system into place to alert mobile equipment operators when they're approaching a trench or other excavation, or when they're operating their equipment adjacent and parallel to the perimeter edge. From inside a miniloader, backhoe, bulldozer, or front-end loader, an operator may not be able to see the edge of the excavation until it's too late. Stop-blocks, barricades, hand signals, and similar systems will work if communicated to all employees and operators. It's also a good idea to slope the surrounding area adjacent to an excavation away (descending) from the perimeter edge.

### Ground-workers/helpers

Keep employees away from digging and lifting equipment. Never allow workers to enter or remain in an excavation when equipment is in use. If they need to move in for a closer look, make sure there's communication between the backhoe or excavator operator and one person on the ground or in the excavation. Don't let the signal calling turn into a free-for-all, with multiple employees giving contradicting signals to the equipment operator. Employees in the trench need to move away from the bucket, digger, or other attachment while it's actively running—and never put themselves or any parts of their bodies between machinery that could easily miscue from operator error, or fail from a hydraulic leak, electrical short, or other breakdown.

### Protective systems

For excavations and utility trenches that are unstable or more than 5 feet deep, use protective systems such as spoils placement, sloping and benching, scaling, shoring, shields or trench boxes, and water management. Unless a soil analysis has been completed, the earth's slope in an excavation or trench must be at least 1-1/2 units of horizontal run to every unit of vertical rise if no other protective systems are in use. OSHA has prepared three decision-making charts (see Figs. 35.1, 35.2, and 35.3) to review protective system options.

1. *Spoils placement.* Keep equipment and the excavated dirt (spoils pile) back at least 2 feet from the edge of the excavation, or more if the space is available. When adequate space isn't available, the soil needs to be placed elsewhere so that it won't slide back unexpectedly into the trench. Veteran hoe operators stress putting the groundbreaking sod and soil crust closest to the excavation's perimeter edge (yet still 2 feet or more away), so that the looser materials can be staged behind the chunked material and are less likely to spill in.

2. *Sloping and benching.* OSHA defines *sloping* or a *sloping system* as a method of protecting employees from cave-ins by excavating to form sides of an excavation that are inclined away from the excavation so as to prevent cave-ins. The angle of incline required to prevent a cave-in varies with differences in such factors as the soil type, environmental conditions of exposure, and application of surcharge loads.

*Benching* or a *benching system* is further defined as a method of protecting employees from cave-ins by excavating the sides of an excavation to form one or a series of horizontal levels or steps, usually with vertical or near-vertical surfaces between levels. A set of OSHA drawings illustrate proper sloping and benching for various soil types (see Figs. 35.4 through 35.8).

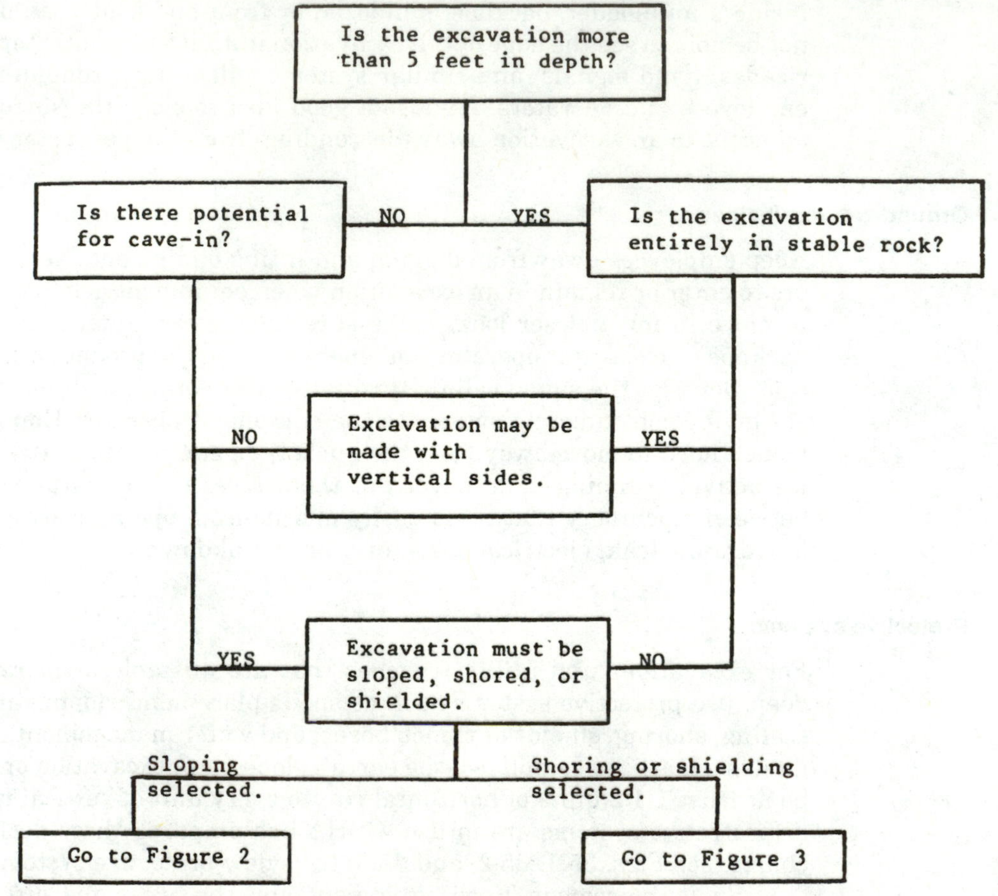


**Figure 35.1** OSHA excavation decision chart.

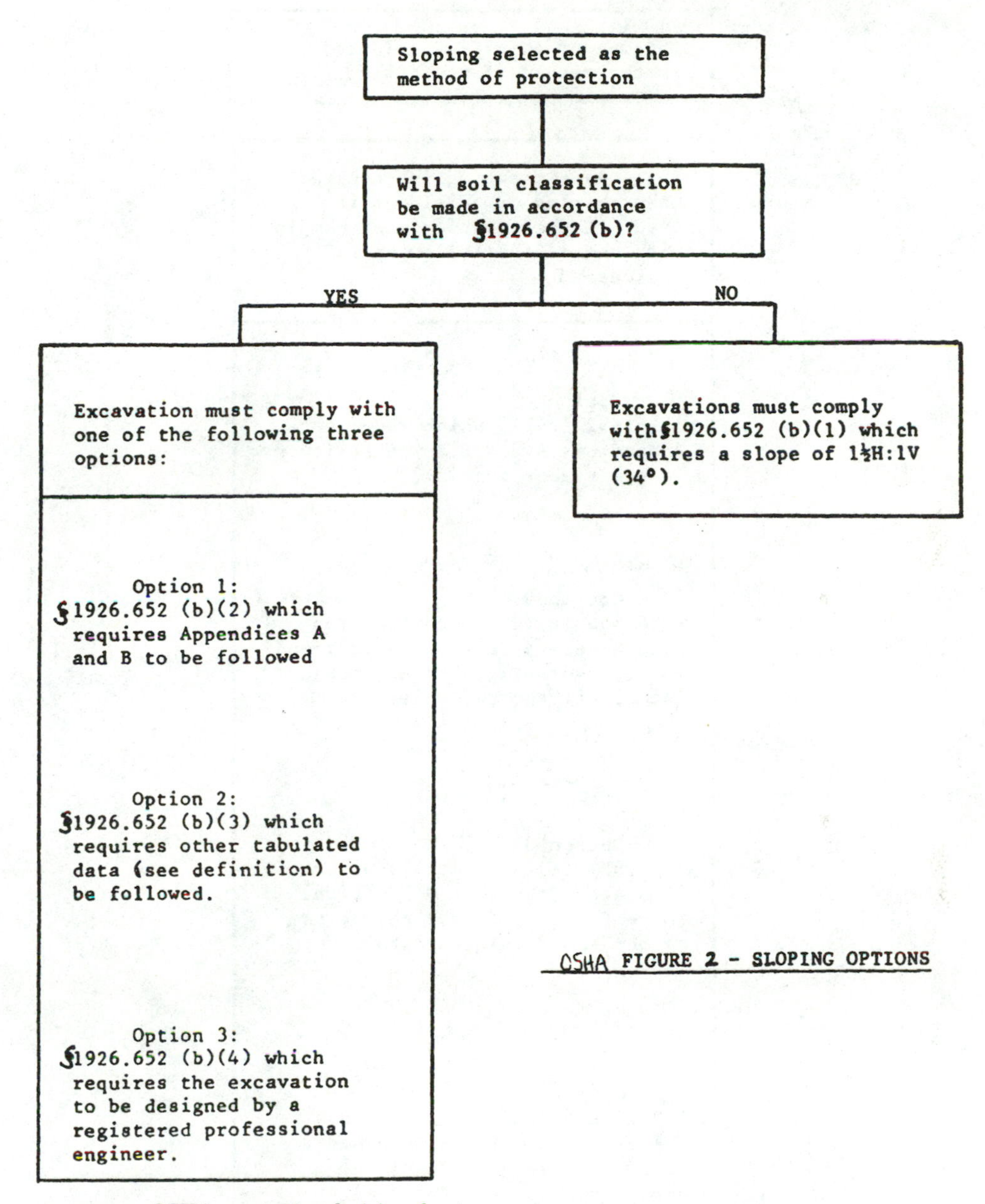


**Figure 35.2** OSHA excavation decision chart.

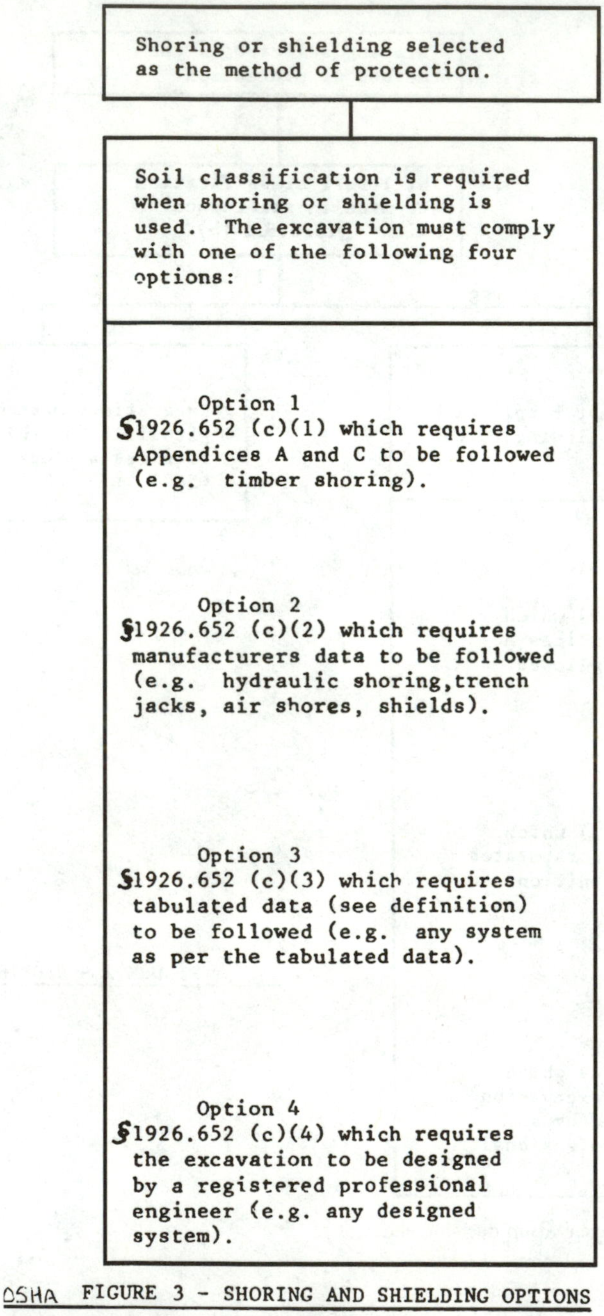

Shoring or shielding selected as the method of protection.

Soil classification is required when shoring or shielding is used. The excavation must comply with one of the following four options:

Option 1
§1926.652 (c)(1) which requires Appendices A and C to be followed (e.g. timber shoring).

Option 2
§1926.652 (c)(2) which requires manufacturers data to be followed (e.g. hydraulic shoring, trench jacks, air shores, shields).

Option 3
§1926.652 (c)(3) which requires tabulated data (see definition) to be followed (e.g. any system as per the tabulated data).

Option 4
§1926.652 (c)(4) which requires the excavation to be designed by a registered professional engineer (e.g. any designed system).

OSHA FIGURE 3 - SHORING AND SHIELDING OPTIONS

**Figure 35.3** OSHA excavation decision chart.

## Slope Configurations

(All slopes stated below are in the horizontal to vertical ratio)

### *Excavations made in Type A soil.*

1. All simple slope excavation 20 feet or less in depth shall have a maximum allowable slope of 3/4:1.

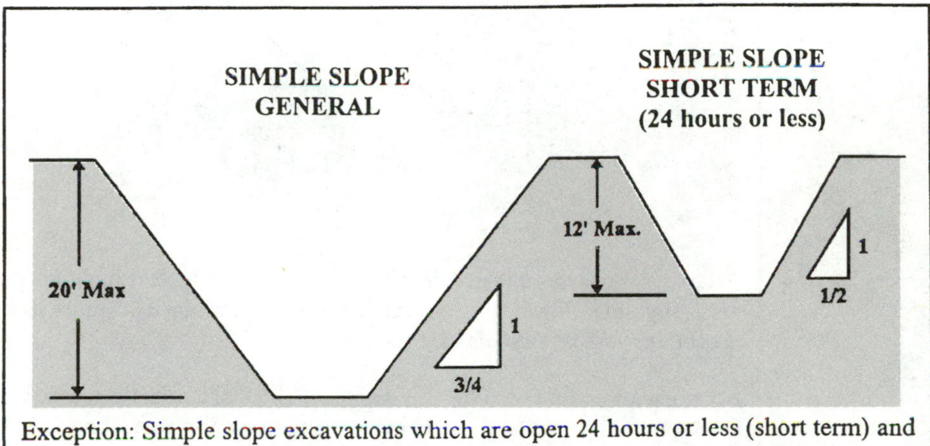


Exception: Simple slope excavations which are open 24 hours or less (short term) and which are 12 feet or less in depth shall have a maximum allowable slope of 1/2:1.

2. All benched excavations 20 feet or less in depth shall have a maximum allowable slope of 3/4 to 1 and maximum bench dimensions as follows:

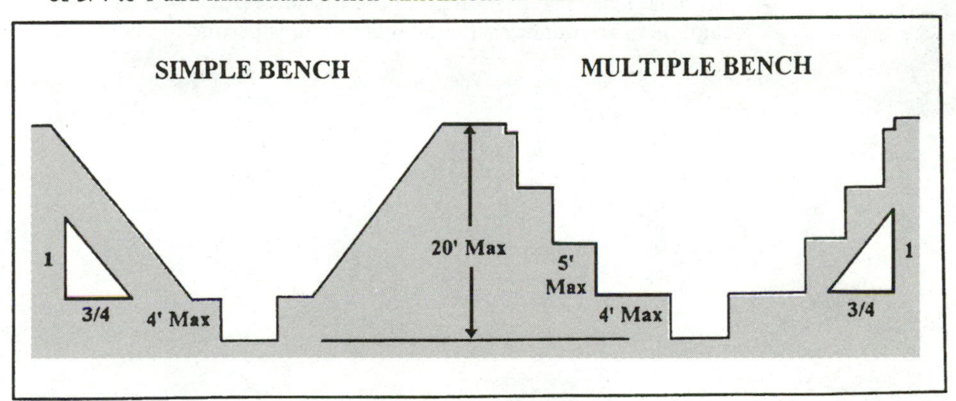


**Figure 35.4** OSHA sloping and benching for type A soil (no. 1).

3. All excavations 8 feet or less in depth which have unsupported vertically sided lower portions shall have a maximum vertical side of 3 1/2 feet.

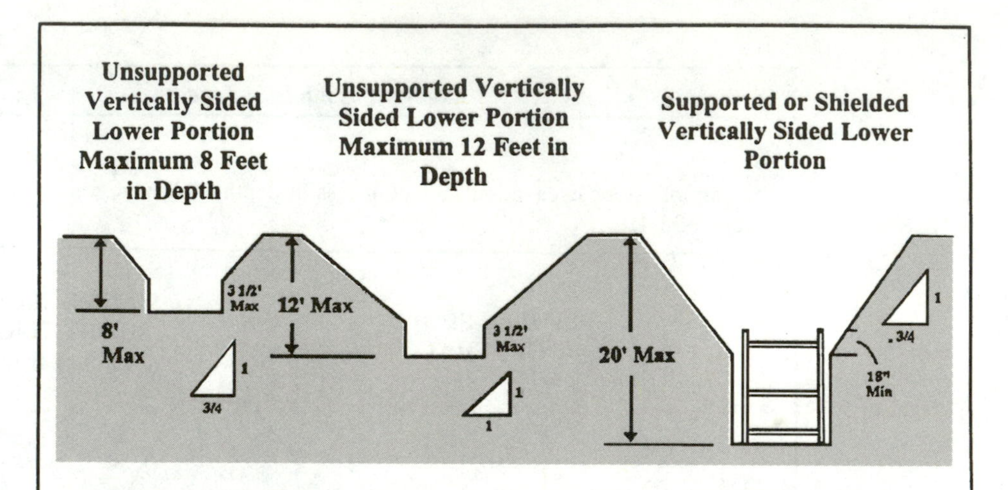


All excavations more than 8 feet but not more than 12 feet in depth with unsupported vertically sided lower portions shall have a maximum allowable slope of 1:1 and a maximum vertical side of 3 1/2 feet.

All excavations 20 feet or less in depth which have vertically sided lower portions that are supported or shielded shall have a maximum allowable slope of 3/4:1. The support or shield system must extend at least 18 inches above the top of the vertical side.

4. All other simple slope, compound slope, and vertically sided lower portion excavations shall be in accordance with the other options permitted under 1926.652(b).

**Figure 35.5** OSHA sloping for type A soil (no. 2).

***Excavations made in Type B soil.***

1. All simple slope excavations 20 feet or less in depth shall have a maximum allowable slope of 1:1.

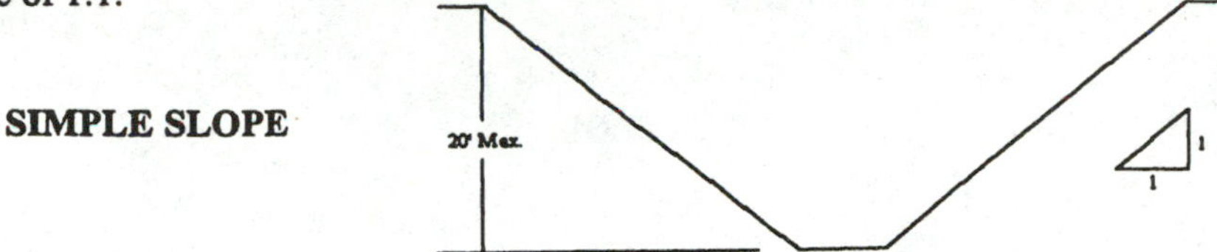


2. All benched excavations 20 feet or less in depth shall have a maximum allowable slope of 1:1 and maximum bench dimensions as follows:

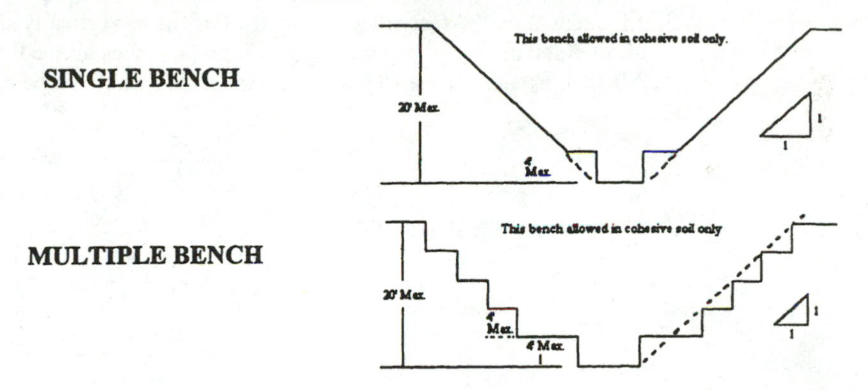


3. All excavations 20 feet or less in depth which have vertically sided lower portions shall be shielded or supported to a height at least 18 inches above the top of the vertical side. All such excavations shall have a maximum allowable slope of 1:1.

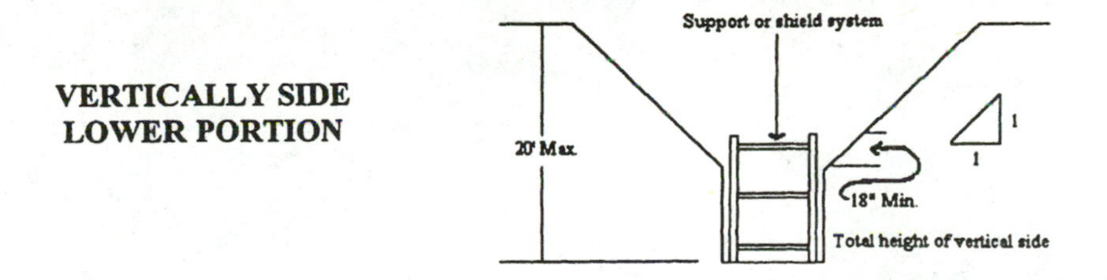


4. All other sloped excavations shall be in accordance with the other options permitted in 1926.652(b).

**Figure 35.6** OSHA sloping and benching for type B soil.

***Excavations made in Type C soil.***

1. All simple slope excavations 20 feet or less in depth shall have a maximum allowable slope of 1 1/2:1.

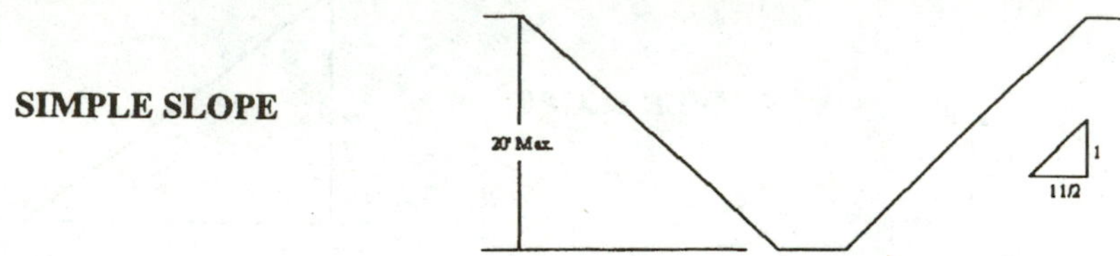


2. All excavations 20 feet or less in depth which have vertically sided lower portions shall be shielded or supported to a height at least 18 inches above the top of the vertical side. All such excavations shall have a maximum allowable slope of 1 1/2:1.

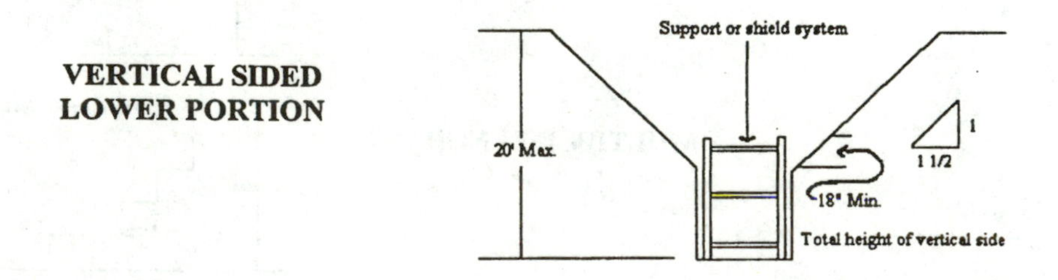


3. All other sloped excavations shall be in accordance with the other options permitted in 1926.652(b).

**Figure 35.7** OSHA sloping for type C soil.

***Excavations made in Layered Soils***

1. All excavations 20 feet or less in depth made in layered soils shall have a maximum allowable slope for each layer as set forth below.

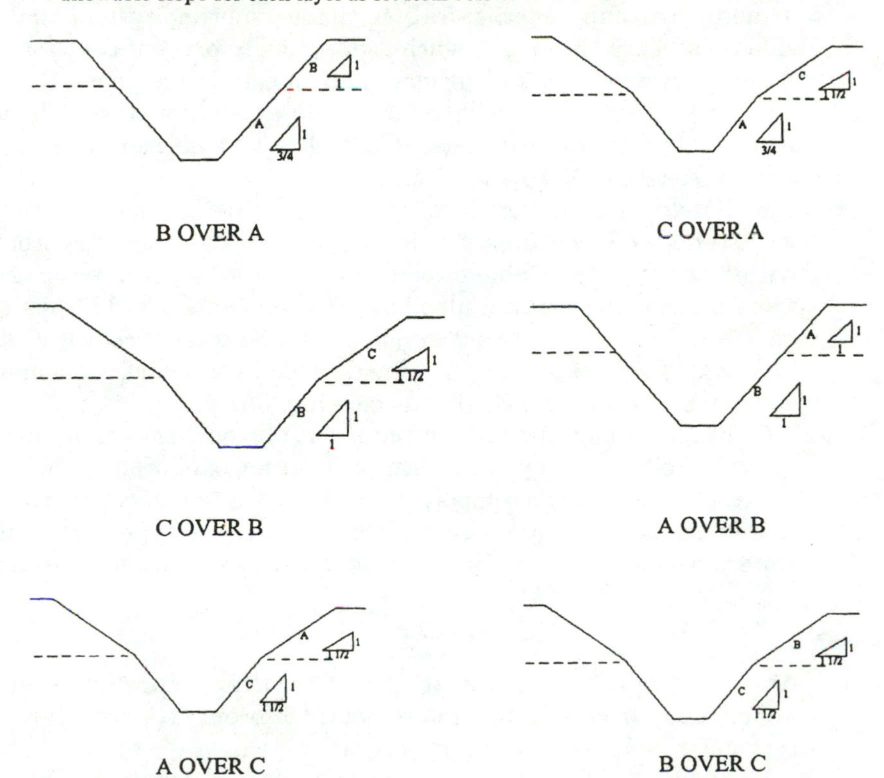


2. All other sloped excavations shall be in accordance with the other options permitted in 1926.652(b).

**Figure 35.8** OSHA sloping for layered soils.

3. *Scaling.* To protect employees from loose soil and rock falling or rolling from an excavation edge or wall, loose material should be scaled away, then protective barricades, shoring, or shields installed as necessary on the wall face to stop and contain additional falling material.

4. *Shoring.* OSHA defines *shoring* or a *shoring system* as a structure such as a metal hydraulic, mechanical, or timber shoring system that supports the sides of an excavation and which is designed to prevent cave-ins (see Fig. 35.9). Shoring systems consist primarily of sheeting members that retain earth against the vertical or inclined faces of an excavation or trench, which are held in place by vertical members or uprights and horizontal members or cross-braces (see Figs. 35.10 and 35.11).

5. *Shields.* A *shield* or *shield system* is defined as a structure that is able to withstand the forces imposed on it by a cave-in and thereby protect employees within the structure. Shields can be permanent structures or can be designed to be portable and moved along as work progresses. Additionally, shields can be either premanufactured or job-built in accordance with OSHA requirements 1926.652(c)(3) or (c)(4). Shields used in trenches are usually referred to as trench boxes or trench shields (see Fig. 35.12).

6. *Water management.* Keep water out of trenches with a pump or drainage system. Following any rainstorm or weather condition in which moisture or water plays a role, the competent person needs to inspect an excavation for soil movement and potential cave-ins. Soils of various types react differently with water; even a minor rainfall can render some excavations or trenches unstable.

## Soil analysis

Anyone familiar with excavating and trenching knows that "all soils aren't alike." That's a classic understatement. Different types of soil require different types of excavation cave-in protection.

Four basic soil types are recognized for excavation purposes. They are "Stable rock," "Type A," "Type B," and "Type C." Stable rock is the strongest, and Type C soil the weakest. A few definitions are in order here, all hailing from OSHA's Appendix A, which tags along the end of 1926.652, in which protective systems for excavations and trenches are discussed. Maximum allowable slopes for excavations in different soil types are presented by OSHA in Fig. 35.13.

*Cohesive soil* means clay (fine-grained soil), or soil with a high clay content, which has cohesive strength. Cohesive soil does not crumble, can be excavated with vertical sideslopes, and is plastic when moist. Cohesive soil is hard to break up when dry, and exhibits significant cohesion when submerged. Cohesive soils include clayey silt, sandy clay, silty clay, clay, and organic clay.

On the other hand, *granular soil* means gravel, sand, or silt (coarse-grained soil) with little or no clay content. Granular soil has no cohesive strength. Some moist granular soils exhibit apparent cohesion. Granular soil cannot be molded when moist and crumbles easily when dry.

*Unconfined compressive strength* means the load per unit area at which a soil will fail in compression. It can be determined by laboratory testing, or

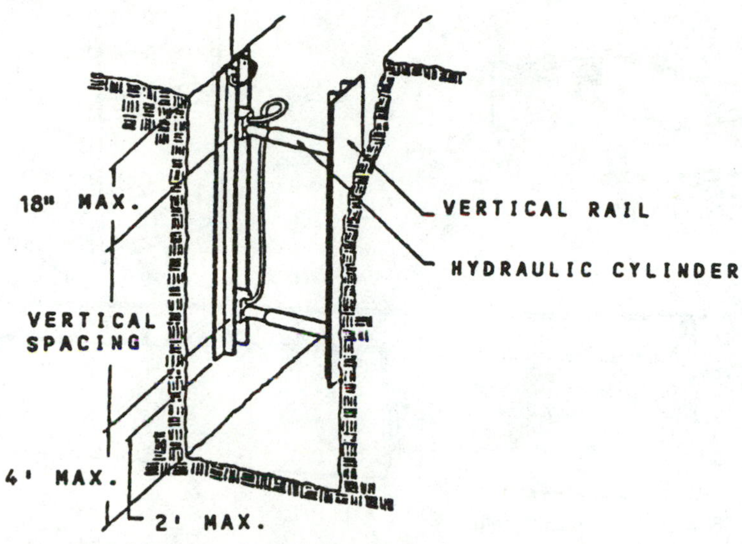


**Figure 35.9** OSHA aluminum hydraulic shoring.

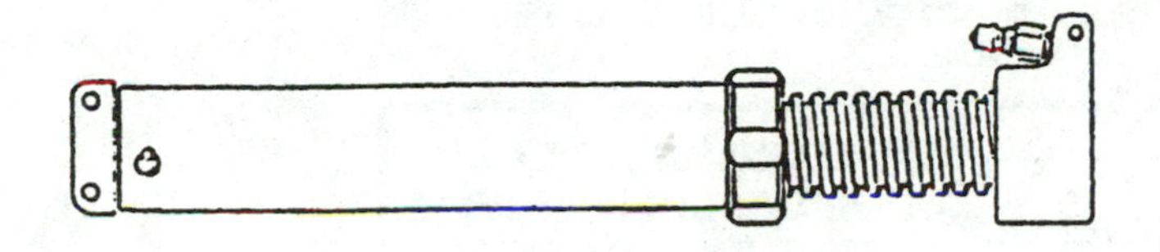

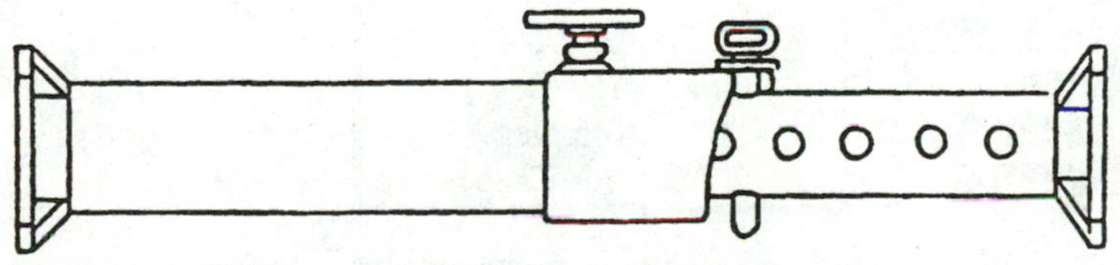

**Figure 35.10** OSHA pneumatic/hydraulic shoring.

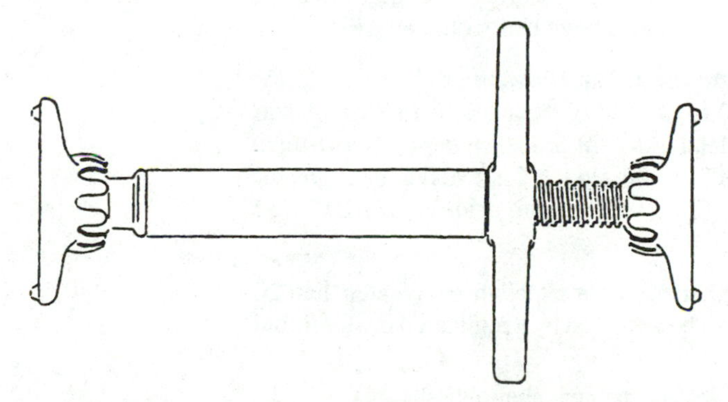

**Figure 35.11** OSHA shoring member.

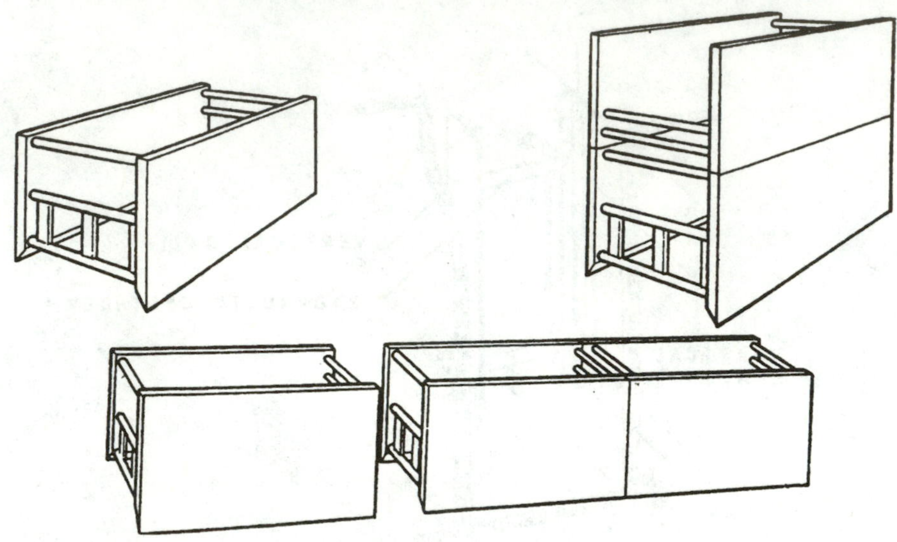

**Figure 35.12** OSHA trench shields.

## Maximum Allowable Slopes

| SOIL OR ROCK TYPE | MAXIMUM ALLOWABLE SLOPES (H:V) [1] FOR EXCAVATIONS LESS THAN 20 FEET DEEP [3] |
|---|---|
| STABLE ROCK | VERTICAL (90 Deg.) |
| TYPE A [2] | 3/4:1 (53 Deg.) |
| TYPE B | 1:1 (45 Deg.) |
| TYPE C | 1 1/2:1 (34 Deg.) |

[1] Numbers shown in parentheses next to maximum allowable slopes are angles expressed in degrees from the horizontal. Angles have been rounded off.

[2] A short-term maximum allowable slope of 1/2H:1V (63 degrees) is allowed in excavations in Type A soil that are 12 feed (3.67 m) or less in depth. Short-term maximum allowable slopes for excavations greater than 12 feet (3.67 m) in depth shall be 3/4H:1V (53 degrees).

[3] Sloping or benching for excavations greater than 20 feet deep shall be designed by a registered professional engineer.

**Figure 35.13** OSHA maximum allowable slopes.

estimated in the field using a pocket penetrometer, by hand-operated shear vanes, by thumb penetration tests, and by other methods.

1. *Stable rock.* Stable rock is natural solid mineral matter that can be excavated with vertical sides and remain intact while exposed. Of course, it's awfully tough to dig—or to cut or blast. Not many home foundations or basements are excavated out of stable rock.

2. *Type A soil.* Type A soils are cohesive soils with an unconfined compressive strength of 1.5 tons per square foot (tsf) or greater. Examples include clay, silty clay, sandy clay, clay loam and, in some cases, silty clay loam and sandy clay loam. Cemented soils such as hardpan and caliche are also considered Type A.

But wait—there's a catch. None of the above soils would still qualify as Type A's if

- The soil is fissured; or
- The soil is subject to vibration from heavy traffic, pile driving, or similar effects; or
- The soil has previously been disturbed; or
- The soil is part of a sloped, layered system where the layers slant into the excavation at an angle.

3. *Type B soil.* Type B soils are cohesive soils with an unconfined compressive strength greater than 0.5 tsf but less than 1.5 tsf; or granular cohesionless soils including angular gravel (similar to crushed rock), silt, silt loam, sandy loam, and in some cases, silty clay loam and sandy clay loam as well as some previously disturbed soils, dry rock that is not stable, steeply sloped layered soils, and the soils above, which—except for their deficiencies—would otherwise qualify as Type A's.

4. *Type C soil.* Type C soils are cohesive soils with an unconfined compressive strength of less than 0.5 tsf. They're granular soils including gravel, sand, and loamy sand, or submerged soil or soil from which water is freely seeping, or worse.

### Determining soil types

OSHA requires that classification of soil types must be based on at least one visual analysis and one manual test. A competent person trained and experienced in soil mechanics must perform the tests. Visual analysis includes an inspection of the entire worksite location, looking for cracks in the ground, evidence of underground disturbances or fixtures, layered soils, or the presence of water, vibration, or any other detriments to a sound excavation. The manual tests can be completed in several ways. Plasticity tests determine that the soil is cohesive if it can be rolled into 1/8-inch by 2-inch-long strands or ropes between the tester's hands without crumbling (as he or she could with clay). If the soil is dry and crumbles on its own or with moderate pressure into crumbs, grains, or fine powder, then it's granular and would not support itself very well in excavations and trenches.

The thumb penetration test can also be used to estimate the unconfined compressive strength of cohesive soils.

Type A soils can be easily indented by the tester's thumb, but can be "run through" or penetrated by the thumb only with an enormous effort. Type C soils are easily poked through by the tester's thumb, and can be molded like a soft clay with minimal hand or finger pressure. Type B soils range somewhere in between the other two types. All thumb penetration tests should be conducted on freshly dug clumps of soil that have not had a chance to dry out.

### Hazardous atmospheres

In some instances, trenches can be considered confined spaces, complete with hazardous atmospheres. OSHA's definition of a hazardous atmosphere is an atmosphere which by reason of being explosive, flammable, poisonous, corrosive, oxidizing, irritating, oxygen deficient, toxic, or otherwise harmful, may cause death, illness, or injury. Potentially hazardous atmospheres must be monitored. Test for oxygen, and if gasoline or diesel equipment is running in the area, for carbon monoxide as well. Depending on the location of the excavation, additional harmful materials can be present, such as hydrogen sulfide and certain explosive gases. If hazardous atmospheres exist, emergency rescue equipment also needs to be nearby, and affected employees must know how to use it.

### Competent persons

Because of the serious nature of the activities surrounding excavations, and the many variables involved, a competent person is needed to plan, supervise, and inspect excavation and trenching operations. For excavation use, a competent person must be trained and experienced in soil analysis, in the use of protective systems for excavations, and thoroughly familiar with OSHA's excavation standard. Again, this work must be done under the direction of a competent person, who understands soils and how they can react to varying conditions. Again, by definition, a *competent person* is an individual who is capable of identifying existing or predictable hazards, or working conditions which are unsanitary, hazardous, or dangerous to employees, and who has authorization to take prompt, corrective measures to eliminate them. The authority to take quick corrective action is a major part of a competent person's responsibilities. It's not enough just to be able to recognize potential hazards; the competent person has to have the authority and capability to act.

Any excavation should be closely monitored. A competent person must conduct daily inspections of the excavation, adjacent areas, and any protective systems that are in place. The competent person should look for situations that could result in cave-ins, or improper heavy equipment operation, or evidence of hazardous atmospheres. He or she should look regularly for changes in the stability of the earth, focusing on water, cracks, vibrations, and the spoils pile. He or she should halt work on the excavation or trench if there is any potential for cave-in, and ensure that the problem is fixed before work is allowed to resume. A major part of the competent person's job must be to correct any hazards before employees enter the excavation or trench at the beginning of the day or shift, as needed throughout the shift, and during or after every rainstorm or incident that could create a hazardous condition in or near the excavation.

### Excavations for residential home foundations

After the foundation walls are constructed, special precautions must be taken to prevent injury from cave-ins in the area between the excavation wall and the foundation wall, *unless:*

- The excavations are made entirely into stable rock.
- The excavations are less than 5 feet deep, and after an inspection by a competent person, are found to be structurally sound, with no indication of a potential cave-in.
- The excavations/trenches are for house foundations or basements (not utility excavations or trenches) having the following characteristics:

  There is no water, no surface-tension cracks, and no other apparent environmental conditions present that jeopardizes the stability of the excavation.

  The excavation or trench is less than 7-1/2 feet in depth, or benching has been installed at a ratio of 2 feet of horizontal travel for every 5 feet or less of vertical height.

  The minimum width of the trench, from the excavation face to the wall formwork at the bottom of the trench, is no less than 2 feet, and preferably wider to give employees more room to maneuver.

  There's no heavy equipment running in the area that causes vibration to the excavation while employees are working within the trench or excavation.

  The excavation plan minimizes the number of employees in the trench and the length of time they spend there.

  The spoil pile is no closer to the trench edge than 2 feet, and preferably no closer to the top edge of the trench than the trench is deep.

- Soil classification types include stable rock, and Types A, B, and C.
- A typical excavation may be benched at a 1:1 slope after the competent person determines the soil type is stable enough for that type of protection. Many residential excavations will be type C soil and will need a slope of 1-1/2:1.

### Landscaping

Site clearing is often one of the first tasks before actual construction takes place. It typically involves work with heavy equipment, tree cutting and grubbing, and soil grading and digging.

### What OSHA says

OSHA comments on numerous activities involved with landscaping and site-clearing activities. Its standards on site clearing are:

> 1926.604 (a) General requirements. (1) Employees engaged in site clearing shall be protected from hazards of irritant and toxic plants and suitably instructed in the first-aid treatment available.
>
> (2) All equipment used in site clearing operations shall be equipped with rollover guards meeting the requirements of this subpart. In addition, rider-operated equipment shall be equipped with an overhead and rear canopy guard meeting the following requirements:
>
> (i) The overhead covering on this canopy structure shall be of not less than 1/8-inch steel plate or 1/4-inch woven wire mesh with openings no greater than 1 inch, or equivalent.
>
> (ii) The opening in the rear of the canopy structure shall be covered with not less than 1/4-inch woven wire mesh with openings no greater than 1 inch.

### Tree cutting

For tree work, make sure that employees are trained to use chain saws, are wearing personal protection, including hard hats, gloves, hearing protection, safety eyewear, chaps or leggings, and safety-toed shoes or boots.

To fell trees and cut branches, certain skills are needed, and trees may also be near power lines, existing homes, and other installations.

What can go wrong?

- A tree or large branch could fall the wrong way.
- A chain saw can slice through a person's limb as quickly as it can cut through a tree.
- Contact with power lines is a very real danger when working with trees.
- Bad weather can make tree work even more hazardous, when electrical storms, high winds, heavy rain, or snowstorms come into play.

Operate and adjust chain saws according to the manufacturer's instructions. The operating manual should be available. A number of additional precautions apply to chain saws:

- Fuel them at least 20 feet from an open flame or ignition source.
- Start the chain saw at least 10 feet away from the fueling area.
- Have the chain saw on the ground or firmly supported, with the brake on, before it's started.
- Circle the handles with the thumbs and fingers of both hands, unless you're given other instructions on holding them.
- Clear paths of brush or other obstacles that may interfere with your cutting.
- Never cut directly overhead with a chain saw.

- Turn off or idle the saw before you move away from a cut tree. Consider that you aren't likely to hear anything else once the saw is fully engaged in its cutting mode.
- Be sure the chain saw is turned off or the brake is on if you have to carry it over slippery or brush-covered surfaces, or if you have to move it more than 50 feet.
- Carry it in a way that prevents contact with the cutting chain or muffler.

Chapter

# 36

# Weather

## Quick Scan

1. Protect exposed skin from hot, summer sun rays.
2. Employees should be able to recognize the signs of heat illness, know how to react to them, and when to seek medical help.
3. During hot weather conditions, employees should rest often, maintain safe salt intake, and stay in the shade whenever possible.
4. Protect eyes from sunlight with sunglasses designed to block out ultraviolet rays.
5. Be aware of cold-weather hazards that require extra care and slower work paces. Ice, snow, and water make extremely slippery surfaces. Recognize the numerous slipping hazards encountered during typical wintry conditions.
6. For cold weather, dress in layers.
7. Allow for extra travel time in wintry conditions, and slow the driving down.
8. Avoid situations in which carbon monoxide gas could collect within a vehicle, garage, basement, trailer, shed, or other shelter.
9. Follow commonsense precautions to limit employee exposure when extreme weather conditions such as lightning, tornado, and other high winds threaten the worksite.

## Working in Hot-Weather Conditions

Heat, a form of energy, is usually considered a good thing. The absence of heat is, of course, cold. People generally prefer to work in warm-to-hot temperatures rather than in cold-to-freezing conditions. But heat in the workplace has its drawbacks. Heat stress—a potentially harmful condition—occurs when the body's natural cooling mechanism breaks down or becomes overworked. Heat stress will likely be a factor when employees are performing strenuous activities in areas where high temperatures, humidity, and radiant heat sources are present.

The hazards related with heat stress, though, are often overlooked because many workers and supervisors have been exposed to heat stress conditions in the past and have weathered the heat before, with no ill effects. Depending on the worksite, heat can radiate from a variety of sources, from a sun-drenched roof covered with a layer of dark roofing paper or shingles, to the inside of an attic that has yet to be vented.

In addition to causing heat-related illnesses, hot environments also increase the potential for accidental injuries. Heat-related injuries can result from fogged safety glasses or goggles, from sweat getting in the eyes, from slippery hands, from dizziness, and from fainting. Mental problems, too, can be associated with hot environments. In hot weather, people tire out faster, are more easily irritated, and are quicker to anger. Poor judgment abounds during hot conditions because it's easier for employees to daydream and lose concentration. All of these hot weather factors result in workers who may not pay as much attention to their tasks as they should.

For argument's sake, a "hot environment" exists when the ambient air temperature is 90°F or greater. High humidity will make high temperatures feel even hotter. But temperature and humidity levels alone do not determine the effects of heat on an individual. Heat stress is also partially dependent on personal factors such as a person's age, weight, level of physical fitness, metabolic rate, use of alcohol or drugs, and any medical conditions, such as high blood pressure that may be present. Alcohol, by the way, is a diuretic, a substance that actually dehydrates the body. Employees who use alcohol to excess within 24 hours of the workday put themselves at considerably greater risk for heat-related disorders.

Exposure of employees to excessive heat at the worksite can bring about medical emergencies in a number of ways. Heat with high humidity tends to tire workers out in a hurry. For some individuals, this may prevent incidents, because the person tires, slows down, drinks liquids, and rests before going ahead at a slower rate. Others, however, may elect to work at the same pace they would have during moderate temperatures, taxing their bodies beyond acceptable heat/work limits.

During periods of rest and mild exercise, the body usually maintains an internal temperature of between 97° and 99°F. But during times of strenuous or extended physical activity in hot temperatures, internal heat created by muscle contraction and flexing combine with outside temperatures and can result in a body temperature of 104° or more. During those times, the regulation of body temperature becomes critical.

### Acclimatization

Individuals who are used to working in hot weather tend to acclimate themselves—they develop increased abilities to sweat more than will individuals who do not frequently work in hot weather. Sweating, of course, and its evaporation, is how people cool themselves. Acclimatization is a process by which our bodies build up a tolerance to heat through gradual work exposure in hot environments. For most people this occurs naturally as temperatures turn warmer from spring to summer. This may not exactly apply to new hires and contractors not used to working in hot temperatures. These individuals may acclimatize over the course of a week, during which they work in a hot environment on successive days for progressively longer periods of time. Consider that a person's heat tolerance may diminish when the person spends time in cooler environments—when on vacation, for instance.

To counteract profuse sweating, continual fluid replenishing is needed to fuel the body's temperature-regulating system. That's why water or thirst-quenching "sports" drinks must be kept at the worksite, especially during warm or hot weather. If water or a water substitute isn't taken, the body tends to shut down its sweating mechanism to conserve fluids. When that happens, nothing helps to regulate the body's internal heat levels. The body temperature starts to rise, and may continue to increase until heat illness results. Dehydration can also come into play. Workers can't trust their own thirst to tell when additional fluids are needed. Experts recommend a minimum fluid intake for someone working in a hot or humid environment of about a cup of water or carbohydrate-electrolyte thirst-quencher every 15 minutes. Water is fine for general use, but the latter thirst-quenching liquids are designed to replace elements of sodium, potassium, and chloride that are needed for maintaining the body's delicate fluid balance. Moreover, muscles contracting and flexing in the heat become "thirsty" for the energy found in carbohydrates. Too, carbohydrates tend to aid the small intestine in its ability to absorb fluid.

Whew! There's a lot more involved with internal cooling than just satisfying a person's thirst.

Dry heat, or high temperatures along with low humidity, can result in more severe heat illnesses than those caused by moist heat. Employees working in dry heat may not recognize exactly how hot they are getting, and are more likely to continue working past their bodies' heat tolerances.

### Heat cramps

Heat cramps are the least serious of the heat illnesses. They come about when employees perform lengthy, active tasks in a hot environment. Depending on the individual, heat cramps can occur in what you may consider little more than a normal temperature. The victim sweats profusely, and tends to drink a lot of water. But the volume of fluids sweated out is greater than those taken in, and along with the sweat go salts and nutrients not replaced by the water. As the sweating continues, severe muscle cramps occur, usually in the legs and abdomen. The victim feels exhausted, and will occasionally collapse and pass out. Heat cramps are often accompanied by dizziness or periods of light-headedness.

First aid for heat cramps includes:

- Moving the patient to a nearby shaded, cool place.
- Providing the victim with water, lightly salted water (one teaspoon salt per quart of water), or thirst-quenchers at the rate of about 1 cup every 15 minutes.
- Applying warm moist towels to the victim's forehead and against cramped muscles.
- Although the above treatments will usually suffice, if symptoms continue, seek medical assistance.

### Heat exhaustion

Heat exhaustion is more serious than heat cramps. It involves an overloading of a person's circulatory system, when the body loses too much fluid and salts while performing strenuous activities. It's a problem typically associated with firefighters, soldiers, athletes, and—guess what?—construction workers. Symptoms and signs of heat exhaustion are rapid, shallow breathing; a weak pulse; cold, clammy, and sometimes pale skin; heavy perspiration; feelings of weakness and dizziness; and unconsciousness.

First aid for heat exhaustion includes:

- Moving the victim to a nearby shaded, cool place.
- Keeping the victim calm.
- Removing enough clothing to help cool the victim, without letting him or her get chilled.
- Fanning the victim.
- If victim is conscious, supply 1 cup water, lightly salted water, or thirst-quencher at a rate of 1 cup every 15 minutes.
- Although heat exhaustion usually responds well to first-aid treatment, if symptoms persist or worsen, seek medical assistance.

### Heat stroke

This is always serious. It requires quick recognition, emergency response, and transportation of the victim to a medical facility—failure to do so in time may result in death. It can happen on a hot, humid day, or during times of hot, dry weather. Heat stroke occurs when the body's heat-regulating systems break down or shut themselves down. Sweating stops and the body can no longer cool itself; instead the victim's body temperature rapidly rises because there's nowhere to get rid of the excess heat, and temperatures rise above 103°F and beyond. Symptoms and signs are deep breathing followed by shallow breaths; a rapid, strong pulse followed by a rapid, weak pulse; dry, hot, often red skin; coma or loss of consciousness; and convulsions or muscular twitching. Sometimes, though, the victim may quickly progress through symptoms associated with heat cramps and heat exhaustion before the signs of heat stroke become apparent.

First aid for heat stroke includes:

- Calling for emergency medical assistance.
- Cooling the victim quickly. Move to shaded, cool area. Wrap victim in wet towels or sheets. Pour cold water on victim's wrappings. Use cold packs or ice packs if available; place under each of victim's armpits, and one against each wrist and ankle, one against the groin, and one on each side of the neck—all places where large amounts of blood flow near the skin. Wait for the EMTs to respond.

Also, watch out for other injuries—such as broken bones or concussions—caused by falls due to unconsciousness.

## Sunburn

Another hazard that's often present at the worksite is the good old sun. From 93 million miles away, the sun showers us in warmth and radiant energy. And to our skin, that's the problem.

There was a time in the not-so-far past when every young home construction worker wanted a "George Hamilton" kind of tan. Some still do, but others have caught on to the fact that to prevent future problems, human skin should be protected from the sun.

What's the difference between then and now? Dermatologists the world over now speak out against the sun's ultraviolet rays as harmful to our skin and eyes. Plus, as a result of subtle changes in the earth's atmosphere, more ultraviolet rays are presently reaching the earth's surface. Studies have also associated ultraviolet light rays with skin damage ranging in severity from sunburn and wrinkling to skin cancer. In short, scientists and physicians are recommending that our skin and eyes be protected from several kinds of ultraviolet rays.

The sun emits three kinds of ultraviolet or UV rays: UVA, UVB, and UVC. Only the first two, UVA and UVB, penetrate the ozone layer and the earth's atmosphere. Both of these can cause serious harm. Exposure to UVB rays is associated with damage to the surface of the skin and can result in sunburn, premature aging of the skin, and skin cancer. Workers who go shirtless are putting themselves at future risk. If that's not harmful enough, UVA rays—which are more deeply penetrating—can damage the underlying tissues that help keep skin firm and wrinkle-free. That's why individuals who have worked under the sun for years and years have that rough, wrinkled, leathery skin. Together, these ultraviolet rays can effectively cause the appearance of prematurely aging skin.

UVB rays are most concentrated in the summertime, especially between 10:00 A.M. and 3:00 P.M.; UVA rays are present year-round, even during cloudy days. At high altitudes, where there's less atmosphere to filter out ultraviolet rays, they're even worse. (That's why skiers and mountain climbers often have shiny red faces, and why you see many of them with sunscreens all over their noses and foreheads.)

### Skin protection

The Sun Protection Factor (SPF) indicates about how much time you can expect to stay in the sun without burning. For example, if it normally takes your skin 30 minutes to turn red in the sun, then a sunscreen with SPF 10 will theoretically allow you to stay 10 times longer in the sun (or 300 minutes) without burning. These are estimates, of course, order-of-magnitude numbers to give you an idea of the amount of protection afforded by a particular sunscreen product. The person's skin type plays a role too; the fairer the skin type, the quicker the burn. And remember that sunblock is necessary on cloudy days, too, because the sun's rays penetrate cloud cover.

Sun Protection Factors range from minimal SPFs of 2, to 15 or more for maximum protection from UVB rays. The low-SPF products are often used by people who tan quickly and never burn, and the high-SPF sunscreens can protect fair-skinned people and others who burn easily. At one time or another, recommendations suggested a SPF of 8 to 14 for those who burn easily, but who tan gradually. Now, more often than not, advice from physicians and sunscreen manufacturers says that a SPF lotion or product of at least 15, with ingredients to block out a broad spectrum of UVA rays, should be used by everyone. Even dark skin which contains more melanin—a chemical which is part of the body's natural defense against the sun—should be protected by a sunscreen with a SPF of 15.

Here's a chart to give approximate times for various skin types:

| Skin type | 1 hr | 1–2 hrs | 3 hrs | 4–5 hrs+ |
|---|---|---|---|---|
| Very fair/extremely sensitive/always burns | SPF 15 | SPF 30 | SPF 30 | SPF 45 |
| Fair/sensitive/tans slowly/burns easily | SPF 15 | SPF 15 | SPF 30 | SPF 45 |
| Fair/tans gradually/usually burns first | SPF 15 | SPF 15 | SPF 15 | SPF 30 |
| Medium/tans well/burns minimally | SPF 8 | SPF 8 | SPF 15 | SPF 30 |
| Dark/tans easily/rarely burns | SPF 4 | SPF 8 | SPF 8 | SPF 15 |

Individuals who have studied the long-term effects of sunlight on skin believe that a significant amount of damage is the result of cumulative incidental exposures, or routine, everyday contact with sunlight. Because younger skin is more delicate and sensitive than older skin, it burns more easily. Studies indicate that much of the sun damage the typical person experiences over a lifetime happens before he or she is 18 years old. It's thought that early exposure could also play a crucial role in the likelihood of developing skin cancer later in life.

All of this points to the importance of protecting workers from the sun. This can be done in a number of ways:

- Working out of the sun as often as possible.

- Wearing hard hats. Hard hat sun shields are also available to extend the brims or cover the neck (see Figs. 36.1, 36.2, and 36.3).
- Wearing white or light-colored long-sleeved shirts and pants will effectively block most harmful rays. A bandanna loosely tied around the neck will also help. Remember that a good deal of ultraviolet radiation can penetrate a sweaty, wet T-shirt.
- A natural suntan helps protect the body from sunburn, but by its very nature a tan is one of the least desirable means of protection. Considering the long-term damage and cancer risks associated with repeated tannings,

**Figure 36.1** Helmet sunshield (no. 1).

**Figure 36.2** Helmet sunshield (no. 2).

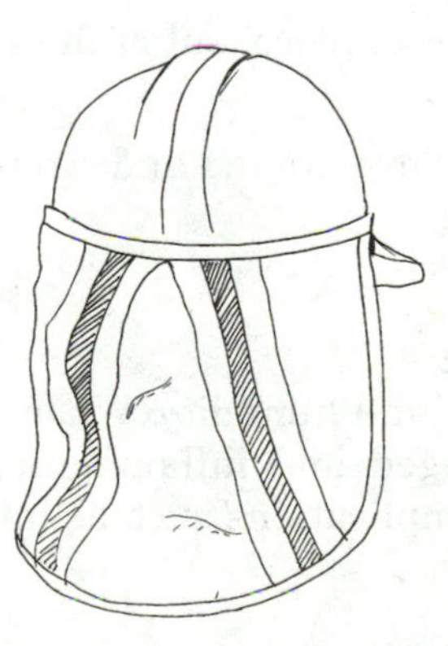

**Figure 36.3** Hard hat neck protector or cape.

individual bouts with sunburn will not be your only problems. And remember that the kind of tan you get "straight from the bottle" will not protect you from the sun's rays either.

- Sunscreens are the most practical products to prevent ultraviolet ray damage. Go for a SPF of 15 or higher. Apply sunscreen at least an hour before going into the sun, on cool, dry skin, and again after perspiring. In general, one application will not be enough for an entire day's activities. When applying the sunscreen, remember to cover all the exposed areas of your body.
- Opaque sunscreens such as zinc oxide or titanium oxide may be used on the nose or lips by people who need to be in the sun for long periods, say, while working on a roof or foundation.

In short, even though overexposure to ultraviolet sun rays is a leading cause of skin afflictions, workers can counteract most of those hazards by using common sense and simple protective measures.

### Protecting your eyes

Although eyes have built-in natural protection against the sun's brightness, glare, and radiation, employees should take additional precautions. Recent research suggests that long-term exposure to the sun's ultraviolet radiation may cause cataracts or other sight-threatening conditions. Concrete, asphalt, chrome, and glass can create unnatural glare situations in the worksite.

Consider that cataract surgery can reduce the eye's natural protection against ultraviolet radiation when the eye's natural lens is removed.

Wearing a good pair of sunglasses during the day will help to protect your eyes and will allow them to adapt more easily to darkness later on.

### Fainting

Fainting can be caused by an inadequate blood supply to the brain. It may in itself be a minor heat-related disorder, but it also indicates that the body is having difficulty adjusting to the heat. A fall caused by fainting can easily injure someone.

- First aid: Have the person lie down in a cool place, out of the sun. Consult a physician.
- Prevention: When in hot temperatures, move around and stretch to improve circulation to reduce the risk of fainting.

### Prickly heat

Prickly heat is a skin rash caused by heat and humidity. When sweat doesn't evaporate, sweat ducts may become clogged and inflammation can result. Severe and prolonged rashes can cause complications such as infection, which should be treated by a physician. Otherwise,

- First aid: apply soothing lotions or powder.
- Prevention: keep the skin as dry as possible. Shower often, and wear fast-drying cotton clothing.

Other hot-weather information to present at tool box meetings includes:

- Eat light, nutritious, preferably unheated meals. Fatty gravies, sauces, and solid foods are more difficult to digest in hot weather.
- Take the heat seriously. Never ignore danger signs such as nausea, dizziness, fatigue, long headaches, and an inability to hold liquids. Instead, get medical help at once.

### Worksite fluid supplies

Again, one of the easiest and most effective preventive measures for heat stress is to constantly replenish the water and salt that's used up by the body's cooling system. Typically, salt is not a problem because most people get enough (and often too much) in their daily food. But fluids are another issue. A person's thirst cannot always be relied upon as a true signal that the body needs water. Instead, fluid intake may require a conscious effort to drink, especially when a person is not thirsty.

Years ago workers took salt tablets along with their drinking water, but that practice is no longer recommended. Any salt tablets or salt water other than *lightly* salted water should only be taken under the supervision of a physician.

Worksite crews find it most convenient when 5- and 10-gallon coolers are used to dispense water or thirst-quenchers (see Fig. 36.4). Thirst-quenchers

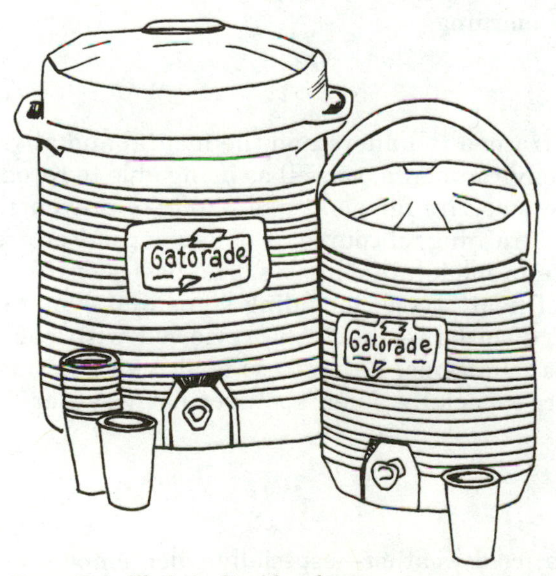


**Figure 36.4** Worksite liquid coolers.

are available in powdered forms, in large packets, which can be mixed on-site with cold water and ice. Like anything, it's better if you survey your crews to learn what they prefer. There are a number of brand-name thirst-quenching drinks, with different flavors. Try several until you hit on a few that the crews like.

Employees working in hot conditions should figure on drinking at least 6 to 8 fluid ounces of water (about a cupful) every 15 or 20 minutes, even if they're not thirsty.

### Work rest periods

A second simple preventive measure for avoiding heat stress is rest. Frequent rest periods should be prescribed for workers performing strenuous duties at the worksite during hot weather conditions.

### Ventilation

Adequate ventilation can be a problem at home construction worksites, especially in basements, garages, upper floors, and attics, where a lack of air flow can create pockets of stale or "dead" air. Consider the use of pedestal or portable fans when employees, or at times, subcontractors, are spending long periods of time working in spaces that lack good ventilation. Fans increase air movement, which helps the body evaporate sweat at increased rates.

### Administrative controls

It can't get much easier than this. An administrative way to help prevent heat stress is simply to schedule hot jobs during the cooler parts of the day—typically early in the morning.

### Training

Employees who are trained to understand the mental and physical hazards associated with hot environments, as well as being able to recognize heat illness signs and symptoms, are far more likely to take preventive measures. The best time for the training, of course, is before exposures are likely. Late spring is a good time in moderate climates. The basics of this chapter—the causes and effects of heat stress, including signs and symptoms and first aid for heat-related disorders—should be reviewed with each crew that's likely to work in hot conditions. Heat stress is also an appropriate tool box safety meeting topic, especially when spells of hot weather fall into work schedules.

## Cold-Weather Safety

It's a simple fact that cold weather—especially when employees are working outside—slows things down. Sleet, snow, and ice, when combined with plywood flooring, muddy grounds, and poor visibility, make walking and operating equipment outdoors difficult and bring about a considerable number of winter-related hazards. Low temperatures, especially when combined with strong winds, can cause serious problems for construction workers whose tasks keep them outdoors for hours at a time. This section keys on hazards peculiar to cold weather, and how those hazards can be guarded against.

### Hypothermia and frostbite

Hypothermia—also called exposure—is a serious loss of body heat. It's usually caused by subfreezing weather, but it can also occur in surprisingly mild temperatures (the forties and fifties, for example), and even when an employee is actively working. The most effective treatment is early recognition, removal of the victim from the cold, and a change into dry and warm clothes. Depending on the individual's tolerance of cold temperatures, if he or she is allowed to continue working, the heat loss can be so great that a victim may no longer be able to produce heat on his or her own for rewarming, and external heat sources will be needed.

Frostbite is a freezing of uncovered or poorly insulated body tissues. Here, the wind-chill factor can be a major cause, but so can damp clothing and wet footwear. Frostbite usually affects fingers, toes, the nose, cheeks, and earlobes. Train employees to keep their extremities insulated and out of the direct wind on subfreezing days. An early sign of frostbite is redness to the cheeks, earlobes, forehead, and fingers. As frostbite progresses, those body parts turn white and numb. First aid for frostbite does *not* include rubbing snow on the affected parts or areas. Instead, gently warm the stricken parts by placing the affected areas against normal-temperature skin or by immersing them in warm (not hot) water.

Much construction work is scheduled so that outdoor tasks can be done during mild to warm temperatures, but when work must be accomplished in cold and subfreezing conditions, employees should dress properly to avoid hypothermia and frostbite. Thermal hard-hat liners (see Fig. 36.5) and cap-style liners (see Figs. 36.6 and 36.7) will help keep employees warm.

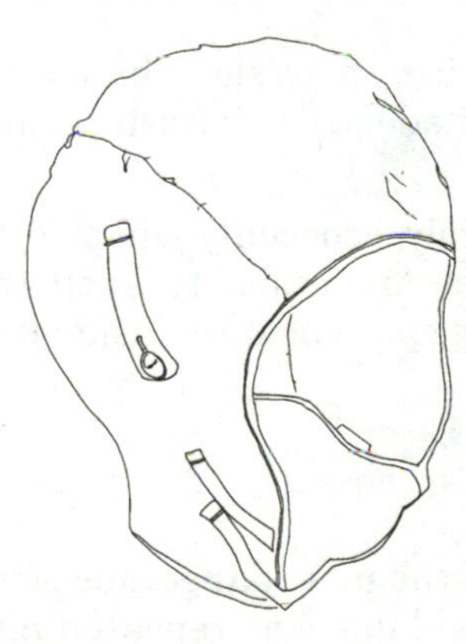

**Figure 36.5** Hard hat winter liner.

**Figure 36.6** Cap-style half liner.

**Figure 36.7** Cap-style full liner.

### Other cold-weather hazards

It's not so much that cold weather itself is a problem; it's rather the effects that cold weather has on the surrounding environment that can cause extra hazards. Employees who work in cold temperatures lose a certain amount of dexterity. And because they're wearing thicker gloves and more layers of clothing, they can't move as quickly. It also takes more energy to perform tasks, and outdoor conditions are more conducive to slipping and falling. The following points should be considered when performing construction work in wintry weather conditions:

- Back injury rates rise, due to the likelihood of snow shoveling, vehicle pushing, falling, and carrying heavy or bulky items improperly.
- Snow and ice collect on walking surfaces. They're especially slippery on temporary wood staircases and steps.
- Treacherous driving conditions include ice, snow, sleet, and fog. Poor visibility, worn tires with poor traction, and inadequate defrosters and windshield wipers make driving more difficult.
- Carbon monoxide poisoning is more likely, especially when some employees spend time eating inside vehicles that are idling to keep the occupants warm, or when employees may be using space heaters inside trailers, sheds, or garages with poor ventilation.

## Lightning

Ever since Ben Franklin pulled off one of the most dangerous scientific experiments in history (the very next two individuals who repeated his experiment were killed by the same activity), lightning has instilled both awe and fear across the globe. Although lightning is not a major hazard on most construction worksites, it's worth mentioning because when it does threaten, the stakes are large.

If worksite crews are outdoors when lightning threatens, have employees:

- Perform inside work, if available. If the structure being worked on is only partially completed, work should be suspended at the site until the threat of lightning is over.
- Sit inside a rubber-tired vehicle. The rubber tires will insulate the vehicle safely, as long as no one touches the vehicle's metal body at any spot.
- Avoid a tent or similar temporary shelter, especially if it is on high ground.
- Never stand beneath a tree or on high ground. Lightning is attracted to the highest object in the area.
- Stay away from overhead wires, and metal fences or poles.
- Avoid carrying umbrellas containing any metal in their construction.

While indoors, have employees:

- Stay away from electrical appliances and plumbing fixtures.
- Stay away from open doors, windows, and fireplaces.

## Tornadoes, Hurricanes, and Other High Winds

These types of extreme weather conditions can come into play during your workweek. Naturally, the likelihood of encountering tornado weather in Oklahoma or Kansas is far greater than experiencing the same conditions in New Jersey. When extreme weather is on the way or threatening, your employees should seek approved shelter—and if that means quitting a partially built structure for the day, so be it. Roofing operations are especially risky in weather extremes, and should be stopped whenever storms, high winds, or other adverse weather conditions threaten or persist.

Remember that with any extreme weather conditions, the weather is not the only hazard. Employees may easily become so preoccupied with lightning storms, high winds, or unbearable heat that they fail to concentrate on what they're supposed to be doing, and are more likely to resort to unsafe behaviors.

Chapter

# 37

# Critters

## Quick Scan

1. Watch out for bees, wasps, ticks, black flies, and other biting insects, snakes, spiders, scorpions, sick and threatening wild animals, and unknown "tame" animals.
2. Any employees who are known to be allergic to bee stings should make that fact known to company supervision, so possible remedies can be arranged for in advance.

## Creatures Great and Small

This is just a tiny chapter, but its contents are worth reviewing with the crews.

Certainly, the possibility of your employees encountering wild or "tame" creatures depends on your worksite locations. Creatures that can harm your employees may either be found outdoors while employees are working on site preparation, landscaping, or other outside construction or maintenance work, or while eating lunch on a backyard picnic table. Some creatures, especially bees and wasps, can be found inside buildings during initial or partial stages of construction. Workers have been bitten, scratched, and stung. They've been sprayed by skunks, punctured by porcupine quills, and savaged by squirrels.

Indeed, strange things have happened between wild creatures and construction worksite employees.

In any event, be wary of local creatures, and train your employees to recognize and avoid any dangerous species they may encounter, including the following:

**Wild animals.** Especially avoid wild animals that aren't acting so wild. This can include small mammals such as raccoons, skunks, opossums, foxes, and squirrels, any of which could be carrying rabies.

**Spiders.** All species of spiders are venomous to a certain extent, but only a few types are really dangerous.

**Scorpions.** These arachnids have a sharp poison claw on the end of their abdomen and will use it at the drop of a hat. The scorpion sting produces intense local signs—acute pain and swelling—that, depending on the physical condition of the victim, can rapidly progress to a serious nature.

**Ticks.** Ticks are tiny blood-sucking arachnids that "inject" saliva and can have a toxic effect on victims. They can carry lyme disease and other maladies. If in a tick-infested area, all employees performing outdoor work should check themselves for ticks after the work is completed. Ticks can be removed with a fine-tipped tweezers if grabbed close to the skin and pulled straight up. Wash the area with soap and water and cleanse with rubbing alcohol. If the tick still can't be cleanly pulled out, medical assistance is needed. If a tick bite causes a red spot, swelling, fever, pain in joints, or flu-like symptoms, have the affected employee report to a physician right away.

**Black flies and mosquitoes.** Believe it or not, these insects can cause major injuries when employees drop whatever they're doing to react to a bite in progress. Best kept off with repellents (see Fig. 37.1) and proper clothing. The repellents should generally be used on both the skin and clothing to be most effective. This is another good reason why employees shouldn't be exposing a lot of skin.

**Bees and wasps.** These probably pose one of the greatest creature hazards to your employees. Their stings produce intense local pain which may result in the victim experiencing shivering, nausea, breathing difficulties, and even unconsciousness. Stings on the face or tongue are especially serious and may cause death by swelling of air passages and asphyxiation. It's likely that

**Figure 37.1** Insect repellent.

employees know if they are extremely sensitive to bee or wasp bites; if they are, it would be wise of them to let their supervisor or crew leader know, so he or she can recognize possible signs and symptoms if a sting and serious medical reaction should occur.

In the case of bee stings, the sting may be removed with tweezers, and various treatments can be applied, including topical or general antihistamine, diluted bicarbonate solutions, and over-the-counter medicines developed for insect stings.

Although practically nothing short of a bee suit will prevent a random bee or wasp sting, there are a number of things that employees can do if they know they're going to work in an infested area. They can:

- Wear light-colored clothing that isn't full of loose voids and baggy sleeve and neck openings.
- Avoid using scented colognes, aftershaves and similar personal hygiene products. Why take a chance by showering with "Peach-Blossom" shampoo? (Sweet foods may also attract bees. Drippings from watermelon and sugary fruit juices have been known to attract bees directly to a sticky forearm.)
- Keep a close eye out for bee nests, and stay away from them. If a ground, tree, or other nest is discovered near to the working area (many bees love to nest in voids between bricks, clapboards, and other siding), it might pay to have the bees professionally exterminated before work continues.

**Snakes.** Although most snakes are harmless, some aren't. If your worksites may feature rattlesnakes, coral snakes, and various vipers or other poisonous species, carry a snakebite kit in your first-aid supplies, and train your employees on how to recognize the poisonous snakes of the area, and how to use the kit.

**"Tame" animals.** Dogs can be anywhere, and any dog can bite. Instruct your employees to never approach a dog they don't know. As a rule, all interaction with unknown pets should be discouraged. If a "drunken" or threatening animal approaches an employee, instruct the employee to either stand still or back slowly away from the animal, speak in a soft, calm tone, and, if available, hold a rolled-up jacket or towel on a lower forearm between their body and the animal.

Chapter

# 38

# Site Security

## Quick Scan

1. Remove valuable tools and materials at the end of the workday, or lock them up securely at the site.
2. Post "No Trespassing," "Danger," "Caution," and other signs carrying specific messages and warnings to discourage unauthorized site entry.
3. Allow no serious hazards to go unguarded or without barricades or other hazard prevention or warning systems.

## When No One Is Watching

This chapter is a brief reminder for your employees to clean up at the end of the day and either remove, barricade, or post potential hazards so that unsuspecting trespassers won't get injured during off hours.

Consider the following points:

- At the end of a workday, it's preferable to completely remove as much equipment and tools from a site as possible.
- Just because your company's employees won't be at the site during off hours doesn't mean that protective devices can be removed or dispensed with. Stairways still need handrails, new glass still needs papers or marks to distinguish it, and debris should not be left scattered about.
- Open trenches and other excavations must be barricaded so that no trespassers—even they shouldn't be on the site—fall into them.
- If possible, illuminate the perimeter of a house or other building under construction, even before all of the walls are up. Lighted areas will discourage all but the boldest interlopers from exploring. Consider motion-detection lamps positioned in key front and back entrance areas.

- Immobilize all pieces of mobile equipment. Park on level ground in illuminated area, if available. Lower buckets, blades, and attachments to the ground. Remove keys from ignitions or place lock-out covers over push-button starters. Lock cabs if possible. A nice big yellow bulldozer, front-end loader, or backhoe—any piece of heavy equipment—has always been a banner attraction to neighborhood youngsters, who love to climb all over a machine.
- Post dangerous sounding signs, and lots of them. You want to convey the concept that the site is hazardous to all unauthorized personnel. "No Trespassing" is not really specific enough to frighten some sightseers. You're not going to scare a practiced thief with typical "Danger" or "No Trespassing" signs, but they will have an effect on neighborhood kids, who may think twice before entering.

  Other sign ideas include messages that say "Restricted Area/Monitored by Video Camera." Some companies have put up dummy video cameras. Then again, some companies with histories of considerable vandalism and theft loss have gone as far as to actually monitor the grounds with real closed-circuit television equipment.
- For some reason—maybe it's psychological—neater sites tend to attract fewer trespassers. Keep piles or stacks of materials orderly; materials or tools scattered about are more likely to be pilfered. Another reason that debris should be removed regularly is that piles of cardboard, scrap wood, plastic, some insulation, and other combustible materials offer tempting targets for youngsters with matches.
- Arrange for the services of someone who lives next door or across the street or down the road to keep an eye on the site at night and on weekends, when your crews aren't there. Supply the watchers with the company's emergency number or let them know how to reach someone in management.
- Remember to burglarproof doors and windows as soon as possible. Exterior doors should have a minimum of single-cylinder deadbolt locks as a first line of defense. Every window in the dwelling should be lockable, and you should pay extra attention to basement windows and garage doors.
- It's happened many times. A contractor takes a several-day or week-long job repairing or installing something in a private home. He leaves his tools there on the site. They're safe, he figures, set up in an attic, basement, family room or locked garage. The homeowners certainly wouldn't *steal* anything. Of course not. But the homeowners might try to *use* the tools while the contractor's gone, and if the homeowners (or their children or grandchildren) aren't properly trained on how to safely use a power saw or portable grinding wheel, at the flip of a switch a serious injury may result.

  Even though it takes extra time, it's better to remove or lock power tools away, or render them inoperable (e.g., with an electric plug lockout device). Doing so will help everyone sleep better.

Chapter

# 39

# Individual Trades and Their Hazards

## Quick Scan

1. Recognize the most frequently encountered hazards to particular trades, then provide initial and refresher training to help prevent those exposures from resulting in worksite injuries and illnesses.
2. Customize training topics somewhat to each trade, so they will be more interesting and applicable to each individual participant.

## A Panorama of Skills and Ills

Learn to recognize and deal with hazards related to individual trades or work types:

### Carpenters and framing crews

In the old days, if you could show me a master carpenter with all of his fingers and thumbs, I'd show you a true anomaly—an individual who was the exception rather than the rule.

Mostly it's the table saw, ripping wood down to width without goggles or safety eyewear, or not using a push stick, and getting the fingers a tad too close for comfort. It's trying to get that one more cut, that extra cut needed before lunch, or while someone is waiting with outstretched hand.

Too, there are head injuries and face injuries from walking into things while preoccupied with the job at hand. During construction, things have a habit of not being there one moment, then being there the next. The environment is constantly changing—that's the nature of construction.

### Electricians

Electricians are exposed to big-time risk. Of the bunch, they tend to be more careful as a class, but even they will let their guards down now and again. Most electricians get into trouble by means of expediency. They've done something a thousand times and expect everything to go just so, and they don't pay close attention to what's happening. There are things that the electrician knows he should do, but instead he relies on his skill and experience to avoid trouble—taking risks he wouldn't think of letting someone else try. Electricians need frequent reminding of the importance of doing things by the book, and *never* taking shortcuts.

In addition to electric shock, electricians are also subject to falls—after being shocked. Burns are another common injury, as are flashes to the eyes if surprise electrical arcing occurs.

### Roofers and siding crews

These trades are practically always up in the air. And since what goes up must come down eventually, they need to make sure that the coming down is a slow, controlled process, rather than quick and abrupt.

Naturally, fall protection is big here. But so is power tool usage, electrical safety, and material-handling hazards. Ladder and scaffold safety is critical, as are slips and falls—which become a lot more dangerous than normal when workers are up a few stories off the ground.

Roofers, as members of any trade where the individuals become familiar with the environment, can develop a feeling of false security and the attitude that they'll instinctively protect themselves from falling by some sixth sense—some mountain-goat-like ability which has gradually developed. This is not true, of course. Roofers must frequently be reminded of the hazards they continuously face.

Subcontractors who are observed taking unnecessary risks must be told to work safely—or be replaced. You don't need the hassle that a serious injury will bring to the worksite and your company.

Siding crews, in addition to being exposed to risks similar to those of roofers, have even greater exposure to sheet metal lacerations and eye injuries.

### Plumbers and pipefitters

If I saved a nickel for every plumber's joke I ever heard, there'd be hundreds of dollars in that bank account. The plumber has one of the most critical trades in all of construction, because nothing will ruin a building like having water and sewage where it shouldn't be, when it shouldn't be.

The hazards to a plumber's trade are legion. There are big-time weights involved—with modern commodes, bathtubs, shower units, water softeners, lengths of pipe—and awkward items that must be manhandled up stairs and over and across unfinished flooring. There are sharp, burry threaded pipe ends to contend with. Plus heated tips of soldering guns, and open flames from burners. Some solders, from materials purchased years ago, still contain lead. Burning and soldering fumes must be carefully avoided.

Awkward positioning is the norm. Hygiene is a big deal, because lots of nasty materials such as pipe dope, sealer, grease, caulk, and similar materials tend to get on or near the hands. There's the powerful-smelling pipe sealer that's used to bond PVC piping.

The knees and back take a beating, and so do the wrists—forever and a day turning pipe wrenches one way or another. Ergonomics definitely comes into play with plumbing tasks.

### Drywall/plaster/trim installers

To outside work crews, the drywall, plasterer, and trim installers have it made. Unfortunately, inside crafts have their hazards too. One is dust. Work with plaster and plasterboard and similar materials day in, day out, and you're talking dust. You're talking foreign bodies in eyes, too. There are razor cuts from sundry sharp instruments, and there are slippery ladder rungs due to spilled plaster and water. The mixes come in heavy containers, so there are material-handling concerns. There's awkward positioning, a lot of over-the-shoulders reaching that tends to tire the shoulders, arms, and wrists. There's the hazard created by the employee's looking upward, instead of where he or she is walking. Individuals have plastered themselves right over edges, down open stairwells, and through unguarded balconies and multilevel floors.

It's tough on the knees, and back, and neck, too.

### Masons

Backbreaking work. Masters of the crouch and awkward position. The "catchers" of the construction trades. Masons deal with a heavy product, with repetitive motions, with dust and sand and mortar and cement. They deal with scaffolding and work platforms. They move a lot of heavy product from level to level, up and down. They accept and handle bulk deliveries. Their work is rough on the hands.

They need strong material-handling skills, and when splitting bricks or blocks, cannot take a chance of forgetting their safety eyewear.

### Flooring installers

They deal with tile, vinyl-flooring types, carpeting, wood, and various other materials. Possibilities for lacerations, for hand injuries, for back strain, shoulder strain, repetitive injuries on the wrist, elbow, and arm tendons. Knee strain. Respiratory hazards, including dust, solvent, and adhesive fumes.

### Painters

Inhalation hazards are present here, with paint and solvent fumes (plus the same materials can make harmful contact with skin). Lead-based coatings can be disturbed and dislodged into the air during stripping, sanding, buffing, and other abrasive methods. Ladder exposures are frequently the cause of painter injuries, as are unplanned contacts with electricity.

Eye injuries must be guarded against. Working with glass is a potential source of lacerations. Neck strains occur from looking upwards, and from constantly raising the arms above shoulder height. Back strain follows. Repetitive-motion injuries are possible, especially to the wrist and parts of the arm and shoulder.

### Mobile equipment operators

Operating machines can be rough on the back. Climbing on and off the equipment is often the most dangerous part of the job. Electrocution is a possibility, especially when something goes wrong—an accidental collision with a utility pole/line, or digging into an underground electrical line or cable—and the operator gets out of the cab, only to act as a grounding mechanism.

Mobile equipment operators are often known to get into a safety rut where they have been doing the same things "for thirty years" and therefore don't believe in "new" safety rules and procedures. Operators like that need to be talked with, in private, until they understand the reasons why they've got to change their behavior.

Part

# 5

# Using Safety to Help Promote Your Business

*The title for Part 5 may, admittedly, sound a bit mercenary. Using safety as a means to a commercial end? Is that fair?*

*It is. It will enable your company to realize additional benefits for becoming a champion of safety within your trade or industry. What's needed today are more companies that operate with the safety of their employees and customers in mind, by investing in the latest user-friendly tools, materials, and methods. That's Chapter 40. The next chapter encourages readers to seek out general safety professionals and safety experts in their fields, personally (they're approachable), and learn from them. No one is out there hiding information on safety. It's available for the asking, and the asking is easy. The last chapter of Part 5, and of the book, discusses how what you learn about safety can be shared with your employees, your customers, and others in your industry and community and beyond.*

*Although Part 5 is just three chapters long, and the chapters are not really "safety chapters" in a conventional way, if you follow some of what they're suggesting, you and your company may lead the companies in your area when it comes to promoting safety.*

Chapter

# 40

# Providing Safety Expertise, Services, and Products

## Quick Scan

1. Review the universe of product types you handle, and offer the safest ones available.
2. Provide safety expertise, services, or products to needy or charitable organizations.
3. Share the knowledge of your craft freely with customers, consumers, and industry peers.

## You Are a Resource

This chapter is a brief plea for you and your company to supply your customers and others with safety expertise, services, or products—relating to whatever you and your company specialize in. Often, safe features can be included instead of "standard" features at little extra cost. In most cases, the benefits will by far outpace the extra expense, and the investment wil be repaid through the well-being of your customers and others, and through the promotional value your company will likely receive.

Since you are making an effort to sharpen the safety skills and resources of yourself and your company, why not share some of what you learn?

The following are examples of "safer" products that, depending on your business, you might offer your customers.

### Fire prevention and protection

- Supply fire extinguishers with fixed hangers in the kitchen, basement, and garage.

- Include smoke and fire alarms, some permanently hard-wired with battery backup.
- Include spark screens or heat-resistant glass doors on fireplaces.
- Include spark arresters on all chimney tops.
- Supply portable escape ladders in bedrooms.
- Include sturdy fire-resistant doors with bedrooms.
- Position windows so that they can be used for escape routes.

### Electrical

- Install only properly grounded electrical systems, with ground-fault circuit interrupters (GFCIs).
- Include plastic safety caps with all receptacles.
- Install higher grades of electrical fixtures, especially well-insulated recessed lighting.

### Walking surfaces

- Install antislip tile, linoleum, or wood flooring.
- Use antislip paint on wood steps.
- Install extra-sturdy handrails on staircases.
- Install heated outdoor steps or walks in cold-temperature locations.
- Choose safety-designed fixtures and installations, such as counter tops with rounded, not sharp corners.
- Discourage the use of painted wood steps for attics and regular living levels apart from basements. Use carpet instead.
- Design hallway and bathroom spaces to accommodate wheelchairs.
- Use texturized concrete on driveways and walks, and texturized sealer on asphalt, to reduce slipping hazards.

### Kitchens

- Smoke and fire alarms and extinguisher.
- Sharp-knife holders in drawers.
- Garbage disposal switches far enough away from sink to prevent opposite-hand injury.
- Extra ventilation for cooking fumes and odors.
- Stoves/ovens with control knobs on top instead of on the front.
- Local fire sprinkler system, such as over the rangetop.
- GFCIs on all kitchen receptacles.
- Good lighting.

### Bathrooms

- GFCIs on all bathroom receptacles.
- Antislip flooring.
- Grab handles near showers and tubs.
- Antislip tub and shower unit floors.
- Install steps with risers on elevated tubs and whirlpools.
- Whirlpool off and on switches should be located far enough away from the tub to prevent anyone in the tub from reaching out to turn on the unit.
- Safety glass on shower doors.
- Easy-to-operate faucets.
- Sun or heat lamps with short-term timers.
- Good ventilation/exhaust fan.
- No-fog mirrors.
- Avoid lighting fixtures over the tub.

### Indoor air and water

- All furnaces vented to the outdoors.
- Extra-quality ventilation and control system.
- Radon detectors.
- Carbon monoxide detectors.
- Products that contain no formaldehyde.
- Extra-quality furnace filtering system.
- Dehumidifier/humidifier units.
- Water filters and purifiers.

### Workshop/basement/garage

- GFCIs installed in all receptacles.
- Good lighting.
- Workshop dust collection system.
- Additional permanent electrical outlets.
- Fire extinguishers.
- Lockout provisions for power equipment.
- Recycling waste bins or containers.
- Texturized, antislip garage floor.
- Garage door opener systems with autoreverse.

- Heating appliances with pilot lights partitioned off from rest of garage or mounted on a concrete or other non-flammable platform at least 18 inches high.
- Safe fold-down garage attic staircase units.
- Home intercom and phone lines.

### Security

- Motion detector lamps installed at entrances.
- Programmable 24-hour wall light switch timers.
- Solar-powered "night lights" to illuminate walkways, poolsides, or decks.
- Video intercoms.
- Burglar alarm system.
- Exterior doors of solid-core wood, fiberglass, or metal-clad construction.
- Door reinforcers and high-security door lock strikes.
- Exterior door hinges positioned inside door.
- Multiview glass peepholes on all exterior doors (allowing a viewer inside to see to the sides, down, and up, not just straight ahead).
- No key-in-knob exterior door handles/locks.
- Deadbolt locks and high-security strike boxes.
- Sliding glass doors with positive locking mechanisms, such as antilift plates.
- Windows with positive locking mechanisms.
- Weather-resistant keyless locks with numeric security pads for operating a nonbeveled dead lock bolt.

### Miscellaneous

- Heat-trace wire on roof edges to prevent dangerous icicles from forming.
- "Smart home" programmable menu-driven appliances and utility systems.
- Automatic controllers for appliances.
- Manufacturer's operating manuals for all new equipment installed.

## Sharing Your Knowledge

Your safety expertise or services would likely be appreciated by numerous organizations in your community, including schools, churches, United Way agencies, and service organizations such as Lions clubs, chamber of commerce roundtables, manufacturer's associations, and similar groups.

For example, if your company does electrical contracting, discuss electrical safety for the homeowner. Talk about how individuals can get themselves in trouble. Explain how using aluminum ladders, and installing (or more likely, taking down) television antennas, gutters and downspouts, and similar fixtures can place an unsuspecting homeowner at risk. Explain the concept of hazardous energy control—how many of us have never been exposed to industrial or construction-type safety training?

The same goes for roofing contractors. Tell how to properly get on and off, and how to walk around on various roof pitches. Explain that someone on the ground, with a good set of binoculars, can accomplish a considerable amount of roof inspecting without ever taking a step on top of the roof itself. Talk about proper roof installations and durable materials—so the homeowner won't have to worry about crawling all over his or her roof a few years later.

Plumbers can talk about modern hot-water controls and the importance of backing off hot-water temperatures from the hottest settings. They can discuss water quality and septic systems and dehumidifiers and humidifiers. They can address slippery bathroom floors and safe bathtub and shower entrances. Energy-efficient heating and cooling units are always an interesting topic.

Home builders can review the latest safety features that can be planned right into new construction. They're able to review basic styles, floor plans, and even specialized homes for families with young children, or for senior citizens.

Carpenters are a big hit when speaking about home workshops or hobby centers. Tool and equipment safety is a strong topic, and so is the proper use of personal protective equipment. These areas lend themselves to lively demonstrations and hands-on participation from the meeting's attendees.

The topics on safety are practically unlimited, and the need for safety expertise and related discussions will always be high and well received.

Participating in these activities will set you and your company far apart from most of your industry's competitors. By getting out in your community and volunteering your services and those of your company, the inevitable result will be a well-deserved reputation for "giving back" and sharing your safety knowledge with your fellow citizens. That, in turn, will likely result in gaining certain amounts of business your company would not have otherwise received.

Chapter

# 41

# Seeking Advice from Safety Experts

## Quick Scan

1. To help further your safety knowledge and education, make the acquaintance of as many safety experts as you can.
2. Attend as many safety conferences, seminars, classes, and other sessions as possible. Arrive early, and mingle with safety experts at refreshment or break time.
3. Keep in contact with as many safety experts as you can.
4. Offer to provide local information related to the expert's field as part of ongoing communications between yourself and the expert.

## Approaching Experts

Experts are like you or me in many ways. With the exception of a relatively small minority of them, they're entirely approachable. Why make contact with experts? For the latest information, trends, and available services regarding the areas about safety that affect your company or interest you most. They can provide you with real-life examples, and with new regulations being proposed or considered. Experts will help you prepare up-to-date materials you'll need for the next chapter, for writing up and presenting "safety successes" your company can share with others.

### Locating experts

You can find experts in sundry ways, including:

- *Through trade associations.* Association officers are often experts in their own rights. Editors of trade publications will know of experts who have been authors or who have been quoted in past articles.

- *Through the Internet.* If you haven't made the connection yet, you probably should. You'll find enormous possibilities out there, mostly free for the asking. OSHA is on the Internet now, and so are many, many more safety, hygiene, and environmental organizations. Through the Internet you can keep up with regulations that may change daily.
- *Through your local library or nearby university library.* Even the smallest libraries nowadays have access to powerful computerized search engines that, within moments, can research huge databases for the experts you need. Librarians will be glad to assist you.
- *Through newspaper and consumer magazine articles.* Safety experts are frequently called upon to comment on home safety and consumer safety issues.
- *Through television magazine news and feature shows.* Locating experts this way takes a certain amount of luck, because no comprehensive listing schedules are so compiled.
- *Through universities and other schools.* Schools and universities are used to supplying experts for articles, speeches, research projects, and even as litigation witnesses. Certain schools are known for their staffs and curriculums in support of safety and industrial hygiene programs. A phone call to the appropriate department secretary will likely lead you to available experts on staff, or visiting or adjunct experts who have part-time relationships with the school.
- *Through speakers' agencies.* There are promotional, media, business, and similar speakers' agencies that specialize in arranging expert speakers who can prepare interesting presentations on practically any topic.

### Preparing to make contact

Nothing will turn off an expert more than stupid questions—or your lack of basic information in the field you're seeking knowledge in. I can remember, years ago, researching an article for a regional magazine about a nationally known wildlife/bird expert who lived a few miles south of Toronto, Ontario. After arranging a meeting at his house for an interview, I and another writer made the six-hour drive and finally found his apartment. He pleasantly greeted us, then ushered us into his den. The first few minutes of the interview went well—until he started inquiring what we knew about the subject matter we were there to interview him about. Indeed, we knew practically nothing, and that became painfully obvious to him. When we couldn't recognize another famous wildlife bird expert he referred to, we lost all credibility with him, and the interview went downhill from there.

Had we done a minimum of research on the expert and what he represented, the interview would have gone quite differently.

The point is, do a little homework on the expert before you contact him or her. Learn a little about the expert's background; if the expert has published articles, essays, or books; if the expert gives lectures, serves on company or association boards, or teaches in university programs or private seminars. If possible, read some of the expert's writing, or speak with some of the expert's friends or associates.

By taking the time to do basic research on the expert and his or her area of expertise, you'll be prepared to carry on a reasonably intelligent conversation when you eventually meet.

### Making contact with experts

There are several main types of experts, including local and national authorities. Local safety experts include ranking members of fire departments, loss-control companies and insurance agencies, police departments, hospitals, physicians' groups, and government agencies (such as OSHA or the EPA or DOT). They can almost always be approached over the phone or in person.

National—or international—experts are more difficult to access. You can try to contact national experts through various ways, including:

1. *By postal mail.* This is the method of contact that puts you and your psyche at least risk. Simply write to the individual and mail the letter to his or her book or magazine publisher, company, association, or group. Nine out of ten times you'll get a reply, but it might take a while coming.
2. *By e-mail.* This is a possibility that is somewhat dependent on the age of the expert and whether in fact the expert has marshalled him or herself into the Internet age. If the expert is Internet literate, it's a great way to make initial contact and to keep in contact.
3. *By fax.* The fax functions like fast postal mail. It's a great way to accelerate feedback response times, especially if the expert belongs to some legitimate company or organization.
4. *By phone.* This method is for "nothing ventured, nothing gained" individuals who can think fast on their feet. It's an immediate way of making contact, but somewhat risky. You could catch the expert at a bad time. You might be misinterpreted. You might, through no fault of your own, make a bad first impression. Perhaps he or she won't like your tone of voice, or the phrasing of a question you ask. Also, considerable preparation needs to be done before cold-calling an expert to avoid saying dumb things or experiencing long pauses when you're at a loss for words.
5. *In person.* This is an excellent way to meet an expert, as long as you don't come across as overbearing, boorish, or threatening. The contact should be kept brief, unless the expert displays interest in lengthening initial communications. There are a number of situations in which you can approach an expert in a natural way:

- *At seminars or classes.* Attend seminars or classes either given by the expert or featuring the expert as a guest speaker.
- *At trade meetings or conventions.* The same goes for trade meetings and conventions, at local, state, and national levels.

At seminars, classes, meetings, and conventions, there are things you can do to increase the likelihood of personal contact and self-introductions.

You can, for example, arrive early. Take a front seat. Often the expert will be there early, too, checking out the room setting, unpacking materials, and seeing if the overhead projector is working or if the microphone is set up. Ask if there's anything you can do to help. Pass out handouts. Tape an extension cord to the carpet. Supply directions to a good Italian restaurant.

- You can try to sit at the same table as the expert for lunch. It's common for an expert at a seminar or class, when it comes time for the in-house refreshments or lunch, to come into the room after most people have been served and are already sitting. Hang loose somewhere in the room until you see the expert collecting his buffet lunch or heading for a table, then go for a spot at the same buffet line or table. The expert is typically very approachable during lunch because it's considered part of the day's activities.

  Participate in whatever conversation results—usually about the morning's topics. If you've done your homework by obtaining and reading one of the expert's books (if there are any), before or right after lunch is a good time to comment on it or get it signed.

  You can make a point to be one of the last ones to leave. Again, ask if you can help pack up, carry items to a car or cab, give the expert a lift to a hotel or airport, or supply information about local restaurants, shopping, or evening entertainment. Sometimes you'll be able to link up with the expert at night. If you get along well with him or her, you've probably made a contact who in future months and years can help with your safety projects.

- *Through meetings arranged by mutual contacts.* The mutual party can be an editor, educator, association director, safety organization, manufacturer, manufacturer's rep, distributor company, government official, or anyone else.

### Reciprocate if possible

Always ask if there's anything you can do for the expert in the future. Perhaps you can send clippings from local newspapers or other publications on topics related to his or her field.

Place the expert's phone number, address, and other means of contact in the media notebook you'll be setting up (described in the next chapter). Include biographical notes for reference.

# Chapter 42

# Presenting Your Successes

## Quick Scan

1. Look for interesting aspects of your business dealing with safety and consumers, or safety and your trade. Choose specific topics to learn more about.
2. Research additional information on your chosen topics.
3. Pay attention to any media exposure your industry gets within your town, city, or region.
4. Identify the individuals, organizations, and publications that deal with the trends, legislation, new products or processes, and news related to your trade or industry.
5. Prepare information for the media, focusing on points of safety dealing with consumers or individuals within your trade group or industry.

## Promoting Safety—and Your Company

Most construction companies do very little to promote themselves. They depend on repeat business and word of mouth. They advertise in the yellow pages of phone directories, in Sunday church bulletins, in local newspapers or area magazines, and by signs painted on their trucks and equipment. They may also have working relationships with one or more real estate companies. Because the contents of this book are concerned with safety, this chapter proposes another means of promotion for your company—a means that will supply, in addition to your company's promotion, valuable information and services to the community.

By promoting safety—no matter what your trade or business—you'll help make a name for your company and you'll be educating consumers by making them more conscious of safe options.

### Prepare yourself

Do some research. Read the trade association publications pertinent to your industry. Be particularly attentive with all safety-related topics. Pay attention to local newspapers, magazines, and public events that focus on any aspect of your trade or industry. Learn to recognize important individuals serving on committees or running your local trade organizations. What kind of information on your industry do consumers care about? Are any interesting local or national trends developing?

### Decide on an angle

By angle I mean an interesting, specific fact, item, or topic about your business and safety that you can explain to the public or to other members of your trade. It should be about information people will find helpful. It doesn't have to be a huge item. It can be very small and specific; in fact, the more specific, the better it's likely to be received.

For example, your company builds houses. As discussed in Chap. 40, numerous safety features can be designed and built into a house—many of which would not be thought of in advance by typical homeowners. Or your company sells and installs windows and doors. There are plenty of safety and security features that can be stressed. Your company pours concrete driveways, sidewalks, and patios. What about nonslip texturized finishes? Proper grading for water runoff? It's likely, no matter what your business is, that you could find and stress some little-known insider information that can make things a bit safer, and perhaps a bit more convenient for consumers.

### Preparing an outline and plan

Simply write a list of the points that you want to discuss or explain. For example, you own, manage, or are employed by a plumbing company. You'd like to discuss the safety aspects of designing and installing a new bathroom. First, you list the points you'd be able to talk about:

- Flooring—ceramic tile, carpeting, wood, vinyl
- Whirlpools
- Vanities
- Tub and sink fixtures
- Windows and ventilation
- Accessories

After considering the possibilities, you narrow it down to your first idea—bathroom floorings.

Sure, ceramic tile is durable and easy to maintain, but tile with a glossy, smooth finish is too slippery for the often-wet conditions that will naturally occur in a bathroom. You want to stress that the best tile will have nonskid glaze surfaces. Maybe explain how floor tiles should have a coefficient of friction (COF) rating of .5 or higher—a rating recognized by the tile industry for expressing slipping resistance. How many people do you think even know that a COF rating system exists? Or what it is and how it works? That, you think, may be an excellent idea for your article. Bathroom floor coverings—from a safety and hygiene angle. It's specific enough to be interesting, and *everyone* knows something about, and uses bathroom floors. Your perspective, as a plumbing expert, brings authority to the piece, and offers perhaps a new way of looking at things.

Next, prepare a simple outline:

Bathroom Floors

A. Carpeting
- Pros and cons
- Recommended types
- Installation methods/maintenance

B. Wood
- Pros and cons
- Recommended types
- Installation methods/maintenance

C. Vinyl/plastic/rubber/similar flooring
- Pros and cons
- Recommended types
- Installation methods/maintenance

D. Ceramic tile
- Pros and cons
- Recommended types (explain slipping coefficient)
- Installation methods/maintenance

### Placing your information

There are many opportunities for placing your ideas. Here are some:

1. Local newspapers. Newspapers are always seeking material for their articles and columns. Manufacturers haven't been the only ones downsizing over the past decade; so have newspapers. They have fewer full-time reporters or staffers and have increasingly relied on part-time and freelance help. Freelance writers are individuals who may work full time elsewhere, and write for various publications "on the side." Newspapers often use them as specialists in their areas of experience—for example, a writer specializing in real estate may have worked in that business and is perfect for contributing investigative and feature articles for a once-a-week real estate supplement put out by the paper. Another freelance specialist may write on business matters, and another, on outdoors topics such as hunting and fishing.

Of course, there are also regular staffers: daily editors, morning editors, Sunday editors, and other editors, depending on the size of the paper.

You need to make contact with the editor whose area of responsibility is most pertinent to your information. The simplest way to make contact here is with a phone call. Explain who you are, and what your idea is. Have the outline of your presentation ready, so that you can review the main points—even if there are only a few of them. Say nothing, of course, about how you expect to receive promotional value from what you're proposing. That's something that will be a side benefit—and only if what you propose is indeed sincerely helpful to the paper's readers.

Chances are that the freelancer or editor will be glad to hear from you; they aren't often approached by concerned members of a building or construction trade, especially members who are volunteering their expertise to help educate consumers.

2. *Local or regional magazines.* These are typically city or area magazines that cover a wide range of topics dealing with their circulation area. Information can be presented here in interview style or in a feature article. Typically, more in-depth coverage is expected for pieces in local magazines, and their lead times are a lot longer than those of newspapers. Many regional publications rely on full-time writers for their articles, but many others turn to freelancers for supplemental pieces. As with newspapers, it's best to approach the local publications first. These may include a Sunday newspaper supplement—which is really put together more like a magazine than a newspaper. Before approaching an editor from a local or regional publication, study the last three or four issues, to get a feel for article lengths and style.

3. *Trade publications.* Trade publications are collectively a huge outlet for writers of specialized information. Trade publications or journals are also known as business, technical, and professional journals. Here are some examples:

*PWC, Painting & Wallcovering Contractor*
*Remodeling News,* for the home remodeling, custom building, and light construction industry
*Security Dealer,* for electronic alarm dealers, and burglary and fire installers
*Aberdeen's Concrete Construction,* a how-to magazine for concrete contractors
*Indiana Builder,* covering residential construction
*Masonry,* covering masonry contracting
*Permanent Buildings & Foundations*
*Roofer Magazine*
*Landscape Design*
*Heating, Plumbing, Air Conditioning*

What's more, practically every trade association has its own publication. Your local library should have a copy of *The Encyclopedia of Associations,* a huge listing of thousands of associations in the United States and all over the world. Your industry is likely to have a number of trade publications looking for materials. Consider the possibilities.

4. *Television shows and segments.* Some localities have educational cable-sponsored shows dealing with consumer information. Nightly news features are also possible, especially if your material can be quickly and graphically shown on narrated tape. Contact the station to find out who you should approach. Be sure to stress that *safety* is your main message.

5. *Home improvement shows/trade association shows.* Some of these are yearly events. Consider preparing a booth with a safety theme.

### On Writing

When a person who doesn't know you reads something that you have written, that piece of writing becomes you. It stands alone as an extension of your intelligence, an example in black and white of how you think.

If your thoughts spill onto the paper haphazardly, so will you be considered disorganized. And if you don't subscribe to rudimentary rules of grammar, you might be thought of as lacking education. Of course, the reverse is also true; if you write clearly, you'll be labeled a clear thinker by those who don't know you (and sometimes, by those who do as well), and, in a sea of wordy, incorrect, unclear sentences and paragraphs you'll emerge like—well, like a fine piece of construction.

But even if you lack English acumen, there's no need to hang your head in shame. Instead, enlist the aid of a relative, friend, assistant, or secretary to proofread your letters and proposals and writeups while you bone up on your grammar and style.

For assistance with your writing style, consult a skinny little book that's still in print, called *The Elements of Style,* by William Strunk, Jr. and E. B. White. Practically everything you need to know about the fundamentals of writing can be found in this clear, brief, entertaining handbook. It explains why and how you should "place yourself in the background," or "write in a way that comes naturally," or "work from a suitable design." It provides a few rules of usage, a handful of principles of composition, and "a list of words and expressions commonly misused."

### Keep a media file

Once you start working with the media, you'll be surprised at how one contact can lead to another. With a little effort, you can build yourself—and your company—into a respected authority on safety in your field. Naturally, that will only continue if you are sincere and you know what you're talking about. Shysters who try to buffalo the media will be found out sooner or later—often with embarrassing results.

Do your homework and be honest, punctual, and sincere in your material presentation, and you can gain for yourself and your company a reputation that few others will ever achieve. It's possible. And that reputation can be a springboard to:

- Radio interviews
- Television interviews
- Giving educational talks for schools and service groups
- Serving on trade association or government-sponsored committees

Keep a file on the interesting media individuals you meet, including names, phone numbers, addresses. Then maintain contact with them.

### A word of hope

If this all sounds too artificial or mercenary to you, please refrain from passing judgment until you try it once or twice. Becoming a safety spokesman for your company, and eventually for your area or region, could lead further, perhaps to a national scale. When this process is set in motion, you may find yourself taking a renewed interest in your career, and you may find that you *like* to be within range of media spotlights, that you *like* to learn more and more about the new products and methods and processes within your industry, and that you *like* to keep in contact with others who are doing the same things.

You may find that by writing up and presenting materials on safety and your industry, you will accomplish many things. Not only will you provide meaningful assistance for consumers and other interested parties, beneficial publicity for your company, and increased professional contacts for yourself and your coworkers—you'll also find that the time spent doing so is extremely rewarding—and lots of fun.

# Index

## ABOUT THE AUTHOR

Dave Heberle is a widely respected environmental safety and health specialist who has written extensively on these subjects. He is coauthor of McGraw-Hill's How to Plan, Contract, and Build Your Own Home, Second Edition, and many other books and magazine articles on home construction, maintenance, real estate, and safety. He has provided safety support, training, and related communications to manufacturing and service organizations for 20 years. He is currently a safety and health specialist with a Fortune 500 company.